Soils and Societies

SOILS AND SOCIETIES

Perspectives from Environmental History

edited by

J.R. McNeill

and

Verena Winiwarter

The White Horse Press

First published 2006 by
The White Horse Press, 1 Strond, Isle of Harris HS5 3UD, UK

Set in 10 point Adobe Garamond
Printed and bound in Great Britain by Biddles Ltd

British Library Cataloguing in Publication Data
A catalogue record for this book is available from the British Library

ISBN 978-1-874267-52-2

The inset illustration on the cover is from Wolf Helmhard von Hohberg, *Georgica curiosa. Das ist: Umständlicher Bericht und klarer Unterricht von dem adelichen Land- und Feldleben*, Nürnberg 1682. Permission to use the copy of the Benedictine Library at Melk, Austria for reproduction is gratefully acknowledged. Photograph © Verena Winiwarter, 2005. Background photograph reproduced courtesy of Professor H.-R. Bork.

Contents

Biographical Notes

EDITORS

J.R. McNeill was born in Chicago and gained diplomas from Swarthmore College and Duke University before joining the History Department and the Walsh School of Foreign Service at Georgetown University in 1985. Since 2003 he has held the Cinco Hermanos Chair in Environmental and International Affairs at Georgetown. Recent books include *Something New Under the Sun: An Environmental History of the 20th-century World* (New York: W.W. Norton/London: Penguin UK, 2000); *The Mountains of the Mediterranean World: An Environmental History* (Cambridge: Cambridge University Press, 1992, reissued in 2003); *The Human Web: A Bird's-eye View of Human History* (New York: W.W. Norton, 2003; co-authored with William H. McNeill); and an edited volume, *Environmental History in the Pacific World* (Aldershot: Ashgate, 2001). He lives an agreeably harried life with his triathlete wife and four exuberant children, and if he had time would enjoy sports, movies and camping.

Verena Winiwarter, who was born and still lives in Vienna, was first trained as a chemical engineer. After years of working in atmospheric research, she earned her M.A. in history and communication sciences in 1991, and her Ph.D. in environmental history at the University of Vienna in 1998. She has been a lecturer (Universitätsdozent) there in Human Ecology since 2003. She currently holds a postdoctoral fellowship in environmental history (APART fellowship) awarded by the Austrian Academy of Sciences at the Institute for Soil Research, University of Natural Resources and Applied Life Sciences, Vienna and at the Faculty for Interdisciplinary Research of Klagenfurt University. Her main research interests within environmental history comprise the micro-history of landscapes, waste, landscape images and soils. She is a founding member of the European Society for Environmental History. From 2001 until 2005 she served as its president. Besides work she enjoys walking and hiking and picnics with her husband and two kids. She has published more than 80 articles and book chapters, and co-authored two CD-ROMs. Recent book publications include *Historische Humanoekologie*, co-edited with Harald Wilfing (Vienna: WUV Facultas, 2003) and the forthcoming *Das Ende der Flaeche*, co-authored with Rolf Peter Sieferle et al. (Köln: Böhlau, 2006). Website: www.iff.ac.at/umweltgeschichte/winiwarter.php

Biographical Notes

AUTHORS

Tim Beach (Ph.D., University of Minnesota, 1989) teaches courses on environmental science, soils, geomorphology, hydrology, natural resources management and environmental archaeology at Georgetown University in Washington, D.C. There he directs Environmental Studies and is Associate Professor of Geography in the Program in Science, Technology, and International Affairs of the School of Foreign Service. In more than thirty field seasons, he has studied soils, geomorphology, geoarcheology, cultural ecology and paleoecology in the Maya region of Central America, the Middle East and the United States. His major research focus has been to study Maya subsistence and long-term human impacts on soils and landscapes of the Maya region. He has published many book chapters and articles in such journals as the *Annals of the Association of American Geographers*, *Geoarchaeology*, *Catena*, *Physical Geography*, *The Professional Geographer*, *The Geographical Review*, *Antiquity*, *Ancient Mesoamerica*, *Latin American Antiquity*, and the *Journals of Archaeological Science*, *Soil and Water Conservation* and *Field Archaeology*.

Hans-Rudolf Bork is a professor of ecosystem analysis and director of the Ecology-Centre of the Christian-Albrechts-University of Kiel, Germany. He received a Dr. rer. nat. (Ph.D.) degree and a Dr. rer. nat. habil. degree at the Technical University of Braunschweig, Germany.

Hans-Rudolf Bork is a generalist in geography, ecosystem research, landscape ecology and landscape development, and a specialist in soil science, geomorphology and hydrology. He is investigating the effects of land use and of extreme weather events on soil formation, soil erosion, water and matter dynamics since the rise of agriculture in Central Europe (Germany, Belgium, Poland), Chile (Atacama, Easter Island, Robinson Crusoe Island), China, Israel, Russia (NW-Siberia), Ecuador (Galápagos) and the USA (Pacific Northwest). Website: www.hans-rudolf-bork.de

Nicholas Dunning grew up on the Island of Oahu in Hawaii, where he developed a fascination with tropical environments and cultural adaptation. He has spent the past 25 years studying the intriguing environment of the Maya Lowlands and the nature of human environmental adaptations, both past and present, in this region. Along the way, he earned B.A. and M.A. degrees from the University of Chicago and a Ph.D. from the University of Minnesota, all in geography. He has authored two books and some 60 articles and book chapters on the Maya world. Currently, Dunning is a Professor of Geography at the University of Cincinnati, where his students affectionately (he hopes) refer to him as 'Dr. Dirt', because of his fascination with soil.

Biographical Notes

Sheryl Luzzadder-Beach (Ph.D., Geography, University of Minnesota) is Associate Professor of Geography and Computational Sciences in the Earth Systems and GeoInformation Sciences Program and the Center for Earth Observing and Space Research, at George Mason University in Fairfax, VA, USA. Her research, publishing and teaching experience includes hydrology and water quality, physical geography, geostatistics, science education, and gender and science. She has conducted fieldwork in hydrology, geoarchaeology and paleoecology in California, Belize, Mexico, Turkey and Syria. Dr Luzzadder-Beach has published chapters in Scott Redford, ***Landscape and the State in Medieval Anatolia: Seljuk Gardens and Pavilions of Alanya, Turkey*** (Oxford: Archaeopress, 2000), Vernon L. Scarborough et al., ***Heterarchy, Political Economy and the Ancient Maya: The Three Rivers Region of the East-Central Yucatan Peninsula*** (Tucson: University of Arizona Press, 2003), A. Gomez-Pompa et al., ***The Lowland Maya Area: Three Millennia at the Human-Wildland Interface*** (Binghampton, NY: Howarth Press, 2003), and Lisa J. Lucerno and Barbara W. Fash, ***Precolumbian Water Management: Ideology, Ritual, and Power*** (Tucson: University of Arizona Press, 2006); and articles in the journals ***The Geographical Review, Annals of the Association of American Geographers, The Professional Geographer, Geological Society of America Bulletin, Catena, Ancient Mesoamerica, Antiquity, Culture and Agriculture, California Geographer*** and ***Environmental Management.***

Andreas Mieth studied Biology, Ecology, Marine Biology and Fisheries Biology in Berlin and Kiel (Germany). He has worked as science-coordinator in the Centre of Biology at Kiel University from 1984 to 2001. Since 2001 he has been the science-coordinator at the Ecology Centre and vice-head of the Department of Ecotechnology and Ecosystem Development. His main research interests are ecosystem and landscape development, in particular the reaction of ecosystems under human impact; technology–ecology interactions; and ecotechnology, e.g. technical use of renewable materials, development of glues based on bio-materials, and ecological sustainability of techniques. His Ph.D. thesis dealt with 'Effects of Land Use on the Development of Landscape and Culture on Easter Island (Chile)'. Publications include A. Mieth and H.-R. Bork, 2005, 'History, Origin and Extent of Soil Erosion on Easter Island (Rapa Nui)' ***Catena*** 63, 244–60; A. Mieth and H.-R. Bork, 2003, 'Diminution and Degradation of Environmental Resources by Prehistoric Land Use on Poike Peninsula, Easter Island (Rapa Nui)', ***Rapa Nui Journal*** 17(1), 34–41

Frank Uekoetter studied history, political science and the social sciences at the universities of Freiburg and Bielefeld in Germany and the Johns Hopkins University in Baltimore, USA. He received his Ph.D. from Bielefeld University in 2001 with a dissertation on the history of air pollution control in Germany and the United States, which was published as ***Von der Rauchplage zur ökologischen Revolution. Eine Geschichte der Luftverschmutzung in Deutschland und den USA 1880–1970*** (Essen: Klartext Verlag, 2003). In 2002, he organised the conference 'Nature Protection in Nazi Germany' under the auspices of the German minister for the environment Jürgen Trittin. He is currently a researcher at the Department of History of Bielefeld University, with a project on the history of agricultural knowledge in the twentieth century. His publications include 'The Frontiers of Environmental History', a special issue of ***Historical Social Research*** (vol. 29, no. 3, 2004), and ***The Green and the Brown: A History of Conservation in Nazi Germany*** (Cambridge and New York: Cambridge University Press, 2006).

Robert S. Shiel lectures on soil management and climatic change at Newcastle University, UK. He is closely involved in research on long-term changes in vegetation and soils, in particular on increasing diversity and improving sustainability in hay meadows (Bardgett, Smith, Shiel, et al. ***Nature***, in press), and manages the 108-year old Palace Leas hay meadow, the world's oldest grazed grassland experiment. Recent completed archaeologically-related research has concentrated on the Mesolithic communities of north east England (Howick) and north west Scotland (Scotland's First Settlers), while publication of results from longer-term projects in Hungary (Upper Tisza Project) and Greece (Boeotia Project) proceeds but slowly, though the fieldwork is finished. At a micro level he has also investigated changes in materials after burial and in the residual evidence of use, particularly on stone tools.

Kate Showers completed her first academic degree in anthropology, but cross-cultural experience suggested that a knowledge of production agriculture would be more useful. Degrees from both Michigan State University and Cornell University resulted in a 1982 Ph.D. from Cornell's Agronomy Department. Her doctoral dissertation, involving two years of fieldwork in Lesotho, southern Africa, included studying the language and culture as well as origins of soil erosion processes, collection of oral environmental histories, and archival environmental history research. After a post-doctoral fellowship in the Department of Theoretical Production Ecology, Agricultural University, Wageningen, the Netherlands and appointments as a research scientist at the Rodale Research Center, Pennsylvania, USA, and as an Academic Visitor at the Social Studies Faculty Centre, Oxford University, she was named Head of Research, Institute of Southern African Studies, National University of Lesotho. This enabled her to continue investigating oral history's potential for obtaining information about historical environmental change. Currently she is Senior Research Associate and Visiting Research Fellow at the Centre for World Environmental History, University of Sussex (Website link: http://www.sussex.ac.uk/development/1-4-5-2-1.html). A synthesis of field work, oral history and archival research concerning Lesotho's landscape history was published as ***Imperial Gullies: Soil Erosion and Conservation in Lesotho*** (Athens: Ohio University Press, 2005) Her interests include the environmental consequences of landscape interventions; social and environmental impact of technologies; soil erosion and conservation; water management; indigenous land use systems; and popular participation, particularly as constituents of environmental history.

Robert J. Wasson is a geomorphologist who has specialised in palaeo-environmental reconstruction in the deserts of Australia and South Asia; the impact of land use on soils and river catchment processes in Australia, New Zealand, India, Indonesia and East Timor; and the development of trans-disciplinary research methods. The river catchment work has been based on a historically-resolved material budget framework in which the major sources, storages and yields of sediment and nutrients are quantified. The results of this approach have been used as the basis for catchment management in many locations in Australia, producing a defensible underpinning for decision makers. When coupled with environmental history, natural resource managers can better understand how landscapes have reached their current state(s).

After a career in CSIRO (Australia), Professor Wasson was recently Director of the Centre for Resource and Environmental Studies at the Australian National University and is now Deputy Vice-Chancellor (Research) at Charles Darwin University in the Australian tropics.

Soils and Societies

1

Soils, Soil Knowledge and Environmental History: An Introduction

J.R. McNeill and Verena Winiwarter

The survival of human communities has always depended, and still depends, on certain environmental goods. Oxygen and fresh water top the list, but soil is not far behind, because for most of history most people got most of their food from plants grown in soils. Our distant forbears who lived by foraging and hunting depended indirectly on soils, as they provided the nutrients necessary for the growth of the fruits, nuts, tubers and, indirectly, the animals, that formed our ancestors' food supply.

When people began the transitions to agriculture (which happened multiple times beginning about 10,000 years ago), their dependence upon soils became more direct and more obvious to them. Agricultural peoples were less mobile than foragers, and more likely to stay put long enough to have significant impacts upon soils, perhaps eroding them, perhaps degrading them, perhaps enriching them. Their religious practice, in many settings, included worship intended to maintain the fertility of soils, something unknown among non-agricultural peoples. The ancient Egyptians, for example, included in their pantheon a male god (Geb or Seb) whose portfolio included the earth in general and Nile silt in particular. More frequently, deities associated with the soil, and its fertility, were female (Winiwarter and Blum, 2006).

With the emergence of urban life some 5,000 years ago, the relationship between humankind and soils shifted once more. Village life had only modest effects upon the distribution of soil nutrients, because although people harvested and ate crops, eventually their wastes or their corpses returned the nutrients to the local environment. But with cities things were different. Farmers exported food to cities, and wastes and corpses never returned, or rarely returned except in small quantities, to the fields whence the food had come. Instead, soil nutrients were systematically drawn from the fields and, after lodging in human bodies for shorter or longer stays, passed into the urban environment, often to be carried away by runoff, local streams and rivers. Henceforth, farmers who served urban markets, or indeed distant markets of any sort, faced a structural problem of long-term nutrient loss. Where soils were deep and rich, the problem might be ignored for generations. Elsewhere, the problem required human ingenuity in the form of soil knowledge and soil management. For millennia, indeed until the nineteenth century, the chief solution to this problem took the form of animal manure. As one Polish nobleman of the sixteenth century put it, 'manure is worth more than a man with a doctorate' (Gostomski 1588). By grazing animals on pasture, scrubland, or in forests, peoples of the Old World could in effect collect nutrients from broad areas and concentrate them on their fields, simply by penning their animals and collecting their manure. Livestock, in addition to their other functions in agro-ecosystems, worked as nutrient gatherers for farmers' fields. In the New World, without large livestock before 1492, farmers lacked manure and had to invest greater labour and ingenuity if they intended to maintain the fertility of their fields. In some areas, human wastes (night soil) supplemented or even replaced animal manure. The use of night soil is very ancient and (until recently) widespread: it is mentioned in Homer's Odyssey and was routine in parts of East Asia and Mesoamerica as well.

The survival, prosperity and power of almost any given farming community rested on its success in resolving the problem of long-term nutrient loss. Egypt was the chief exception. There the waters of the annual Nile flood brought a regular nutrient subsidy in the form of a film of silt left behind by receding floodwaters on the banks of the Nile. It came from the volcanic (and nutrient-rich) mountain landscapes of Ethiopia. In effect, the Nile did the job that domestic animals (or carts of night soil) did elsewhere. Less thoroughly and less reliably, floodwaters in Mesopotamia and from the Indus, the Ganges, the Irrawaddy, the Mekong, the Amazon and elsewhere served in the same capacity. Everywhere, long-term economic trajectories, the ebb and flow of political power, the waxing and waning of populations, rested on the successful management of soil nutrients. For want of nitrogen, many a kingdom was lost.

This is a fundamental truth which historians, even environmental historians, have scarcely recognised. Some inspired amateur polymaths and historically minded soil scientists have probed the subject of human–soil relationship through time (Hyams 1952; Lowdermilk 1953; Hillel 1991). Geographically inclined historians have often pointed out the significance of soil factors upon human affairs, but without acknowledging that soils have their own histories. Historians typically regard soils as fixed features of the environment and therefore background considerations to agricultural and economic history. The great contribution of environmental historians over the past generation is to remind others that environments are not mere backdrops to the dramas of history, but participants in their own right, interacting with all the others. This book aims to extend that logic to soils in particular, to present them as entities with histories.

Soils have their own histories, both natural and human. Geological substrate, climate and vegetation shape their natural histories, which in turn shape their human histories. The human histories of soil are both material and intellectual. What happens with soils and the societies that use them is a matter of both biogeochemistry and culture. What people believe about soils influences (although it does not necessarily determine) what they do with them, whether they conserve and nurture them, whether they abuse and abandon them. What people understand – and misunderstand – about soils is thus a necessary part of any history of the nexus between soil and society. Hence this book accords considerable attention to soil knowledge and its maintenance, transmission and impacts, from the ancient Roman agronomic writers to Sotho farmers and German agricultural experts.

This book also devotes considerable attention to soil erosion. Since the 1930s soil erosion has (to fluctuating degrees) occupied many a fine scholarly mind (e.g. Neboit 1983; Blaikie and Brookfield 1985). Erosion's corollary, soil deposition, has occupied rather fewer minds. But both processes have helped shape human affairs wherever food came from the soil. By and large scholarly discussion of soil movement in history is conducted between two poles, which we may call Malthusian and Boserupian. The Malthusian position relates soil erosion to population pressure, growing food demand, deforestation and the cultivation of marginal, highly erodible lands. The Boserupian pole is occupied by those who see soil stability deriving from active efforts at conservation and enhancement, which in the past were normally labour-intensive (e.g., building and maintaining terraces or manuring) and thus more easily undertaken in thickly populated landscapes. Either position might be correct in a given situation. But of course population is not the only variable that affects patterns of erosion and deposition, often not even the crucial one. Technologies, agricultural customs, climate change, livestock behaviour and many other factors shape soil erosion histories. On different time scales, different variables may dominate.

For these and other reasons, this book includes studies on different time scales. It is organised roughly according to the length of time under consideration. Papers deal with millennial, secular or decadal time scales. Soil creep, which typically carries small amounts of soil downslope annually, is normally a negligible concern at decadal time scales, but sometimes a decisive one on millennial scales. Conversely, government policies may be quite influential on decadal scales, but of negligible impact on millennial scales (because they, like the governments behind them, are almost never maintained long enough).

Fernand Braudel, a geographically inclined historian (although he was not interested in soil change or soil processes), divided historical time into three parts in his great work on the Mediterranean world (Braudel 1972). He wrote of ***l'histoire événementielle***, of ***la conjuncture***, and of ***la longue durée***. His categories are of shorter duration than ours, and less arbitrary: ***la conjuncture*** for example was connected to the 50- or 55-year Kondratieff cycles sometimes used by economic historians. But in (faintly) echoing Braudel with our arrangement of time, beginning with longer scales and moving to shorter ones, we are merely trying to make clear, as he was, that different processes appear salient depending on the depth of historical vision.

Spatial scale and geographic location are also variables that affect both soil processes and what looks important about them. Chapters in this book range in spatial scale from the local to the continental, from Easter Island to all of Africa, with many stops in between. Every inhabited continent is represented, and most of the great historic centres of human population. As a result, the book deals with a wide array of cultures, political economies and landscapes. It is a step towards a world environmental history of soils.

Lastly, this book is an interdisciplinary collaboration. Any approach to the study of humankind and soils requires the methods and insights of multiple scholarly disciplines. The arrangement of inquiry and knowledge into discrete disciplines, which owes a little to Aristotle and lot to German universities of the nineteenth century, has been an enormously rewarding strategy for at least 150 years. But it surely hides as much as it reveals. And no scholarly discipline is adequately equipped (nor any single scholar as far as we know) to cope with the range of methods, perspectives and data that is appropriate to the subject of the linkages between humankind and soils through history.

The history of soils is perhaps the most neglected subject within environmental history (McNeill and Winiwarter 2004). With this book we hope to address that neglect. We also hope to provide a long-term perspective on soil-society relations that may be helpful in putting soils more securely on the agendas of environmental policy. Soil ecosystems lie, albeit uncharismatically, at the root of human existence. But modern behaviour strongly suggests there is much as yet not understood about soils and their links to society, and much that is understood that is imperfectly practised. Soils are a very slowly renewable resource, and the use we make of them, today as in much of the past, is often quite unsustainable. Understanding this in all its dimensions, including its historical ones, is a step on the way to putting the soil–society relationship on a sounder basis.

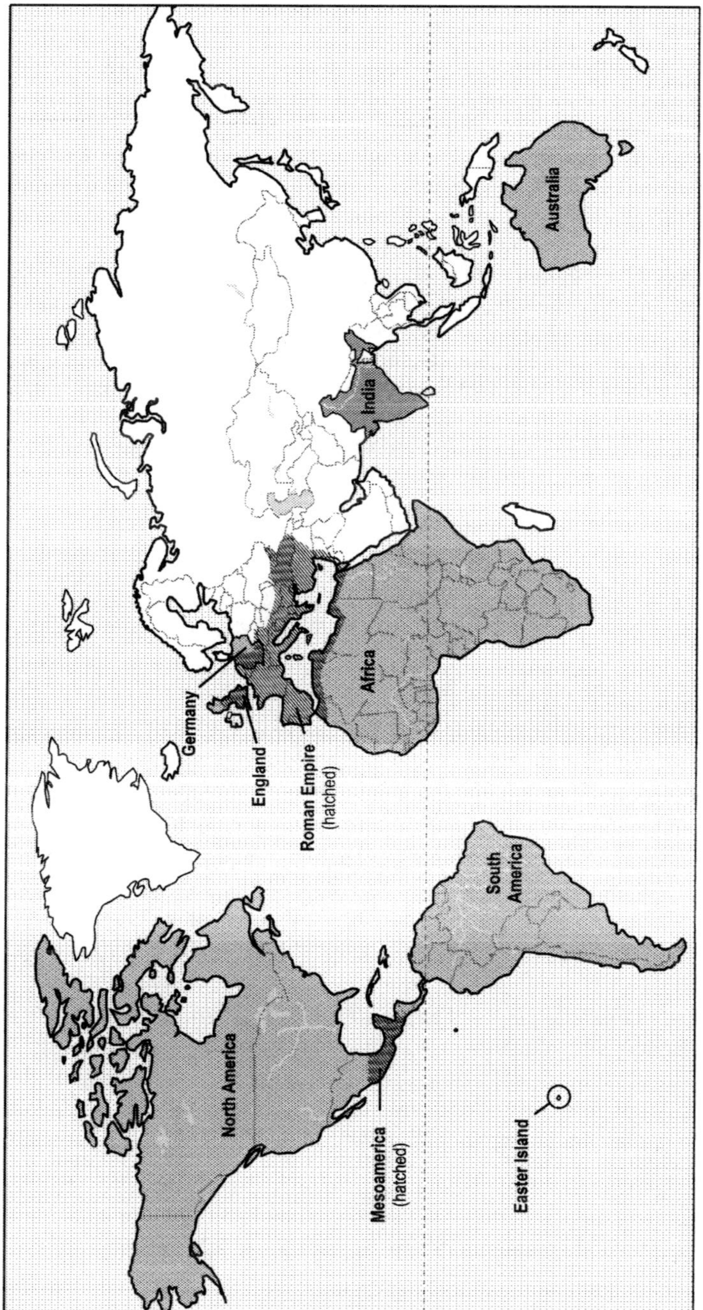

Figure 1.1. Location map of the regions treated in this book

ACKNOWLEDGEMENTS

This book is the outcome of a collaborative effort over several years. First, we would like to thank all authors for their work in summarising and presenting knowledge from different parts of the world in the context of this book.

We are very grateful to the Breuninger Foundation, Stuttgart (Germany) for their generosity in hosting a workshop on Wasan Island in August 2003. Most of the papers of this volume were presented and discussed there. We would in particular like to thank Helga Breuninger and her team for their hospitality and interest in soil matters.

It was a great pleasure to co-operate with Andrew and Alison Johnson of The White Horse Press, who were instrumental in turning the idea of an interdisciplinary environmental history of human–soil interactions on a worldwide scale into reality.

Verena Winiwarter acknowledges financial support for her work in contributing to and editing of this volume from the Austrian Academy of Sciences, APART fellowship program, Grant 10916, 'Precious Dirt Underfoot. A History of Soil Knowledge'.

Production of maps for the Introduction and the chapters on Mesoamerica was made possible by a grant from Georgetown University's Walsh School of Foreign Service.

REFERENCES

Blaikie, Piers and Harold Brookfield. 1985. *Land Degradation and Society*. London: Routledge.

Braudel, Fernand. 1972. *The Mediterranean and the Mediterranean World in the Age of Philip II*. New York: Harper & Row, 2 vols.

Gostomski, Anzelm. 1588 [1951]. *Gospodarstwo*. Wroclaw: Wydawn. Zakladu Narodowego im. Ossoli´nskich, 1951.

Hillel, Daniel. 1991. *Out of the Earth: Civilization and the Life of the Soil*. Berkeley: University of California Press.

Hyams, Edward. 1952. *Soil and Civilization*. London: Thames and Hudson.

Lowdermilk, W.C. 1953. *Conquest of the Land through Seven Thousand Years*, Agricultural Information Bulletin no. 99. Washington DC: US Department of Agriculture.

McNeill, J.R. and V. Winiwarter. 2004. 'Breaking the Sod: Humankind, History and Soil'. *Science* **304**: 1627–9.

Neboit, René. 1983. *L'homme et l'érosion*. Clermont-Ferrand: Publications de la Faculté des Lettres de Clermont II.

Winiwarter, Verena and Winfried E.H. Blum. 2006. 'Souls and Soils: A Survey of Worldviews'. In Benno Warkentin et al. (eds), *Footprints in the Soil: People and Ideas in Soil History* (Amsterdam: Elsevier), pp. 107–22.

2

An Introduction to Soil Nutrient Flows

Robert S. Shiel

Until the widespread availability of 'artificial' fertilisers in the mid-twentieth century AD, the growth of crops and forages was frequently limited by an inadequate amount of nutrients in the soil (Cooke 1975). The appropriate management of soil, therefore, was closely linked to ensuring a sustainable supply of plant nutrients. In examining how well this was achieved, it is, however, inappropriate not to consider other soil factors, many of which also contribute to plant growth and health. The soil is an holistic material, and alteration of one characteristic invariably influences numerous others. For example, much of the nitrogen taken up by plants comes from the soil organic matter (humus), but the amount of this converted to available forms is influenced by soil wetness, temperature, cultivation, acidity and the sequence of crops produced (Cooke 1967). Thus, it is not just the amendment of soil with manures and fertilisers that affects the supply of nitrogen. In addition, environmental conditions – excess or shortage of water, temperature or light – as well as the presence of toxins, pests, weeds and diseases, all affect the efficiency of the plant's metabolism and therefore its ability to take up the nutrients that are in an available form. This behaviour was recognised at an early date and can be described as an example of Liebig's (1843) Law of the Minimum – which in its broadest interpretation states that the growth that occurs depends on the requirement that is available to the plant in the smallest amount.

The problem of maintaining soil nutrient supplies also depends on the demand for agricultural products. This demand has, other than at exceptional periods such as during the Black Death of the mid-1300s AD, tended to increase with the passage of time (Thirsk 1997). To achieve greater food production, either more land must be cultivated or the existing area must be cultivated more intensively. As any use of land for food production leads to reduction of overall soil fertility – particularly as the humus content decreases, and the extent and rate of this change depends on the intensity of the use (Jenkinson 1988) – either solution to the problem of increasing food output leads to deterioration of some soil properties. Such fertility deterioration invariably results in slower growth of plants and, by leaving the soil surface exposed for longer, frequently increases soil erosion (McGregor et al. 1999), and this, by removing the nutrient-rich portion of the soil preferentially (Stoltenburg and White 1953), has an even greater negative effect on fertility. In the absence of modern fertilisers, declining fertility reduces crop productivity per unit area and, if the supply of land is limited, can have serious consequences. A shortage of nutrients leads to use of a narrow range of crops, which can be grown under such conditions, over a large proportion of the total area. This narrowing of the food base can result in catastrophic regional failures – as happened in Western Europe with potato crops in the middle nineteenth century AD (Burton 1948; Bergman 1967). Where land is not in short supply, degradation of soil at one site invariably leads to clearing of virgin land elsewhere, and this may, in its turn, suffer the same fate; such a shifting cultivation system can only persist so long as there is a fresh supply of unoccupied land (Nye and Greenland 1960).

The properties of the land in its virgin state depend on the material from which it has formed (the parent material) and the extent of 'ageing' to which it has been exposed. The 'ageing' is the result of chemical weathering of the soil mineral particles and the loss of the potential plant nutrients, either by leaching or by removal in the harvested crop. The extent of this weathering is one of the fundamental factors used in determining the soil orders in the United States system of soil classification (USDA 1999). The parent material from which the soil forms can vary from siliceous sands through loessic or alluvial silts to nutrient-rich clays, and depends very much on the underlying geology and processes of erosion and deposition in the Pleistocene and Holocene (White 1997). The north and west of Europe has been repeatedly affected by glacial episodes, the last of which finished 12,000 years ago, and which, to a variable extent, have replaced ancient, highly weathered soils with glacial till (boulder clay), outwash sands or loess. While the distribution of boulder clay largely follows the area glaciated, the loess and outwash sands extend far to the south of the area covered by the ice sheets (Flint 1971). Even in the peripheral areas to the south of the glaciation, the abrupt climatic change, particularly in rainfall, at the end of the ice age caused widespread erosion and the establishment of alluvial fans (Borsy 1990). Where the surface was not disturbed by the ice sheets (Fitzpatrick 1963) then the residual soils are often very acidic and low in plant nutrients and, like aged plants and animals, tend to have a small productive potential. In these areas of very weathered and impoverished soils it is in flood plains or where erosion has exposed fresh material that the soils are likely to be most fertile. This is not to say that all of the young soils are fertile, for those with a high content of silica sand contain few nutrients and evolve rapidly to podzols (Paton et al. 1976) and have poor ability to retain nutrients or water, in addition to being very acidic. Thus, Europe can be viewed as a patchwork of soils, often with variation over short distances, between those youthful soils that are naturally productive to those which require large amounts of inputs to achieve a useful product. The proportion of these contrasting 'types' will originally have differed greatly between localities but this pattern has been affected by subsequent exploitation of the land by farmers; their management may have improved conditions in some, but in many other places has made the soil less productive.

An analogy with human ageing can help to appreciate the utility of soil. A child has little productive ability and requires assistance, a juvenile tends to become over enthusiastic, but an adult reaches a peak of productivity. As maturity progresses, the range of work that can be done decreases, while in old age considerable assistance is again needed. If we imagine an area of freshly exposed boulder clay at the end of the ice age, this has a good content of mineral nutrients but lacks any system of pores through which water, air and roots can move, and it contains little if any of the essential nutrient nitrogen. The first plants to grow on such material have to cope with a poor rooting environment and the lack of nitrogen – the plants are Stress Tolerators (Grime et al. 1988) and are likely to include nitrogen-fixing legumes. The growth of these plants improves the conditions in the soil and the Stress Tolerators are frequently replaced progressively by the more vigorous Competitor plants that could not tolerate the original conditions. In the juvenile stage, the conditions may be too 'rich' for some species, but as the conditions become more stable a maximum of productivity is reached. Throughout all this time nutrients are being lost by leaching, and the soil progressively becomes more acidic and depleted in nutrients such as potassium (White 1997). Plants which can tolerate these oligotrophic conditions – a different group of Stress Tolerators (Peet 1992) – tend to replace the Competitors, and growth again becomes restricted to a narrow group of slower growing species. The period of time over which these changes occur depends very much on the parent material, the climate and the topography. On a level sandy site with a moderate rainfall, the whole sequence can be completed in a few thousand years (Paton et al. 1976). As agriculture spread across Europe, beginning from areas far to the south of those influenced by the direct effects of the glacial episodes (Renfrew 1987), the 'young' soils in most of the continent had gone through the initial stages of formation, usually to the adult stage before they were used. The main exceptions were in alluvial soils in river valleys and where erosion occurred on hill slopes due to human activity. In these cases, the soil forming in the alluvium contains much organic matter from the eroded soils and frequently it stabilises rapidly into a fertile productive soil profile while the eroded hillsides are rejuvenated – sometimes to the infant stage!

Although crops and animals for food, fibre or fuel can be produced at all of these soil development stages, some are better suited than others to each stage. If species suited to an earlier stage or one which has not yet been reached is desired, then the soil must be amended by the user to provide the appropriate conditions. Examining first the selection of the most appropriate soil types, we have to consider the crop species available. The cereals that are the backbone of modern food production systems tend to grow best on relatively mature soils. If the nitrogen supply is excessive, the plants grow too tall and fall over (lodge) making the crop difficult to harvest and also prone to attack by predators and by fungal diseases in the damp conditions near the soil surface. Modern wheat suffers less from this problem because the varieties grown today are short and do not lodge so easily – this has allowed farmers to increase the nitrogen supply and hence the yield (Percival 1974). On more mature soils that have become acidic then rye and oats can be grown (MAFF 1979); there is also a form of barley, bere (Jarman 1996), that is tolerant of soil acidity. Acidic soils, however, tend to have relatively poor nitrogen (Alexander 1961) and phosphorus (Wild 1988) availability, so ensuring an adequate nutrient supply under such conditions is not easy. Many plants grown for fruit grow better in soils with a somewhat restricted nitrogen supply (MAFF 2000) – strawberries for example will produce masses of foliage in fertile soil, but the fruit is hard to find and tends to rot in the moist atmosphere created by the rampant foliage. Most legumes prefer soil with a high pH and good supply of available phosphorus but frequently compete poorly with more competitive species if the nitrogen supply is good (Cooke 1975); the answer is therefore to grow legumes after crops which have removed much nitrogen from the soil unless the soil contains relatively little organic matter.

This in effect has introduced the concept of crop rotation as a means of manipulating the nutrient supply to advantage the species that we desire. Rotation will also reduce soil erosion (Stoltenburg and White 1953) and, by providing a diverse range of crops, will reduce the risk of a catastrophic failure. While a rotation has the objective of returning the soil properties at the end of the cycle to the state that they were in at the beginning, and as such returns most of the nutrients removed in animal feeds via the manure, the sale of crops for human consumption in distant urban areas results in local loss. In the case of nitrogen these losses can be made good by fixation from legumes within the rotation but other nutrients will be depleted unless an external source is available. Recycling of urban wastes has only ever benefited the areas close to the towns and is limited today by a variety of concerns involving disease organisms and toxins in the wastes (Lampkin 1990). Other management methods available to the farmer operate to modify the conditions in the soil to render it more similar to those in which the species desired is most likely to succeed; these techniques include drainage, liming, manuring, marling and, in the extreme, warping (Russell 1915). The increased use of 'artificial' fertilisers since the Second World War has arguably solved the nutrient management problem, but by removing the need for the rotation of crops and the production of animal manures has exacerbated the other soil problems that are related to nutrient management. Thus, we have more severe erosion, decreasing soil organic matter, and associated with this poorer soil structure, increased drainage and soil management problems

(MAFF 1970), increased droughtiness and leaching of nitrates into water supplies (Addiscott et al. 1991). To continue the analogy with human ageing used above, liming, manuring and warping can be related to rejuvenation, rotation of use between crops and grass can be related to taking exercise or a recreational change, while the repeated growing of extractive crops, particularly with the use of large amounts of fertiliser, could be related to self abuse!

Recently, there has been widespread and increasing interest in organic farming (Lampkin 1990) which emphasises the use of crop rotation and maintenance of soil organic matter, and uses natural resources, such as limestone and phosphate rock, to increase the nutrient content. Although this is argued by its proponents to greatly improve the quality and sustainability of the soil, it is only returning the land use system to one that was common in the early years of the twentieth century AD (Fream 1919). It remains therefore to be seen how well the use of such rotations, whether in the nineteenth or twenty-first centuries, can satisfactorily ameliorate soils and maintain them in a sustainable condition while providing large amounts of high quality food.

REFERENCES

Addiscott, T.M., D.S. Powlson and A.P. Whitmore. 1991. *Farming, Fertilizers and the Nitrate Problem.* Oxford: C.A.B. International.

Alexander, M. 1961. *Introduction to Soil Microbiology.* New York: Wiley.

Bergman, M. 1967. 'The Potato Blight in the Netherlands and its Social Consequences (1845–1847)'. *International Review of Social History*, **12**: 391–431.

Borsy, Z. 1990. 'Evolution of the Alluvial Fans of the Alfold'. In A.H. Tachocki and M. Church (eds) *Alluvial Fans: A Field Approach* (Chichester: Wiley), pp. 229–46.

Burton, W.G. 1948. *The Potato.* London: Chapman and Hall.

Cooke, G.W. 1967. *The Control of Soil Fertility.* London: Crosby Lockwood.

Cooke, G.W. 1975. *Fertilizing for Maximum Yield.* London: Granada.

Flint, R.F. 1971. *Glacial and Quaternary Geology.* New York: Wiley.

Fitzpatrick, E.A. 1963. 'Deeply Weathered Rock in Scotland, Its Occurrence, Age, and Contribution to the Soils'. *Journal of Soil Science*, **14**: 33–43.

Fream, W. 1919. *Elements of Agriculture*, ed. J.R. Ainsworth-Davis. London: John Murray (10th edn).

Grime, J.P., J.G. Hodgson and R. Hunt. 1988. *Comparative Plant Ecology: A Functional Approach to Common British Species.* London: Unwin Hyman.

Jarman, R.J. 1996. 'Bere Barley: A Living Link with 8th Century?' *Plant Varieties and Seeds*, **9**: 191–6.

Jenkinson, D.S. 1988. 'Soil Organic Matter and its Dynamics'. In A. Wild (ed.), *Russell's Soil Conditions and Plant Growth,* 11th edn. (Harlow: Longman), pp. 564–607.

Lampkin, N. 1990. *Organic Farming.* Ipswich: Farming Press.

Liebig, J. 1843. *Chemistry in its Applications to Agriculture and Physiology*, edited from the manuscript of the author by Lyon Playfair, 3rd edn. London: Taylor and Walton.

MAFF (Ministry of Agriculture, Fisheries and Food). 1970. *Modern Farming and the Soil.* London: HMSO.

MAFF. 1979. *Fertiliser Recommendations.* London: HMSO.

MAFF. 2000. *Fertiliser Recommendations for Agricultural and Horticultural Crops* (RB209). Norwich: The Stationery Office.

McGregor, K.C., R.F. Cullum and C.K. Mutchler. 1999. 'Long Term Management Effects on Runoff, Erosion, and Crop Production'. *Transactions of the ASAE*, **42**: 99–105.

Nye, P.H. and D.J. Greenland. 1960. *The Soil under Shifting Cultivation,* Technical Communication 51. Farnham Royal: Commonwealth Agricultural Bureau.

Paton, T.R., P.B. Mitchell, D. Adamson and R.A. Buchanan. 1976. 'Speed of Podzolisation'. *Nature*, **260**: 601–2.

Peet R.K. 1992. 'Community Structure and Ecosystem Function'. In D.C. Glenn-Lewin, R.K. Peet and T.T. Veblen (eds), *Plant Succession* (London: Chapman and Hall), pp. 103–51.

Percival, John. 1974. *The Wheat Plant: A Monograph.* London: Duckworth.

Renfrew, C. 1987. *Archaeology and Language: The Puzzle of Indo-European Origins.* London: Cape.

Russell, E.J. 1915. *Soil Conditions and Plant Growth.* London: Longmans, Green and Co.

Stoltenburg, N.L. and J.L. White. 1953. 'Selective Loss of Plant Nutrients by Erosion'. *Proceedings of the American Society of Soil Science*, **17**: 406–10.

Thirsk, J. 1997. *Alternative Agriculture: A History From the Black Death to the Present Day.* Oxford: Oxford University Press,

USDA. 1999. *Soil Taxonomy: A Basic System of Soil Classification for Making and Interpreting Soil Surveys,* 2nd edn. Washington, DC: US Dept. of Agriculture, Natural Resources Conservation Service.

Wild, A. 1988. 'Plant Nutrients in Soil: Phosphate'. In: A. Wild (ed.), *Russell's Soil Conditions and Plant Growth* (Harlow: Longman), 11th edn, pp. 695–742.

White, R.E. 1997. 'Soil Formation'. In *Principles and Practice of Soil Science* (Oxford: Blackwell), 3rd edn.

3

Exploitation and Conservation of Soil in the 3000-Year Agricultural and Forestry History of South Asia

R.J. Wasson

INTRODUCTION

Themes and Sources

For most of the 3000 years for which a written record exists, exploitation of soil, vegetation, water and animals has dominated literate ideas about and relationships with the non-human world in South Asia. There is also a conservation ethic in this literature, focused largely on maintaining soil fertility, good water quality, and a mix of land covers that included forests. The major purpose of conservation was human wellbeing, but the largely spiritual literature calls upon people to look after the Earth as a gift of god.

The literature surveyed here is mostly concerned with soil in agriculture. Soil and land classification, land capability assessment, methods of preparing soil for planting, techniques for ploughing and soil fertilisation appear from the earliest times. While soil is treated as a discrete entity when the technicalities of agriculture are being considered, more usually soil is seen to be connected to landforms, erosion, flooding, sedimentation, vegetation, land use, provision of water, water quality, and human health. The world, it seems, was viewed as an interconnected whole.

Soil was also used for building houses, as a canvas for paintings in temples, as a way of providing essential salts for animals and people alike, and as a source of some minerals. Elsewhere there are some references to soil minerals and soil for health, but in the surveyed literature these other uses are not mentioned.

The earliest manuscripts were written in Sanskrit, and are mostly spiritual texts. Some are strongly secular with a focus on law and management of the state, and there are two Sanskrit treatises concerned solely with agriculture. In the seventeenth century, during the time of Mughal rule, a rich literature written in Persian focused on management of the rural economy largely for revenue. There is also a treatise devoted to agriculture. During the British period (c. 1760–1947), the relevant literature includes documents designed to steer the management of the rural economy, again largely for revenue. Scientific papers in the newly developed tradition of western science also appear, along with the observations of travellers from various European countries.

Almost all of the pre-British literature refers to what is now Bangladesh, northern India and northern Pakistan, with few accounts of central and southern India. For continuity, only those documents from the British period that refer to the same areas as those described in earlier documents have been used, with one exception.

The degree to which this literature is normative or reflects agricultural practice is not easy to determine. The Brahmanical tradition of stating what ought to be suggests that at least some of the Sanskrit literature is normative. But this literature also contains a lot of detail about methods for tilling soils, collecting water, land capability, and also relationships between soil, land, water and vegetation. It is difficult to interpret this as anything but a record of common observations and practice (cf. Elvin 1993 for China). Some of the literature provides methods for predicting rainfall, and a calendar for agricultural activities that is astrologically based. Given the significance of astrology in modern Hindu society, and its early beginnings, it is likely that this literature reflects practice. There are however many passages that are essentially spiritual, in which the farmer is urged to follow good ways.

From the thirteenth century onward a large proportion of South Asia's population was Muslim. The Muslim literature appears to reflect agricultural practice and also to describe commonly used methods of soil classification and land capability assessment. One exception is the only solely agricultural treatise of the seventeenth century, which contains much of a scholarly nature that may have been the musings of the writer rather than a reflection of widespread beliefs and practice.

The literature of European travellers and of the British East India Company and British government bureaucracy reflects practice and classification, suitably biased by the cultural lenses of the authors and their often commercial purposes. Even the authors of the scientific literature serve their employers, and this material must be read in the light of its purpose. Yet the British material is easier to interpret than much that came before, simply because it is closer to our modern experience and it is rich and abundant, providing many voices and views that can be cross-checked. The audience for this literature is also clear.

For the local and in the main pre-European literature, the audience is not clear. The earliest literature was almost certainly a codification of oral tradition. The Vedas were, and still are, recited for religious purposes well after they had been written down: a result of illiteracy and adherence to long tradition as illustrated by, for example, the family name Trivedi for people who could recite three Vedas.

Were the agricultural treatises written for bureaucrats, literate farmers, or for students? Were they committed to memory and recited by farmers? They may have occupied in South Asia the same place in society as the agricultural almanacs of Europe and North America, read by a few and disseminated by word of mouth.

The themes of this chapter are in summary: soils in agriculture, soil classification and land capability, soil degradation, and the overarching concepts of exploitation and conservation. The narrow range of conservation issues is also explored, and some insights provided by a comparison with the environmental history of China. Some attention is also paid to human responses to soil changes, mostly from the British period for which documentation exists. Finally, the 'modern' concept of a pre-British balanced agroecosystem is examined.

Before dealing with the written evidence, the chapter begins with a brief and necessarily inconclusive summary of the relationship between hunter-gatherers and soil. This leads into the evidence from early agriculture and the rise of urban centres during the Harappan period. While the Harappans were literate, their writing has not been deciphered. We must wait until the Vedas, starting around 1000 BC, for the first decipherable written accounts.

THE EVIDENCE

Hunter-Gatherers

The place of soil in the worldview of hunter-gatherer societies is only known from modern examples. In South Asia, hunter-gatherer peoples are now very rare, and take part in cultivation and the market economy to varying degrees. There is therefore almost nothing known of the place of soil in the pre-agricultural worldview of such people. We therefore turn to archaeological records, almost solely from Pakistan, northern India and Bangladesh (Figure 3.1).

The Harappans

The appearance in the Indus River valley of the Harappan Civilisation c. 3200 BC, marked by '…extensive, well-planned cities…' (Kenoyer 2003), had precursors in the beginnings of agriculture in Baluchistan (in what is now western Pakistan) c. 7000 BC (Possehl 1997). Shinde (2002) sees the south-west Asian agro-pastoral system of wheat, barley, cattle, sheep and goats reaching Baluchistan by c. 7000 BC. Small villages appear with stone-banked fields, mud brick houses, and rituals indicated by figurines and burial. Soil was clearly central to the agricultural life of these people, at a time when climate was highly variable but becoming wetter

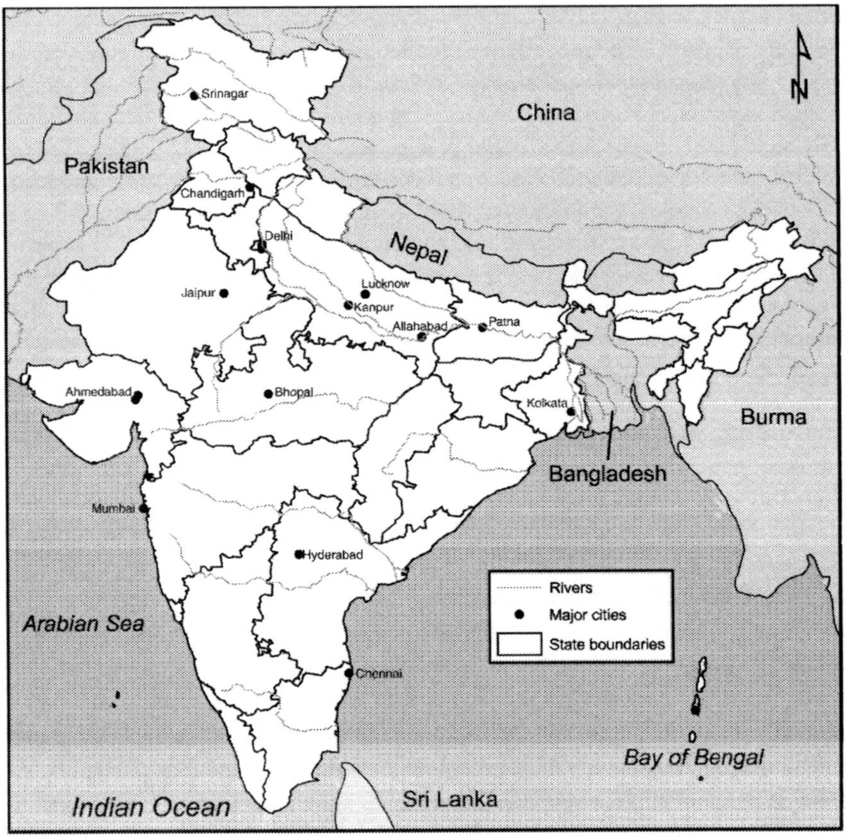

Figure 3.1. Map of South Asia

(Singh et al. 1990). But we can only infer a system of cultivation, moisture retention and erosion control afforded by the terraced fields, and possibly fallowing. There is nothing in the words of the people of their time.

The earliest villages appear along the Ravi River in the Indus valley c. 3300 BC, according to Kenoyer, (2003). These people herded cattle and cultivated wheat, barley, legumes and sesame and are sometimes described as representing the Early Harappan or Early Indus period of settlement (Possehl 2004). Between 2800 and 2600 BC the city of Harappa thrived economically and technologically. Exquisite luxury goods such as beads and glazed pottery were produced, along with fine figurines painted with what appear to be representations of woven fabric. Merchants used seals of clay, and writing appeared. The determination of value by weight developed, possibly for taxation or tribute.

Agriculture and grazing supported the economy of early Harappa, along the Ravi River floodplain, but trade was clearly an important source of wealth based on the high quality technological achievements of Harappan craftsmen. Granaries in the larger cities evidence a surplus of agricultural production. Brick buildings and the use of wood in construction show that forests were also being used, presumably in the vicinity of the cities. From 3200 to 1900 BC, cities scattered over the Indus Valley and to the east were nestled in a landscape of villages with cultivated fields and grazing. The cities appear to have been abandoned c. 1900 BC, an 'event' described by many archaeologists and historians either as 'collapse' or 'eclipse' of civilisation. Wheeler (1947) believed that a massacre had occurred at the great Harappan city of Mohenjodaro, and that the Harappan civilisation had collapsed as a result of invasion by the Aryans. He drew from the Rigveda, the oldest of the Vedas written in Sanskrit with a largely spiritual and moral philosophical content. Wheeler viewed the Rigveda as history, and saw in the destruction of ninety forts by Indra, the Aryan warrior as ***puramdava*** (fort destroyer), documentary support for an Aryan invasion.

Other explanations of the demise of the Harappan cities included climate change or damming of the Indus by a fault (Raikes 1965) that resulted in cities along the river being engulfed by mud. The latter is implausible (Wasson 1984). Singh (1971) found evidence of climatic drying at about the time of abandonment of the cities from pollen preserved in salt lakes in the Thar Desert. The science of climate change in the region has subsequently been strengthened by Singh et al. (1990) and Wasson et al. (1984). But the timing of the change has been questioned by a detailed study of the geochemistry of Lake Lunkaransar, also in the Thar (Enzel et al. 1999). Perhaps the most damaging problem for the argument that climate change led to the Harappan collapse is that the climate during the period of greatest vitality was not very different from that of today. Also, the cities were on rivers, and the agriculture benefited from annual floods which continued even in a drier climate because the rivers rise in the Himalayas.

Fairservis (1967) used estimates of the population size and calorific need of the city of Mohenjodaro to calculate the area required for cultivation, the quantities of fodder needed for cattle, and the amount of timber needed to fire the bricks for the city and to keep its cooking fires alight. He concludes that the Harappan city was in '... a precarious economic situation (which was) a significant reason for the downfall of one of the world's earliest civilisations'. Too many demands were being made on the soil and water resources of the area, and in addition the removal of gallery forest (the trees along the river margins) would have exacerbated flooding, according to Fairservis, and would have made agriculture difficult.

Leaving aside all of the discrepancies and implausibility of the various explanations, a more considered view of this civilisation has recently appeared. The idea that cities such as Harappa were suddenly abandoned is not tenable. The breakdown in maintenance of drains and city walls in the period 1900 to 1300 BC suggests that the ruling elites could not control the city (Kenoyer 2003). Trade also suffered, and so did the crafts. In the absence of decipherable records, the decline of the Harappans remains an open question.

The Ganges Valley and the Harappans

The growth of Harappan settlements in the Ganges River plain, specifically the Ganges-Yamuna Doab (interfluve), and in Gujarat may have resulted in a society that could no longer communicate across its full breadth. It had become many separate entities. Climate had got worse as well, and one of the major rivers of the Indus valley, the Ghaggar-Hakra, had shifted and then dried up, perhaps because of tectonically induced beheading. Maybe also the cities had placed stress on the natural resource base. But it is not possible either to disentangle one putative cause from the others, or to ascribe the change of Harappan society to one cause alone. The post-urban Harappan society was vibrant and growing, especially in the eastern part of Punjab, Haryana, northern Rajasthan and western Uttar Pradesh at the time of the so-called collapse of the civilisation (Possehl 1997).

The centre of Harappan settlement shifted progressively from Baluchistan and the Indus River valley south-east to Gujarat and east and north-east to the Ganges River valley over a period of over 5,600 years. During this period, settled agriculture developed alongside pastoralism and forest exploitation, then cities developed and were gradually abandoned. When the pattern of cultural change is seen in its entirety, exploitation and over-exploitation of natural resources can only be one factor in these changes.

Kenoyer (2003) claims that by 1300 to 1000 BC, 'a new social order characterised by a distinctive ideology and language, began to emerge in the northern Indus Valley and the Ganges River region to the east'. He draws upon the ancient literature of the Vedas, and the Mahabharata and Ramayana epics to describe the new social order as one of competing polities practising Vedic religion and speaking an Indo-Aryan language. This view has its origins in the idea of Aryans displacing non-Aryans.

The archaeological evidence shows that in the Ganges Valley wild rice was being used as early as 9000–8000 BC at sites near Allahabad (Agrawal 2002), but reliable evidence from pollen of settled agriculture for most of the valley is first dated by the radiocarbon method to about 6000 BP (Before Present, which is equivalent to about 5000 BC) at Sarai-Nahar-Rai (Gupta 1976). Vishnu-Mittre (1972) claims evidence of agriculture at Chirand in Bihar 5000–6000 BP (4000–5000 BC), but the chronologic basis of this claim is weak. The estimate of 5000 BC in the Ganges

Valley for settled agriculture can be compared with dates between c. 3500 and 2000 BP (2000 to 0 BC) in Southern India (Shinde 2002).

The evidence for a progressive shift of the Harappan culture from west to east and then the continuing evolution of agriculture in the Ganges Valley is clear. Urbanisation reappears in South Asia in the Ganges Valley c. 500 BC (Agrawal 2002; Erdosy 1998) some 4500 years after settled agriculture developed in this region. The mix of cultivation and pastoralism is also clear from the archaeological record during the entire period of 8,500 years, as seen earlier. This land use pattern is also reflected in the Vedas.

The Vedas and Epics

The Vedas are part of the classical literature of India, which during the so-called Vedic Period also included the Brahmanas and Upanishads. These texts are essentially religious and their dates are uncertain. They are believed to be written versions of oral traditions that possibly lasted for centuries. Basham (1971) observes in frustration that the Vedic hymns '…tell us little of the great events of the time'.

They do however provide insights into agriculture, and soil and water resources of northern India. In the Rigveda, Taittiriya Samhita, Khadira Grihya Sutra, Gobhila Grihya Sutra, Kaushitaki Brahmana, the Artharvaveda Samhita, Shatapatha Brahmana, Apastamba Grihya Sutra, the White Yajurveda, and the Yajurvediya Maitrayani Samhita there are numerous but brief references that allow a sketch of agricultural practices (Saxena 1976; Nene and Sadhale 1997).

Vedic agriculture was developed as land was cleared by cutting or by burning. Irrigation channels (***surmi***) and wells (***kupa*** or ***avata***) were used, water being lifted by wooden buckets (***ksha***) with the help of pulleys made of stone (***ashmachakra***) to be poured into the irrigation system. Floodplain agriculture was also practised, after the summer flood period. Multiple cropping occurred each year of rice and barley, and a fallow system also existed, supplemented by manuring. Possibly crops of barley, rice, beans and sesame were grown in one year in wetter places. The main operations were ploughing (***krishantah***), sowing (***vapantah***), reaping (***lunanta***) and harvesting (***mrinantah***). Small and large wooden ploughs (***langgala*** or ***sira***) were used along with sickles (***datra***), rollers (***matya***), winnowing baskets (***sthivi***) and spades (***khanitra***). ***Sira*** were drawn by six, eight, twelve and sometimes twenty-four oxen.

Land was classified as waste (***ushara***), especially when saline, pastureland (***gocharalvrajalghoshtha***) and cultivated (***karsha***). A distinction in the last class was between furrowed (***sitya***) and ploughed (***halya***) (Thapar 1984). Cows were grazed in forest and barley fields, and the value of clean pond water for cattle was recognised.

Ancient Indian texts also contain many comments about water and landforms (Tripathi 1990). The Rigveda recognises velocity differences among rivers, and a simple classification of topography (mountains, clefts, stony places, habitable places, lakes, badlands). This is mirrored in the Artharvaveda, along with the idea that rivers rising in snow-capped mountains will be perennial. The epic Ramayana contains (or repeats) a simple landform classification, noting in addition sandy riverbanks and the dense and hard nature of the land watered by the Ganges; presumably a reference to clays on floodplains some distance from the river channel. The erodibility of sandy riverbanks is noted along with the erosive action on mountains of intense rain. Rivers that move by breaking their banks (avulsion in modern terms) are distinguished from those that overflow their banks in the Ashtadhyayi, the major work of the great Sanskrit grammarian Panini (Vasu 1962). Perennial rivers are also identified as those that deposit fertile soil.

The Vishnu Purana, apparently contemporaneous with the epics (Basham 1971), contains a classification of soil: black, white or yellowish, blue or red, yellow, gravelly, hilly or bouldery, and golden. In the Brihat Samhita soil colour is said to be an indicator of water quality: '... that pebbly and sandy soil containing copper makes water astringent. Brown-coloured soil gives rise to alkaline water, pale white soil (makes) salt water and blue-coloured soil makes water pure and sweet' (Tripathi 1990).

Differences in soil and rock type with depth are recognised in the Brihat Samhita, particularly yellow soil overlaying rock in which a large amount of water will be found. The rock is said to occur at about one half a ***purusha***, that is, about one metre. Termite mounds are also viewed as an indicator of groundwater, especially in dry regions. The rock might be Kankar (calcium carbonate nodules) or else the location might be on the fringes of the Gangetic Plain where rock is at or near the surface.

The link between water quality, vegetation cover, agriculture and the differences in soil is described in the Bhava Prakasha, but with little that is explicit. Poor water quality is caused by decaying vegetation, stagnation and lack of penetration by sunlight. The health impacts of different kinds of water arising in catchments with different types of vegetation are also recorded in the Bhava Prakasha, apparently derived from several medical (Ayurvedic) texts.

This array of observations and advice contains classifications of the natural world, for human purposes, and some of the linkages between landforms, soils, vegetation, water flow and quality, river behaviour and agriculture. They indicate a deep and abiding interest in agriculture and soil.

When were these texts written and to what period and area do they relate? Almost all are believed to be the codification of long oral traditions, and many have borrowed from earlier writings. The Rigveda is believed to have been codified c. 1000 BC (Basham 1971; Possehl 1997) but that is uncertain, as is the date for all of these texts. The Bhava Prakasha is clearly derivative and could be as late as the sixteenth century AD (Tripathi 1990). The Artharvaveda is dated at about 800 BC, and the Vishnu Purana to the first century AD (Tharpar 1984). The Grihya Sutras probably date from the second half of the first millennium BC (Thapar 1984), the White Yajurveda to about 800 BC, the Brahmanas to AD 1300–1400, the Ramayana to 400–500 BC (Fosse 1997), and the Ashtadhyayi to the end of the fourth century BC (Basham 1971). These texts therefore span about 2600 years.

It is generally accepted that the Rigveda is the oldest text at about 1000 BC, although this can never be known with certainty (Leach 1990). Nonetheless, analysis of its vocabulary and content relative to other texts supports the view that it is the oldest, even if the date is uncertain (Fosse 1997). Agriculture at this time took account of soil type, as would be expected, and ideas had developed about rivers and their floodplains. Half a millennium later the Ramayana records ideas about catchment processes that affect water quality.

The earliest literature refers to both deserts and wetter country, names locations that some scholars believe can be identified, and provides some evidence of an eastward migration of people. These pieces of evidence have been woven together with archaeological finds to reconstruct a migration of agricultural people along the Ganges Valley into the forests and swamps of Uttar Pradesh and Bihar (Thapar 1984; Sharma 1996). The internal consistency of the argument supports the idea that the texts were written after the de-urbanisation of the Harappan civilisation, although it is possible that some of the early literature draws upon the Harappan oral tradition. It seems highly unlikely that any of the early literature describes Aryan people who overran the Harappans.

A major social change occurred in the first millennium BC: states appeared in the Ganges Valley. The change was the result of many factors (Thapar 1984). The migration of people along the Ganges Valley provided readily available new land, presumably because the people in the forest were few in number and unable to resist the migrants. Therefore, conflict did not arise because the Vedic people were not crowded, nor was there rapid demographic increase. The Vedic people were not strongly socially stratified, and external conflict was limited, hence no necessity for states existed. The rise of rice agriculture and irrigation probably had some effect on social stratification, because landowners differed from labourers. Internal tension caused by this stratification may have played a part in state formation. When the head of a household (the ***grihapati***) was transformed into a landowner and participant in trade, and the ***shudra*** into a peasant cultivator and artisan, stratification and tension increased. ***Shudras***, the lowest of the four classes (priest, ***brahmin***; warrior, ***kshatriya***; peasant, ***vaishya***; and ***shudra***) is a labourer who is somehow subservient in the Rigveda (Thapar 1984).

Intensive use of the soil for agriculture that was partly dependent upon irrigation in the first millennium BC contributed to social stratification embodied in the ***varna*** system of classes. Growing wealth and the development of cities in the second half of the first millennium BC required defence. And so the states were born, to reduce conflict within a socially stratified system and to defend cities.

Sharma (1996) adds an element to this account of state and *varna* formation: iron. With the development of agriculture in the Ganges Valley, iron tools became more common. Iron appears in the mid-Ganges plain c. 800 BC, and in the eastern plain 1200–600 BC. Sharma believes that iron was first used in war and hunting, and only from c. 600 BC was iron used by farmers and artisans. The clay soils of Magadha (south of the Ganges, east of the Son River, and west of the Hazar Tuki River, bounded to the south by the Vindhyan Hills and Chotanagpur Plateau) and the plains north of the Ganges in Bihar and eastern Uttar Pradesh were too hard for wooden ploughs, it is claimed. The *Manusmriti*, a compilation of Hindu law composed in its final form in the second or third century AD (Basham 1971) advises against the use of iron-tipped ploughs because they injure the Earth and its creatures. This appears to be reaction to a technology that was more intrusive than others, and may reflect a forest dweller's view of nature.

Iron allowed the cultivation of larger areas, especially for wet rice, and therefore greater food production. The greater complexity of society, and particularly the *varna* system, could thereby be maintained. The postulated increasingly intensive use of the soil is therefore seen as a contributor to state and *varna* formation.

Makkan Lal (1986) counters Sharma's argument from a detailed study of Kanpur District on the southern bank of the Ganges. Lal focused on the argument that iron tools were needed to clear the 'thick monsoon forest' of the mid-Ganges Valley to allow cultivation. Lal sees no change in iron technology from the Painted Grey Ware period (ending c. 500 BC) to the Northern Black Polished Ware (NBPW) period (500–160 BC) when agriculture spread, cities arose and states formed. During the NBPW period, most settlements were small with fewer than 500 people. Many were along river banks where soils are coarse–textured and easy to plough, and others were near lakes where forest was often absent. Most agriculture could therefore be carried out without forest clearing. Lal concludes that urbanisation, and by inference state and *varna* formation, was the result of social and economic forces that generated an economic surplus. This idea is an echo of the traditional view of the rise of cities in the Near East, and has none of the sophistication of Thapar's explanation involving many factors that produced social stratification, tension, defence and the need for a state. As with the decline of Harappan cities, the reasons for rise of states in the Gangetic Plain remains a mystery.

The Bhagavadgita and Albiruni

Albiruni (Abu-Rihan Muhammad) travelled to India with Mahmud of Ghazni early in the eleventh century AD, the first deep penetration into South Asia by Turks (Sharma 1983). Albiruni was a gifted man who turned his attention to many fields of study. By AD 1030, he had completed an 'Enquiry into India' (*Tahkik-i-Hind*), which is an investigation into Hinduism. He paid particular attention to the Bhagavadgita, the greatest devotional book of Hinduism and part of the Mahabharata. But it appears that his translation and transcription of the Gita is not faithful to the original, and includes many of his own ideas and ruminations (Sharma 1983).

There is a particularly fascinating section in the *Tahkik-i-Hind* that deserves notice, particularly given current interest in the concept of sustainability:

> The life of the world depends upon sowing and procreating. Both processes increase in the course of time and this increase is unlimited, whilst the world is limited. When a class of plants or animals does not increase any more in its structure, and its peculiar kind is established as a species of its own, when each individual of it does not simply come into existence and perish, but besides procreates a being like itself or several together, and not only once but several times, then this will as a single species of plants or animals occupy the earth and spread itself and its kind over as much territory as it can find.
>
> The agriculturalist selects his corn, letting grow as much as he requires, and tearing out the remainder. The forester leaves those branches, which he perceives to be excellent whilst he cuts away all others. The bees kill those of their kind who only eat, but do not work in their beehive.
>
> Nature proceeds in a similar way; however, it does not distinguish for its action under all circumstance one and the same. It allows the leaves and fruits of the trees to perish, thus preventing them from realising that result which they are intended to produce in the economy of nature. It removes them so as to make room for others.
>
> If thus the earth is ruined, or is near to be ruined, by having too many inhabitants its ruler – for it has a ruler, and his all-embracing care is apparent in every single particle of it – sends it a messenger for the purpose of reducing the too great number and of cutting away all that is evil.

This translation by Sachau (quoted by Sharma 1983) contains the idea of unbounded human increase, and the self-limiting feedbacks of natural ecosystems; very 'modern' concepts. It introduces the 'ruler', the deity, who sends a messenger to put matters right if ruin is suspected. The concepts are largely Islamic, according to Sharma, but the messenger is Vasudeva, a Hindu invention. Sharma also observes that in the Artharvaveda, the Earth is described as 'Not overcrowded by the crowd of Manu's sons'. Manu are the children of the many gods (*deva*) created by Brahma in the great creation myth of Hinduism (Prime 2002). And the *Manusmriti* is a list of rules for right behaviour, as we saw earlier.

The idea that the natural increase of people should not dominate the Earth, and that ploughing with iron-tips is destructive, is probably a forest-dweller's view of the world. As cultivators encroached on forests, destroying plants and animals, and ripping open the soil (note that the word *krishi*, or agriculture, also has the meaning 'to tear, to cause pain, to torture'; Sadhale 1999), oral traditions took up an ecological view that resonate with many modern environmentalist groups that see the forests as sacred. These oral traditions, codified in the ***Manusmriti***, along with other ancient texts, continue to energise modern-day environmental activitists (Prime 2002).

These proto-ecological ideas had no discernible impact on either the Hindu or Islamic (Mughal) rulers who came later, as we will see.

The Arthashastra and Krishi-Parashara

Early in the history of Indian state formation, the ***Arthashastra*** (a 'Treatise on Polity'), was written reputedly by Kautilya (also called Canakya or Visnugupta) a minister at the court of Chandragupta Maurya (Basham 1971). The treatise gives detailed instructions on the operation and control of the state, the organisation of the economy, and the conduct of war. According to Kangle (1972), ***artha*** is the sustenance of people's livelihood (***vritti***); that is, the ***Arthashastra*** is the science by which material wellbeing is secured. It is regarded as one of the three goals of human existence, the others being ***dharma*** (duty) and ***kama*** (desire).

The date of the ***Arthashastra*** is doubtful, as is the name of its author. It may have been compiled at the Mauryan court, perhaps at Pataliputra (modern Patna in Bihar) on the banks of the Ganges. It is certainly pre-Guptan according to Basham (1971), which places it somewhere between 300 BC and AD 300.

Soils appear in the ***Arthashastra*** as part of the state economy. The state is sustained by revenue from agriculture (***krishi***), cattle-rearing (***pashupalya***) and trade (***vanijya***). The state receives grain, cattle, money, labourers and other products. The state can control its people and enemies by using revenues to maintain an army. A state cannot exist without territory or country (***janapada***). The ideal country should be easy to defend, and should contain agricultural land, mines, forests, pastures and trade-routes, and should be populated by hard-working agriculturalists and men of the lower ***varnas***. The emphasis is on livelihood and exploitation, and when selecting locations for new settlements a ruler is advised to seek abundant water, and land that yields crops without much rain and is suitable for grain crops.

Settlement of unoccupied land (***sunyanivesa***) is an important activity of the state, providing evidence that there was plenty of unoccupied land at this time. Slaves and labourers in state service prepare the new territory for cultivation.

The ***Arthashastra*** also includes a basic land classification. Land unsuited for cultivation is said to have a weakness (***bhumicchidra***) and should be used for pasture or forest, the latter divided between use for forest products (including base metals) and elephants. The idea of soil fertility is included along with a basic topographic subdivision (Tripathi 1990). On the royal farms, at least, soil fertility was maintained by adding cow manure.

The text notes that irrigation works (***setubandha***) are an important activity of the state. Two kinds are recognised, one based on tanks and wells (***sahodaka setu***) and one on reservoirs (***aharyodaka setu***). A well-organised irrigation system at the time of Chandragupta Maurya is attested to by a Greek visitor to Pataliputra, Megasthenes (Basham 1971). Megasthenes' writings have not survived, but were used by later Greek and Latin authors.

Megasthenes travelled along the Ganges Valley, and was impressed by the extent of irrigation, which enabled two crops to be harvested each year. He was also impressed by the rich soil of India, perhaps by comparison with the drylands further west.

The account in the *Arthashastra* of agriculture, rivers and soil emphasises the production of food and the sustenance of the state; that is, exploitation of natural resources. There is little detail of agricultural technique, but there is much said about supervision and administration of agriculture and taxation. Megasthenes' account simply confirms the *Arthashastra*'s description of irrigation, and makes a 300 BC date for Kautilya's treatise plausible if not definite.

At about the same time as the *Arthashastra*, the *Krishi* by Parashara was written (Sadhale 1999). The Vedic distinction between cultivated and uncultivated land remains in the *Krishi*. Cultivated land is classified according to whether it is river-fed or rain-fed, the latter lying higher than the former and able to produce only one crop each year. The higher ground, known as ***bagara bhumi***, is dry and cultivation is not possible without irrigation. The river-fed land, ***nadimatrika***, was highly fertile and produced up to three crops a year. This land was excellent for paddy, which another and earlier Chinese traveller, Xuen Zang, wrote about very favourably between AD 629 and 645.

The low-lying land was also known as ***kacchabhumi*** or ***kaccha*** meaning riverside land or land contiguous to the water. This area consists of fresh alluvium, only available for agriculture during the dry season when cucurbits in particular are currently grown with the aid of cow manure. The ***kachabhumi*** also grows good pasture.

The limitations of soil properties on cultivation were well recognised at this time. This appears particularly to be the case with forest areas where fields were scarce and ploughing was difficult in 'hard, iron-black soil' (Choudhary 1971, pp. 114–15). It might be supposed that by the twelfth century forests remained on land that was not good for cultivation.

Slash-and-burn agriculture, or ***jhum***, was also described in the *Krishi*. The spade was the means of breaking the soil in the limited area of cultivation in forests. Elsewhere a plough was used, apparently the same as that used in many villages in northern India today. Soil was ploughed as many as five times before sowing, and the remaining clods broken by harrow.

Fertilisation follows the prescription of the *Arthashastra*, and the *Krishi* gives detailed instructions for the use of cow manure. Paddy in particular needed the application of dried, powdered manure.

Paddy was extensive in Bengal, Assam and north Bihar, with wells, ponds and tanks supplying irrigation water. Sugar-cane was also widely grown in Bengal, Bihar, Uttar Pradesh and further west. There are also records of wheat and barley in Bengal.

Widespread paddy, irrigation systems and instructions in the *Krishi* combine to suggest an intensive agricultural system in which soil preparation was exacting.

Unfortunately the identity and purpose of Parashara, the text's author, are unknown, although the first passage says that the *Krishi* is '... for the benefit of farmers' (Sadhale 1999). It is clear from the text that the author was an authority on astrology, cattle husbandry and meteorology, as well as agriculture. Sadhale claims that the *Krishi* is like a farmer's almanac, containing information arranged by season and month. But how many farmers could read the *Krishi*? The language is simple, possibly enabling its memorisation and recitation. It is remarkable in one particular aspect; it is solely concerned with rain-fed agriculture, suggesting that it drew on experience in the better-watered regions, such as Bengal. But Nene (in Sadhale 1999) suggests that it refers to northern Pakistan, on the basis of the crops mentioned in the text.

Of the 243 verses in the *Krishi*, 117 have an astrological component (Nene in Sadhale 1999). Sixty-nine relate to prediction of rainfall. Also, some 27 verses deal with cattle, showing the importance of pastoralism in the rural economies. Ploughing by cattle is to be strictly controlled by astrology, as are many other aspects of agriculture. Manuring by dried cattle dung is clearly described, and crops will yield little without it. Other topics covered include water harvesting and water retention by bunding, weeding, plant protection against disease, threshing, construction of ploughs, and many festivals and ceremonies that should accompany parts of the agricultural cycle.

Early Medieval Northern India

Some 700 years after Megasthenes' stay at the Mauryan court, Fa Xian, a Chinese monk who visited India in search of authentic copies of Buddhist scriptures, reported a gentler and more humane society than that recorded in the Vedas (Basham 1971). Buddhism and Jainism had helped in this process, and Hinduism had replaced the old sacrificial Brahmanism. Fa Xian reported that most respectable people were vegetarian, and that only low castes and untouchables ate meat. This shift in dietary pressures must have had some effect on the agricultural system, presumably by reducing the size of cattle holdings and areas devoted to pasture. We might imagine that larger areas of soil were put under the plough.

This was the time of the Gupta Empire, a time of peace and creativity. Around AD 450 a marauding group from Central Asia arrived on the scene and the Gupta Empire was gone by AD 550. The political unity of the Guptas did not return. A '... drab story of endemic warfare between rival dynasties' (Basham 1971, p. 71) followed, until the beginnings of incursions by Arabs and Turks starting in AD 712. Between AD 1001 and 1027, Mahmud raided deep into northern India as seen earlier. He annexed Sind and Punjab. The remainder of northern India was independent, but divided between three chief kings who were at war: Prthvijara Cahamana, Jayaccandra Gahadavala and Paramardideva Candella. After one failed attempt in 1191, the Muslim invaders from Afghanistan overcame resistance as far east as Bengal by 1203. Nepal, Assam, Kashmir and Orissa maintained their independence, cut off from the Ganges Plain by natural boundaries. Until the eighteenth century, Muslim rulers dominated northern India.

During this troubled time, the *Kashyapiyakrisisukti* was written. This is an agricultural treatise, written mostly between AD 800 and 700 although there may be later interpolations (Wojtilla 1985). The author, Kashyapa, was probably one of a group of Brahmins of the Vaikhanasa school, followers of Lord Vishnu. Most of the writings of the Vaikhanasa school were of a socio-religious kind and their works were designed as a code of conduct, intended for other Brahmins, and for government and village officials (Ayachit 2002).

The *Kashyapiyakrisisukti* is a detailed account of methods of cultivation of a wide range of crops, both irrigated and unirrigated. While highly practical, the text is set within a larger purpose. The production of grains and other vegetation is said to be '...the sole purpose for (the) highest fulfilment of the Earth'. It stresses support for agriculture by the ruler, particularly identifying suitable land, building reservoirs and planting trees on their banks, constructing canals, making seed available, donating land to 'weaker' people, arranging markets, standardising weights and measures, and afforestation. This list reminds us of the *Arthashastra*.

Once again the concepts of land capability appear. The importance of good land is emphasised because it '... yields good results to everyone, confers good health on the entire family, and causes growth of money, cattle, and grain'. The ruler should appoint knowledgeable people (interestingly from any caste) to assess land for cropping. Good land should be: devoid of stones and bones; a pliant (plastic) clay, very unctuous (greasy) with reddish and black hue, and glossy with water; neither too deep nor too shallow; conducive to speedy seedling emergence; easy to plough and cultivate; water absorbent, and replete with beneficial organisms such as earthworms; and devoid of thorns and cow dung, thickly set and compact, and heavy when it was lifted. This account reveals a combination of a physical and biological concept of soil.

Verses I 51 to I 54 (in the enumeration of Ayachit 2003) provide methods for examining soil, some of which would be familiar to a modern soil scientist. At a point, a hole is dug and the effects of the digging observed repeatedly. The characteristics of the soil of note are its colour and uniformity of colour, the '... drops of water ...' (probably its turbidity), taste, fluidity, and its stickiness.

Kashyapiyakrisisukti charges the ruler with many responsibilities. Among them are planting of trees on reservoir banks, establishment of gardens, and propagation of forestland on the edge of existing forests '... or on the mountain slopes ...' (II 141). Again we see the Brahmanical tradition of stating what ought to be, rather than explaining why such afforestation is required. But the reference to afforestation of mountain slopes is unique in the pre-British period. The text mentions many places in northern India, parts of central India (Madhya Pradesh), Nepal, the Indus Valley in modern Pakistan, and the deltas of the Godavari and Krishna rivers in Andhra Pradesh. Within this region there are many mountain slopes which may have needed repair, or could simply have been places more useful for forest and forest products than the plains which were mainly used for cropping.

Although it is clear that Kashyapa had knowledge of a large part of North India, Wojtilla (1985) believes that the author's main experience was in South India in the deltas of the Krishna and Godavari rivers. This is supported by the inference of assured rainfall, more than one rice crop each year, the breed of cattle described, and the description of good soils for cultivation as clayey, red and black with good density, among other evidence (Nene in Ayachit 2002).

Finally, Kashyapa provides detailed accounts of manuring using cow dung, goat dung, compost and tendrils of creepers (as green manure). Nene (in Ayachit 2002) notes that there is no mention of dead animals as manure, a reflection of the author's Brahmin tradition. Also, in this text and all others by Hindus there is no mention of the use of night soil as a fertiliser, by contrast with treatises from Europe, China, and the recommendations of a Muslim agricultural text (*Nushka Dar Fanni-Falahat*) from the seventeenth century.

The Mughal Period

From AD 1556–1719 the Mughal Empire exercised control of almost all of modern India, Pakistan, Bangladesh, Afghanistan and a small part of Tibet (Richards 1993). This was an empire of exquisite sensibilities, expressed in art, music and poetry. It was also immensely wealthy. The Mughals developed a bureaucracy that administered an innovative land revenue system, coinage and military organisation. The organisation of functions at the centre was duplicated in the provinces.

Soils and Revenue

For our purposes, the elite's role in land revenue collection is most important. They received salary assignments (*jagirs*) from land revenue, and the entire empire depended upon these revenues. The centre received revenues that were not used for essential provincial expenses. Under the Afghan ruler Sher Shah Sur, who became the ruler of northern India in 1540 and died in 1545, the state extracted cash from the peasants calculated from expected crop yields. Careful cadastral survey of land and crops was the basis of this system. Sher Shah believed that the state should encourage agricultural productivity, and cultivators who cleared forest and woodlands received exemptions or reduced taxes. Those who either increased the cultivable area or sunk wells received similar relief, so there was an incentive for increased exploitation. During Akbar's reign (1556–1605), taxes were in the form of produce, taking into account variations between regions in yields.

The land and soil classification used by the Mughals was superimposed onto classifications that have their roots in the Vedas. In north Gorakhpur District, due north of Varanasi on the Gangetic Plain, the East India Company employee Francis Buchanan Hamilton described the 'native' classification of land and soil, and land uses, in the eighteenth century (summarised by Bhargava 1999) (Table 3.1).

Table 3.1. Soil and Land Classification in Gorakhpur district

Soil Types	
Balua	White sands near rivers, low productivity but can be productive with levelling and further alluvial and organic deposition as in the *diara*
Dhoosi	Sandy, sterile soil with a reddish tinge, of low productivity
Dorus	Soft, ash coloured, absorbent, clay-rich fertile soil with abundant earthworms
Muttear	Loam of light colour, cracks in the hot season
Land and Soil Types	
Banjar	Cultivable wastes or never cultivated, requiring little tillage
Bangar	Dark brown soil like *dorus* of high productivity, on high ground secure from inundation, fallowed after 3 years
Kuchara bangar	Inundated *bangar*
Khadir bangar	Irregular surface, retains moisture
Humwar bangar	Level surface
Bhat	Low-lying, clay-rich, good moisture retention, needs fallowing more often than *bangar*
Chowur bhat	Low, marshy land, inferior productivity, fallowed every 3 to 4 seasons
Chowrear bhat	Higher than *chowur bhat*, and of higher productivity

Source: Hamilton summarised in Bhargava 1999

The soil types are straightforward, showing a clear local correlation between soil and crops. *Dorus* was used for everything except paddy, unless it occurred where water collects. *Muttear* was used for all crops, while *balua* was used only for wheat, millet and arhar. ***Khadir bangar*** was used for barley, lentils, rice and peas, while ***humwar bangar*** was used for wheat and sugar-cane. ***Chowrear bhat*** was useful for sugar-cane and corn.

Many of the District's farmers were highly mobile, cropping ***banjar*** for up to 3 years before the fertility was reduced and they moved on. This pre-fallow period was also the time when the lowest rents were charged by landowners, and so there was an incentive for migration. In the most northerly part of the area, this migratory agricultural system was not tamed by the Mughals, and only changed as a result of the British Settlement (see Hill 1997, for Purnea District).

In the sedentary agricultural areas, soils and land use were similar to an infield-outfield system (Bhargava 1999). Land within 400 or 500 m of a village, called ***gongyer*** or ***goind***, was well manured and irrigated annually, producing two crops a year. Beyond ***goind*** was ***madhyam***, ***bich*** or ***meeano***. This less managed land produced only one crop a year. The most remote and lowest lands, or the outlying fields, were called ***pallo***, which received neither manure nor irrigation. Winter rice and pulses were grown in ***pallo***. Generally ***muttear*** occurred in ***goind***, ***dorus*** in ***madhyam*** or ***meeano***, and ***balua*** occurred in ***pallo*** near rivers. While the pre-Mughal land and soil classification matched land use to land type (and therefore capability), the land classification used by the Mughals was based on production. The capability of the land to produce was not explicitly surveyed, simply the outputs. The detailed records upon which revenues were based come to us in their most complete form in the ***Ain-i-Akbari*** compiled by Abul Fazl and completed in 1598 (Blochmann, 1872). Other documents from the late seventeenth and eighteenth centuries provide further revenue data and other insights (Habib 1982; Moosvi 1987).

The ***Ain*** used a threeway classification of land: tilled (***kishta***) or cultivated (***mazru'a***); and waste (***uftada***) in the sub-categories of cultivable waste and uncultivable waste (Trivedi 1998). The crop yields are given for land under continuous or near-continuous cultivation (***polaj*** and ***parauti*** respectively), and for each crop three classes of yield are recorded: high (***gazida***, ***a'la***), middle (***miyana***) and low (***zubun***). It is not clear on what basis these three classes are distinguished (Moosvi 1987).

Trivedi (1998) calculated the area of forest at the time of the ***Ain*** for the ***subas*** of Agra, Awadh, Delhi and Lahore (see Habib 1982 for locations). Near-natural vegetation, in many cases forest, covered over 50 per cent of the provinces of Agra and Delhi, 64 per cent in Awadh and 72 per cent in Lahore. The area of cultivation ranged from 14 per cent to 28 per cent in these ***subas***, and the area of uncultivable waste ranged from 47 per cent in Agra to 89 per cent in Allahabad. A reasonable fraction of the land designated uncultivable waste probably included soils recognised at the time as being uncultivable or barren. There was therefore a concept of land capability at this time.

Moosvi (1987) makes the important point that uncultivable land is not a rigid concept. Fertility and capability for cultivation is only absolutely limiting on rock or where salinity is extreme. As population increases, inferior land is cultivated, assisted perhaps by irrigation. The area of cultivable waste therefore should increase through time, which Moosvi demonstrates by comparison of the *Ain* data with those from the British figures of 1909–10.

The social structure of power, seen by Elvin (1993) as the most important factor in Chinese environmental history, was also a major control in Mughal times. The *jagirdars* were almost completely controlled by the Emperor, including the methods of assessment and collection of land revenue (Habib 1963). The creation and collection of surplus created Mughal wealth, and the demand for revenue increased with time. Vast wealth and grinding poverty sat side by side as a result, in the words of the Dutchman Pelsaert (quoted by Habib 1963).

Rotation of *jagir* every 3 to 4 years meant that long-term planning by *jagirdars* was impossible, and their interests were best served by extracting maximum revenue as fast as possible. The Frenchman Bernier observed (also quoted by Habib) of *jagirdar* behaviour: 'Let me draw from the soil all the money we can though the peasant should starve or abscond and we should leave it, when commanded to quit, a dreary wilderness.'

The testimony of Pelsaert and Bernier is supported by other European observers of the seventeenth century, and by local writers. So great were the depredations of at least some of the *jagirdars* that whole villages were deserted, land lay uncultivated because the oxen had been sold to pay *jagir*, and one observer reported that many areas '… have become forests infested by tigers and lions' (Habib 1963, p. 325). During this period there was some extension of cultivation into the *tarai* (the swampy area on the Gangetic Plain near the Himalayas) and the delta in Bengal. This increase of cultivation may have offset the area of abandonment, and so the cultivated area in Aurengzeb's time exceeded that in the *Ain* despite abandonment. The relentless Mughal quest for revenue inspired soil abuse that, in turn, required that new soils be brought into use.

While the original sources need to be further consulted, we can infer that soils over large areas in the seventeenth century were not cultivated or manured, and some were re-vegetated presumably by thorny scrub of the kind that can now be found on poorly managed land. The migration of peasants, to escape unreasonable revenue demands, led to the farming of new areas. A peasant moving to a new area might even obtain assistance to cultivate, a practice eventually outlawed (Habib 1963, p. 329).

Some areas of uncultivated soil therefore were cultivated not as a direct result of Mughal concessions but as a result of oppression. Areas previously cultivated were abandoned, sometimes to be re-occupied. But the extent of areas involved and the impacts on soils are not explicitly known (Raychaudhuri and Habib 1982).

A Mughal Agricultural Text: Nushka Dar Fanni-Falahat

The son of the Emperor Shahjahan, Dara Shikoh, wrote *Nushka Dar Fanni-Falahat* (the Art of Agriculture), transcribed in 1693 (Akbar 2000). Dara Shikoh was a scholar who was as concerned with understanding Hinduism and its similarities with Islam as he was interested in agriculture. Although little is known about him, his status and scholarly inclinations suggest that his insights into agriculture came from government officials rather than directly from farmers.

Most of the text is devoted to methods of cultivating and propagating a large number of plants and collecting seed. The suitability of soils for particular plants is clearly set out (see Table 3.2). The descriptions of suitable soils combine ideas of soil, land and land capability classification. Nene points out that the use of soil for vines where they have previously grown suggests that the microflora was appropriate to subdue root pathogens in the new vines. The same may also apply to carrots. Melons are resistant to salinity, but most soils will do. The planting of palm trees in saline soil reflects their tolerance or maybe they need salts, according to Nene. Dara Shikoh provided instructions for planting different crops. Pits dug for olive trees should be left open for a year or burnt, presumably to kill pathogens. Burnt cow bones and dung are also recommended as fertilisers for trees. Apart from dung, and bones, salt and nitre (saltpetre), vine sap, eggs, olive leaf-sap, pig dung, human urine, night soil and sheep blood are all mentioned. As a Muslim, Dara Shikoh could recommend the use of animal remains, night soil and human urine, in contrast to the Hindu texts.

The range of plants discussed in the text, and some of the agricultural practices suggest that Dara Shikoh was not solely reliant upon South Asian experience. Information from West Asia and the Mediterranean appears to have been included, although this may indicate the use of plant species and agricultural methods in South Asia derived from further west. Mughal India was in constant touch with Iran and the Ottoman realm, so crops or information – or both – could easily have come from afar.

Saltpetre

Soil was not only the basis of agriculture in Mughal India, but also supplied another product: saltpetre. This is a naturally occurring chemical compound of potassium and nitrate (KNO_3), although sometimes the name saltpetre is applied to sodium nitrate ($NaNO_3$). Indian saltpetre is mainly the potassium compound (Sarkar 1975).

The *Ain* records saltpetre 'mines' along the Ganges and to its north scattered about on the alluvial plains of rivers in Bihar (Habib 1982). It also occurred south of the Ganges. Other major sources were north of Agra on the Doab, and north of Ahmedabad in Gujarat.

Table 3.2. Suitability of soil for different plants

Plant	Description
Palm Tree	saline soil
Vines	'clean and pure' where vines have been grown before
Fig	'strong and humid but away from water'
Olive	'low-lying red soil which is not 'humid'
Pomegranate	'dry and gritty soil'
Peach	'wet soil'; 'on the banks of canals or near them'
Damson	'stray and fertile soil'
Mulberry	prepare soil well, mix in sand
Walnut	'soil should be very fertile'
Sissoo	'grown in saline as well as in non-saline soils. In saline soils it will be white and in non-saline soils it will be black in colour'
Oranges	'wet and fertile soil, prepared properly with manure'
Wheat	'clayey soil, of black or yellow colour'; soil prepared well by several ploughings by four pair of bullocks
Melon	moist, clean and saline soil; 'if the soil is bad, a mixture of manure, sand and sugarcane leaves will do'
Carrot	'if old soil is mixed in carrot's soils....carrots will become long'

Source: Nene in Akbar 2000

Saltpetre was used by the Mughal elite as the oxidising component of gunpowder, for cooling water, and as fertiliser (Akbar 2000; Sarkar 1975). When the British and Dutch began to trade in the Ganges Valley in the early seventeenth century, saltpetre was one of the goods of interest to them. The Dutchman Pelsaert and the Englishman Peter Mundy have left records of the source and method of production of saltpetre. It was prepared from three types of earth: black, yellow and white. But the best saltpetre was said to come from black earth, being free of salt or brackishness (Sarkar 1975).

The entire focus of the Mughals was revenue and exploitation, so soils were considered only as a contributor to the yield of different kinds of land. The judgement of barren, particularly saline, lands as unproductive is also a reflection of the focus on revenue. Saltpetre production from soil had a commercial and military purpose.

Sri Lanka

A brief account of agricultural development in the northern dry zone lowlands of Sri Lanka is now warranted, because here we see the same features encountered in northern India. The dry zone is 'the cradle of Sinhalese Civilisation', where the development of tank irrigation touched almost every river valley (Ismail 1995). From the fifth century BC to about the first century AD, settlers spread up the main river valleys such as the Malwatlu-oya and the Mahaweli Ganga. Tanks allowed the population to occupy all of the dry zone by the end of this period, coincident with the spread of Buddhism.

Even more tank building occurred between the second and the fifth century AD during a period free from foreign invasion. The first of the great tank-building Kings, Vasabha (AD 65–109), appears in the contemporary source the ***Mahavamsa*** (the Chronicles, Geiger 1934). The author(s) of the Chronicles attribute to Vasabha an objective for his tank and canal constriction: to 'make the land more fruitful'. These words are a gentler version of the motivations of the Mughals and the author of the ***Arthashastra***, but there is a common purpose.

But there appears to have been a further purpose, that of developing the ability to repulse foreign invaders particularly from what is now India. In the ***Culavamsa***, the later part of the ***Mahavamsa***, we learn of the King Sena II (AD 851–885) that 'From this time onward he made the Island hard to subdue by the foe and made it increase in wealth like land of the Uttarakurus' (Geiger 1953). The Uttarakurus were a mythical people who lived in a utopian land (Ismail 1995). So the ability to marshal armies, fund their activities and feed them, depended upon a strong agricultural economy that in the dry zone required extensive irrigation. All of this in the presence of Buddhism.

The British and Post-Independence Periods

The eighteenth century in India saw the demise of the Mughals and the rise of foreign interests, especially British. Under the emperor Aurengzeb, the Mughal Empire had united the greater part of India. But repeated wars, struggles for the throne, decline of the bureaucracy, incursions by Hindu powers, the rise of the independent states of Oudh, Orissa and Bengal, and the intervention of several European commercial interests led to the disintegration of Mughal administration (Sarkar 1975). Out of the ruins grew a new political system, first dominated by the British East India Company and then in the nineteenth century by the British Government.

The Rise of Science

From the early seventeenth century until the twentieth, it is the accounts of foreigners that tell us about the soils of South Asia. The local voice is only heard again as the local inhabitants are trained in Western science, for this is the period when this worldview arose and soil science, geology and geomorphology developed in Western Europe.

James Hutton, the son of a respected Edinburgh merchant, had his *Theory of the Earth* published in 1795. The idea that soil was formed by the rotting of rock first appeared in this work. This was a substantive challenge to the prevailing view that the Earth was created for man, and so the idea that it could rot, and erode, was untenable to the faithful. The so-called 'denudation dilemma' was solved by Hutton by supposing that erosion and soil formation were in balance, so erosion did not remove all soil. The habitability of the Earth was thereby maintained (Davies 1968).

Soil science had its origins in agricultural studies and in earth science and chemistry. In the late nineteenth century, Russian scientists established many of the principles by which soils are still explained. The idea of soil forming processes, and soil zonalism in which soil is formed over time by the action of climate and organisms on parent material (rocks and sediments) conditioned by relief were firmly established in the USA by 1938 (Paton et al. 1995).

Prior to the first scientific accounts in the early to mid-nineteenth century, the only written accounts from South Asia in which soil plays a role are those of European travellers, missionaries and East India Company men. Early travel accounts say next to nothing about soils (Wessels 1924; Archer 1980) One very informative account does however exist from the eighteenth century. A keen observer, mapmaker and Surveyor-General of Bengal was the Englishman James Rennell (1742–1839). In a paper of 1781, Rennell provided a thoroughly modern account of the behaviour of sediment in meandering and straight rivers. His observations of the Ganges and Burrampooter (Brahmaputra) rivers are mostly morphological and sedimentological. He notes how the rivers are '... enriching of adjacent lands ...' (p 89); presumably a reference to fresh sedimentation on floodplains. He recounts how a meandering river removes sediment from its outer near-vertical bank and deposits sediment on inner shelving banks. He describes how banks more than 30 feet (~9m) high in the dry season are swept away during floods, especially where the soil is loose, causing a change in the course of the river. Rennell lived in Bengal for eleven years, during which time he observed changes to both the Jellinghy River and the Ganges. The latter shifted one bank the breadth of an English mile and a half (~2.4 km) in nine years. This is an astonishing rate of bank erosion of ~270 m/yr, although Rennell notes that this is an extreme rate and the more usual is between ~130 and 160 m/yr. Rennell goes on to give an account of water flow in a river bend, and of the origin of channel bifurcation. The meandering habit of the Ganges is attributed to the looseness of the soil.

In his account of the delta of the Ganges-Brahmaputra, Rennell describes the turbidity plume in the Bay of Bengal, produced by fine-grained sediments in suspension in water, which extended a distance of twenty leagues (roughly 100 km) into the sea. The deposition of mud and organic matter and sand produces banks that are just below the surface of the sea, and one day will rise above the sea to be cultivated. Rennell's account also includes a plausible physical explanation of river channel islands in the Ganges, which after only a few years accumulate enough 'mould' to be cultivable; that is, deposited mud and organic matter that provides a natural fertiliser for annual crops. This system of opportunistic agriculture is recorded in the Vedas, and in the ***Krishi-Parashara***, and continues to the present in the flooded or ***diara*** lands along most rivers of northern India. Today, cow manure is used as fertiliser, and the people who cultivate this rudimentary soil have little else to depend upon, unlike ***diara*** cultivators of earlier times who probably had additional land on floodplains and elsewhere, given that the pressure on land was lower.

Rennell, writing at the same time as James Hutton, was formulating his geologic and geomorphic ideas, which combined keen observation and an understanding of fluid mechanics and sedimentology to account for the behaviour of the Ganges River in its lower reaches. Stratification of sediment exposed in eroded river banks and in the delta is accounted for by appealing to differences in the 'respective gravities' of sand and finer particles producing strata of different texture depending upon the stage of flood. He provides an estimate of the suspended sediment concentration in the Ganges at its height, wherein a glass of water contains 'one part in four of mud'. The concentration would have been less than 25 per cent of course, because the mud contains water. Finally, in describing the hydrology of the Ganges, Rennell draws upon the observations by Count Buffon to explain levees on the Ganges as the result of 'precipitation of mud' as the water 'purifies itself as it flows over the plain'.

Rennell's paper, communicated to the Royal Society of London by the great botanist Joseph Banks, must rank as one of the earliest scientific accounts of a river and its sediments and soils anywhere in the world.

Other writers were more interested in the political and economic aspects of India. Francis Buchanan Hamilton was one of the most travelled and productive writers who worked for the East India Company (Allen 2002). His journey into Nepal in 1802–03, from Bihar to Kathmandu, resulted in an account of the landscape, people, agriculture and politics of the mountain kingdom (Hamilton, 1819). He described the rocks, using whatever knowledge he had of the young science of geology, and listed the rock types found in the stream gravels. He described forests, cultivated areas, and the oddities of soil such as a 'singular black ferruginous earth' on the banks of the Kosi River that was eaten by elephants and used by the local people for ink. But we learn little else, except that cultivation is restricted to valley floors where soils are fertile, water is available, and gradient workable.

Buchanan, as he was known, also wrote extensively of Bihar as it was in 1811–12. But once again the usual accounts are given of soil fertility variations, crops and irrigation systems without the scientific analysis that was developing in the ranks of British residents in India. But Buchanan's purpose was to obtain information relevant to the Company, and this was mostly political and commercial.

About 1820, Major General Sir Alexander Walker, originally an employee of the East India Company, authored a detailed account of agriculture in various parts of India (Walker 1997). Soils appear in this account in a number of ways. Attempts to introduce English ploughs and other agricultural implements to the Malabar area failed because the Mahrattah (Maratha) people found them to be too heavy, thereby fatiguing their oxen. The soils were too dense for these ploughs, and Walker celebrates the different kinds of ploughs, both drill and common, that Indian cultivators use for different sorts of seeds and soils. The English generally blamed the Mahrattahs' prejudices, sloth and obstinacy for the failure. But Walker saw the matter differently. The Indians already had a workable and adaptable system for cultivation. They did not need English implements.

Walker described the 'native' soil classification of Malabar, noting that soil fertility without water is of little value, especially in a land where for half the year the soil is hard and cohesive. In Malabar, soil is classified into three sorts:

- *Pasheemah Koor* – the highest quality soil consisting of rich clay. The quality of the soil is discovered by digging a hole a yard (~1 m) deep and wide. If the soil is the right kind, then not all of the excavated soil will be needed to fill the hole.

- *Rashee Pasheemah Koor* – this is of medium quality and will exactly fill the hole. This soil is not as adhesive as that of the highest quality, the word *rashee* meaning a mixture of earth and sand; and today it would probably be called loam.

- *Rashee Koor* – the poorest soil that will not fill the hole, consisting of loose sand.

Walker was astounded to discover that this method of determining quality should accord with methods and ideas recommended by Europeans such as Lord Kairns and Sir H. Davy, despite his otherwise favourable views of Indian agriculturists. The same technique was used in the Roman Mediterranean (See Chapter 7)

There is one further reference by Walker to soil fertility from Bengal. Quoting a Mr Colebrooke, Walker observes that the extraordinary fertility of the soil of Bengal is a disincentive to anything but subsistence agriculture because of the ease of producing crops. By contrast, in Scotland nature provided little benefit to the farmer and as a result some of the finest agricultural improvements have occurred. Walker does not tell us if such improvements have occurred in those parts of India where nature has not been as kind as in Bengal.

When problems appeared after British interventions in the 'native' irrigation system, at a time when Western science and methods of observation were developing, a new relationship with soil was born.

In the nineteenth century, the Western Yamuna Canal system was built in what is now Haryana and Punjab States between 1819 and 1886 to boost production and alleviate the poverty of the local people through irrigation (Islam 1997). The British were more concerned with the former. Salt efflorescence (***reh***) appeared in the lands near the canal, causing hardship for the local people. This triggered investigations, inquiries and attempted solutions that have continued to today (Whitcombe 1972).

By the early twentieth century, after many decades of canal construction by the British, it was not only salinity (***reh***) that was of concern. Willcocks (1930) called for a total reappraisal of the effects of river embankments and canal systems on agricultural productivity, soil quality and malaria. Anticipating concerns raised in the 1970s, Willcocks argued that embankments 'congested' rivers and canals, and embankments stopped the natural replenishment of alluvial soils by flood-borne sediment. By trapping water on floodplains much longer than occurred naturally they also increased the incidence of malaria.

From the Himalayas to the Sea

Almost all of the discussion thus far has concerned the Indo-Gangetic Plain. The soils and agricultural systems described are developed in alluvium of various ages and elevations above the modern rivers. References in the ancient literature to places that may have been in the foothills of the Himalayas or the Vindhyan Hills are extremely rare. The writers were mostly concerned with the most productive lands, which are not in the mountains. This emphasis continued into medieval times and the Mughal Period, and of course the British had a major focus on the Plains because of the greater revenues that could be achieved from irrigation and the difficulty of political control in the mountains.

There was however another focus in British and later in Indian minds concerned with soil. This focus is part of what has become known as the 'Himalayan Dilemma' in an influential work by Ives and Messerli (1989). In 1975 Eckholm set out a catastrophic view of Himalayan degradation of forest cover and soils that led to widespread and deadly flooding downstream on the Indo-Gangetic Plain. Flooding was seen as the direct consequence of deforestation and soil erosion that led to more rapid runoff and sedimentation of rivers and canals. The idea that river embankments, themselves designed to limit flood damage, could exacerbate flooding by trapping water on floodplains, as argued by Willcocks in 1930, did not significantly enter the debate until the 1970s. That much of the soil erosion in the Himalayas is natural, from landslides triggered by earthquakes and by glacial outwash, was highlighted by Ives and Messerli: 'At the macro-scale, however, natural processes will reduce the role of human interventions to insignificant proportions' (1989, p. 145). Yet many elements of the debate began much earlier, in the 1860s.

The volatile mix of scientific forestry, peasant needs and uprisings, the British Government's desire for revenue, uncertain and distorted science, and the convenience of blaming floods and sedimentation in the Ganges Valley on peasants in the mountains, produced a debate of high importance and major uncertainty (Thompson et al. 1986).

In the mid-nineteenth century, the Forest Departments of Punjab and Himachal Pradesh developed what Saberwal (1999) describes as an 'alarmist discourse on environmental degradation' over one and a half centuries, partly in response to threats to its bureaucratic power from the Revenue Department and partly in response to criticism of its forest and soil management. Saberwal argues that the Forest Department was forced into alarmist rhetoric to legitimise its authority. Recently better forest management and reforestation has been called for by both national and international bodies to alleviate the suffering of millions of people downstream, based largely upon this alarmist rhetoric.

Uncertainty and distortion of natural science has played a major role in this discourse from the 1860s, involving: a hydrological view of forest soils as a giant sponge that has a dramatic effect upon the runoff rate and therefore river flow; the idea that forests also dramatically impede runoff and soil erosion across the entire Himalaya; the notion that minor forest use by peasants had a huge effect on soil erosion and runoff; the use of small-scale measurements and observations from which to construct explanations and solutions at the mountain-range scale; the belief that increasing forest cover will reduce sedimentation and flooding dramatically; ignorance of world hydrological literature that demonstrates the relatively small effect of forest cover on runoff in terrains like that of the Himalaya; and an almost complete disinterest in natural erosion, and methods for determining its rate relative to that induced by land use.

Saberwal demonstrates convincingly that science was used in an exercise of power. The dismissal of the 'non-scientific' views of local people allowed contrary opinions to be discarded. Bureaucratic interests helped to shape both the environmental rhetoric and science. The spatial patchiness and heterogeneity of scientific investigations evoke concerns about the appropriateness of the scientific enterprise, allowing for science to be both wrong and political.

The debate continues, even while major reforestation occurs. India still blames Nepal for floods, but does nothing about the embankments on its rivers. And science is only now wrestling with data collection and analysis at appropriate spatial and temporal scales. Meanwhile, the Chipko ('tree huggers') environmental and social movement has had major impact at the village level, without substantive engagement with mainstream science to find mutually acceptable understandings and solutions. Non-Government Organisations have used whatever science and opinion exists to push their case for reforestation, and government inquiries continue.

Recent scientific research (Wasson et al., in prep.), based on a whole-of-catchment approach in the Upper Ganges has shown that the huge sediment transport event of 1970 (which triggered much of the Chipko activism) originated in an area of deforestation of the Southern Himalayan Front at the boundary between the Lower and Higher Himalayas. About 70 per cent of landslides in this area came from deforested hill-slopes. The highest percentage of Lower Himalayan sediment in river sediments deposited over the last 6300 years along the Alaknanda River, a major tributary of the Ganges, occurred in the 1970 deposit. And this occurred just after the most significant period of deforestation in the 1950s and 1960s. It can therefore be concluded that deforestation has increased erosion in this part of the Himalayas, but there is as yet no evidence that the rivers of the Gangetic Plain have aggraded as a result, thereby exacerbating flooding.

Even more convincing results come from Sikkim where Starkel and Basu (2000) have shown that landsliding, channel aggradation and flooding both within the mountains and in the mountain piedmont were all increased by deforestation. As in the Upper Ganges, major erosion of deforested hill-slopes occurred during isolated extreme rainfalls.

Indian Soil Science

The scientific study of soil by employees of Indian Forest departments were not in the main local empirical investigations. Rather, they drew upon results from the USA and Europe to tell a story consistent with their intended management and bureaucratic objectives. In parallel, empirically driven soil science was developing in India in support of agriculture.

The history of Indian agriculture during the British period focused on exports (Kanwar 1997). Emphasis was given to jute, cotton, spices, sugar-cane, tobacco, tea and coffee, not to food crops. The 1943 Bengal famine and food shortages during the Second World War launched an intensification strategy to grow more food. There was little science in this strategy, a condition that continued well after independence in 1947. Despite a Royal Commission in the late 1920s which found that the fertility of India's soils could only support subsistence agriculture and emphasised the need for research, little of substance happened until the late 1950s. Kanwar even claims that the soil fertility and manuring experiments of the 1940s set back Indian soil science because they were poorly designed and incorrectly interpreted.

Analysis of the relationship between nutrient availability, fertiliser needs and crop yields on soils all over India laid the foundations for the modern era of soil science in the service of agriculture. Yet the relationships between plant physiology and soil properties were not explored, and still have not been in systematic ways. Science has developed in a disconnected way. It was only in the mid-1960s that a need for integrated research involving soil physics, chemistry, microbiology, pedology, land use planning, soil conservation, soil management and soil testing was perceived by land managers. The first nationwide survey of soil degradation was published in 1994 by Sehgal and Abrol (eg. Velayuthum et al. 2002).

The separation between the science of soil for agriculture and for understanding soil erosion and sedimentation remains, although foresters working in the Himalayas and a handful of scientists concerned with catchment management have come closest to making the links.

A Balanced Agroecosystem

Gadgil (2001) and Prime (2002) among others have pictured the pre-British village-based agroecosystem as one in balance. Prime quotes extensively from Banwari, editor of *Jansata*, a Hindi daily newspaper published in Delhi, to support his case. Banwari also published a book on forest culture (1992), in which he depicts the Hindu idea of the world as a forest, drawing on the Ramayana and Mahabharata epics that are rich in descriptions of the forests of the first millennium BC. The intrusion of agriculture broke up the natural forest cover, so that gradually a three-way classification of forest emerged: ***tapovan***, where sages seek truth; ***shrivan*** which provides prosperity; and ***mahavan***, or the great natural forest where all species find shelter. Once again we find resonances with the three pillars of sustainability; economy, society (dominantly spirituality) and environment. The preservation of these forest types makes a village a full entity, married to the idea of the ***panchavati***, symbolising earth, water, fire, air and ether (Banwari 1992).

Gadgil, one of India's most insightful modern ecologists, provides less detail than Prime and Banwari, and essentially no scientific evidence for the postulated balanced agroecosystems. Examples of such scientific evidence comes from the Himalayas in the hands of P.S. Ramakrishnan (pers.comm) and K.G. Saxena (pers. comm), but nothing is known from the plains perhaps because these pre-modern agroecosystems no longer exist to be studied. Bhargava (1999) sees no evidence for equilibrium in the pre-British village agroecosystem. Peasants could see no end to the natural vegetation, and were not natural conservators of the environment, according to Bhargava. And the Mughal agricultural system scarcely favoured balanced agroecosystems.

The descriptions of pre-British agroecosystems of the Gangetic Plain, and the accounts by old villagers that this author has been privileged to hear, certainly indicate a much greater forest cover, local water harvesting systems of high water quality, a nutrient recycling system that maintained soil fertility and limited problems of salinity and sodicity prior to the advance of canal irrigation, waterlogging and salinisation.

In the absence of extant examples of these agroecosystems we must turn to reconstructions based upon pollen, pond sedimentation and archaeology. Unfortunately, only a little has been done on the Gangetic Plain along these lines, but a beginning has been made.

In the area between Lucknow and Kanpur, in Uttar Pradesh, the interfluves (land between rivers) are studded with ponds, which, in most cases, are fragments of former channels. The ponds are underlain by river sand that has been dated by luminescence techniques to between 13,000 and 6000 BC. When the channels were abandoned, ponds began to form around 6000–4000 BC first as clearwater ponds rich in gastropods, then as mud-floored features with varying quantities of water (Srivastava et al. 2003; Agarwal et al. 1992; Sharma et al. 2001; Gupta 1976; Chauhan et al. 1991)

The accumulated muds and fine sands in the ponds record soil erosion in the surrounding landscape. At Misa Tal, near Gosainganj in Lucknow District, luminescence dating has been used to estimate sedimentation rates through time (Wasson et al., in prep). The area that drains into the pond is only 7.26 km^2 in area, and consists of low-relief and low-gradient agricultural land with little or no forest. The soils are silty loams with some clay enrichment in the subsoil (B horizon) and, in places, calcareous subsoils that have formed within the last 10,000 years. The Misa Tal catchment lies within the area of most of the geopolitical developments since agriculture began nearby 6,000 years ago. The site was part of the Mauryan Empire, the Gupta Empire, the Mughal Empire and the British Empire. We might therefore expect a gradually increasing rate of sedimentation, reflecting soil erosion as agriculture spread and intensified.

Surprisingly, this is not what is found. While there is a low rate of sedimentation from about 5000 BC, after that date, at the time of the ***Arthashastra***, the rate rises to a peak of about 90–120 t/km^2/yr at the same time as the climate becomes wetter. The rate then falls slightly but then appears to rise again during the last 1000 years. Modern rates vary between 940 and 1410 t/km^2/yr.

Not all of the soil eroded from the catchment of Misa Tal stays in the lake, because of loss through drains to the Gomti River. If we conservatively estimate the trap efficiency of the pond at 50 per cent, the modern rate is between 8 and 13 times higher than the rate during the last 1500 years.

The villagers of the Misa Tal catchment removed groves of trees and increased the area of cultivation in the post-1950 push for food self-sufficiency. In that time they have watched the pond fill with sediment, and the amount of water standing in the pond each year get smaller. The modern erosion rate appears to have increased as the land has been even more heavily exploited.

This record does not support Gadgil and Prime. But many more sites need to be investigated to test this conclusion.

Change and Responses

We can imagine but cannot know the detailed nature of the responses by people to changes to soils in ancient India. Manuring was almost certainly a response to declining fertility, but nothing is recorded about the time taken to respond and the failed experiments before successful methods were found. Similarly, *diara* cultivation must have been a trial and error method, but nothing is recorded. It was only in the British period that records exist to allow some analysis of changes and responses.

Saberwal (1999) has summarised the literature beginning in the 1860s on the relationship between forests, climate, runoff, floods, erosion and sedimentation both in India and elsewhere, as already discussed. Based on the opinions and conclusions contained in an extensive bibliography of papers published in the journal *Indian Forestry*, Saberwal summarised the prevailing ideas which are here presented diagrammatically (Figure 3.2). This conceptualisation is similar to the Theory of Himalayan Environmental Degradation (Ives and Messerli 1989, Fisher 1990; Thompson et al. 1986).

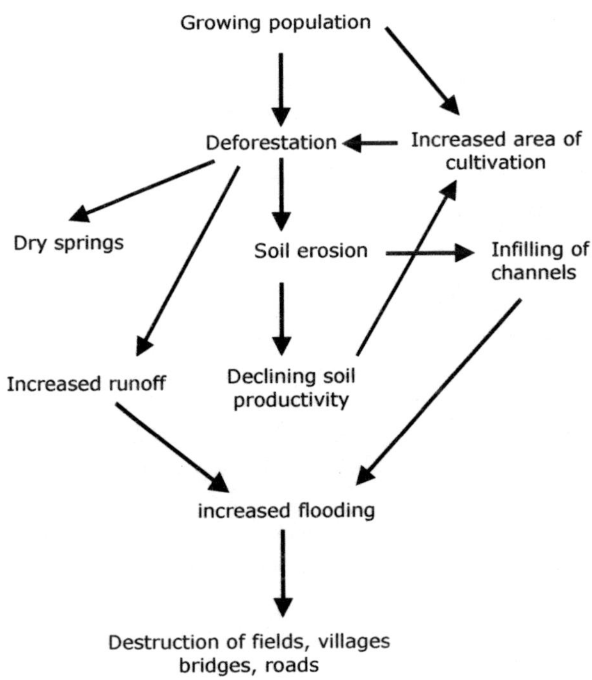

Figure 3.2. The theory of Himalayan environmental degradation

There is little evidence of massive recent deforestation in the Himalayas, but there is evidence of recent small-scale deforestation and depasturing of domestic stock in forests against the will of various State Forest Departments. There is however evidence of steady deforestation over the last two centuries, especially since the introduction of commercial forestry by the British and its acceleration between the 1950s and 1990s, after which it largely came to an end.

The myths about the link between deforestation and floods centre on the ability of forest soils to absorb a large fraction of monsoon rainfall and therefore act as a buffer to runoff and stream-flow. There is little evidence in support of any soil being able to buffer substantial rainfalls, especially when saturated during the monsoon. There is however evidence that springs dry up after deforestation, and return upon reforestation.

Modest increases in the area of cultivation have occurred in the Himalaya. Agricultural terraces were seen as the villain, but in many cases may reduce erosion rather than increase it. Because of terracing and manuring, and the small increase in cultivated land, there is little evidence that soil productivity has declined substantially, and so the feedback to a further increase in the area of cultivation is suspect.

Increased flooding is believed to be the result of both increased runoff following deforestation and sedimentation of rivers which reduces their capacity. Yet there is little evidence of increased flooding as measured by volumes of water flow each monsoon, and there is a poor link between deforestation and increased runoff. There is however evidence that the area flooded has increased over the last few decades (Centre for Science and Environment 1991), a change that could be the result of embankments.

There is little but anecdotal evidence of sedimentation of rivers, and even if this was happening the link to human-induced soil erosion in the Himalayas is extremely weak. Wasson (2003) has constructed a sediment budget for the Ganges-Brahmaputra system, but concluded that existing measurements of soil erosion and sediment transport in rivers cannot be used to identify with any certainty the role of land use separate from natural processes. Basin-wide estimates of erosion rates are needed, and the only obvious way to derive them is to use geochemical tracers. Initial results reported in Wasson (2003) suggest that the High Himalaya is the main source of sediment, where human impacts are slight. More recent results show that significant quantities of sediment also come from the Lesser Himalaya, where human agency is significant. From this, the role of land use can be inferred and then quantified. The current policies, which emphasise reforestation and have their origins in the battles for bureaucratic supremacy in the late nineteenth century, may need to be re-assessed.

Exploitation and Conservation of Soil in South Asia

SUMMARY

For most of the 3000 years of documentary records of northern India, soils have been seen as part of the production system. The focus has been on simple classification of soils for cropping and pastoralism, and for revenue calculation. At times the links between soil and water quality, sedimentation and river change have been noted but these links were secondary to the view of soil as a sustaining resource for agriculture. At no time prior to the modern era does the role of soil in ecosystems receive much attention, and it only appears during the British period as part of a forest production system or as a buffer to both runoff and flooding. Environmental problems such as *reh* and soil impoverishment receive little attention in the surviving sources, except as a brake on crop yield, until the post-Independence period.

The idea of a pre-British equilibrium agroecosystem on the plains of northern India is part of the nationalist reaction to the colonial legacy of the British. The lost agro-ecosystems are items of reverence to be recreated. The slight available evidence suggests that, if there was an equilibrium, it existed for 2000 years from well before the Mughals to nearly the present day. As gauged by erosion rate estimates, it has been the post-Independence drive for food-self-sufficiency that has dramatically upset the putative equilibrium, not the pressure of British rule on agriculture.

Conservation for most of the period was focused on soil fertility and forest resources. This is not surprising, given that most of the literature describes conditions on the plains where these issues, along with water supply, were of paramount importance. The geographically wide-ranging *Kashyapiyakrisisukti* mentions a role for the ruler to reafforest mountain slopes, the only pre-British reference to this idea. While it is not clear from the brief account by Kashyapa why such afforestation is desirable, it is possible that his experience (or accounts from others) of the mountainous lands adjacent to the Indo-Gangetic Plains, and in central India, where hills rise abruptly from plains, was like that of China. Elvin (1993) documents many cases in eastern China, where hills rise abruptly from plains, of deforestation leading to massive erosion and flooding. The need to protect adjacent areas of cultivation appears to have driven a call for conservation of the hill-slopes. If the literature of South Asia were not dominated by the plains, similar reflections might have occurred prior to the British period when interest in the commercial opportunities of the Himalayas, in particular, drew their attention to the erosion and flooding hazards of over-clearing.

While it has been shown that much of what passed for science in the debate over Himalayan forestry was driven by the imperatives of bureaucratic power, nevertheless many of the British were genuinely concerned about the impact on the cultivated plains of land use in the mountains. This concern, and the debate, continues to the present. Here, at least, we get a more complete view of the response to soil changes, and of the many factors that contribute to the response. In the earlier literature, the response of cultivators to the impost of the Mughal revenue collectors is visible, with apparently large-scale migration as the major response and recolonisation of once cultivated soils by native vegetation and animals.

Manuring is the most pervasive response to soil fertility decline, with every text describing techniques and sources of manure. That manuring is an early response is clear from its appearance in the Vedas.

Exploitation of soil is deep in Indian culture and society, with the *Kashyapiyakrisisukti* of some 1300 years ago setting the tone. The earth is fulfilled to the highest degree by the production of grains and other vegetation for human use. It is therefore no surprise that India is struggling to include conservation issues in the thinking and actions of its rural population, that are broader than those that directly sustain life in agroecosystems.

This chapter has depended upon existing translations of works in Sanskrit and Persian for the earlier literature. Most of the translators have been concerned with social, economic and religious issues, and a re-reading of these works in the original languages may draw out fresh insights into natural phenomena, including soil, and the human relationship to them. This might best be done by expert linguists working with soil scientists, geomorphologists, ecologists and historians. Field surveys of current agricultural practice, enlightened by such a re-reading, may allow comparisons with ancient practices and views of soil, showing where traditional views still exist and where they have been supplanted.

The major new area of research is of the kind begun at Misa Tal. The sedimentary record in lakes, rivers, tanks and reservoirs is a rich archive containing information that can be compared with the documentary record. Calibration of the pollen record in modern agroecosystems, such as has been pioneered by European scientists, would permit a more quantitative reconstruction of land cover. From this the state and treatment of soils could be inferred. Here is another multi-disciplinary research field, involving palaeoecologists, earth scientists, agricultural scientists, historians and linguists.

Such new research could not only aid our understanding of how people have related to soil, and how this relationship has and has not evolved, but also how the practices and attitudes of the past inform current opinion, and how present practice might be improved.

ACKNOWLEDGEMENTS

Dr. K Proust and Dr. Y.L. Nene for critical comments, McComas Taylor for checking the orthography, and Margo Davies, Melinda Wall and Joanne Pinter for assistance with the manuscript. Merrilyn Wasson for teaching me so much about India.

REFERENCES

Agarwal, A.K., M.H. Rizvi, I.B. Singh, A. Kumar and S. Chandra. 1992. 'Carbonate Deposits in Ganga Plain'. In I.B. Singh (ed.), *Gangetic Plain: Terra Incognita* (Geology Dept., Lucknow University), pp. 35–44.

Agrawal, D. P. 2002. 'The Earliest Agriculture and Pottery in South Asia'. In Y. Yasuda (ed.), *The Origins of Pottery and Agriculture* (New Delhi: Roli Books).

Akbar, R. (trans.) 2000. 'Nuskha Dar Fanni-Falahat (The Art of Agriculture)'. *Agri-History Bulletin,* **3** (Secunderbad, India: Asian Agri-History Foundation).

Allen, C. 2002. *The Buddha and the Sahibs*. London: John Murray.

Archer, M. 1980. *Early Views of India: The Picturesque Journeys of Thomas and William Daniell, 1786–1794: The Complete Aquatints*. London: Thames and Hudson.

Ayachit, S.M. (trans.) 2002. 'Kashyapiyakrishisukti (A Treatise on Agriculture by Kashyapa)'. *Agri-History Bulletin,* **4** (Secunderbad, India: Asian Agri-History Foundation).

Banwari. 1992. *Pancavati: Indian Approach to Environment*. Delhi: Shri Vinayaka Publications.

Basham, A.L. 1971. *The Wonder That Was India: A Survey of the Culture of the Indian Sub-Continent before the Coming of the Muslims* (includes information on Sanskrit literature in 'Language and Literature'). New York: Hawthorne.

Bhargava, M. 1999. *State, Society and Ecology: Gorakhpur in Transition 1750–1830*. New Delhi: Manohar.

Blochmann, H. (ed.) 1872. *A'in-i Akbari*. Calcutta, 2 vols. Osnabruck, Biblio Verlag, 1985.

Buchanan, F. 1936. *An Account of the Districts of Bihar and Patna in 1811–1812*. Patna: Bihar and Orissa Research Society.

Chauhan, M.S., A. Khandelwal, S.K. Bera and H.P. Gupta. 1991. 'Palynology of Kathauta Tal, Chinhat, Lucknow'. *Geophytology,* **21**: 191–4.

Centre for Science and Environment. 1991. *Floods, Flood Plains and Environmental Myths*. New Delhi: Centre for Science and Environment (*State of India's Environment: A Citizens' Report*, no 3).

Choudhary, A.K. 1971. *Early Medieval Village in North-Eastern India (AD 600–1200)*. Calcutta: Punthi Pustak, Patna University.

Davies, G.L. 1968. *The Earth in Decay*. London: MacDonald Technical & Scientific.

Eckholm, E.P. 1975. 'The Deterioration of Mountain Environments'. *Science,* **189**: 764–70.

Elvin, M. (1993). 'Three Thousand Years of Unsustainable Growth: China's Environment from Archaic Times to the Present'. *East Asia History,* **6**: 7–46.

Erdosy, G. 1998. 'The Prelude to Urbanization: Ethnicity and the Rise of Late Vedic Chiefdoms'. In F.R. Allchin (ed.), *The Archaeology of Early Historic South Asia: The Emergence of Cities and States* (Cambridge: Cambridge University Press), pp. 75–98.

Enzel, Y., L.L. Ely, S. Mishra, R. Ramesh, R. Amit, B. Lazar, S.N. Rajaguru, V.R. Baker and A. Sandler. 1999. 'High-Resolution Holocene Environmental Changes in the Thar Desert, Northwestern India'. *Science,* **284**: 125–8.

Fairservis, W.A. 1967. 'The Origin, Character and Decline of an Early Civilization'. *Novistates,* **2302**: 1–48.

Fisher, R.J. 1990. 'The Himalayan Dilemma: Finding the Human Face'. *Pacific Viewpoint,* **31**(1): 69–76.

Fosse, L.M. 1997. *The Crux of Chronology in Sanskrit Literature: Statistics and Indology, a Study of Method,* Acta Humaniora 21. Oslo: Scandinavian University Press.

Gadgil, M. 2001. *Ecological Journeys. The Science and Politics of Conservation in India.* New Delhi: Permanent Black.

Geiger, W. (trans.) 1934. *Mahavamsa or the Great Chronicle.* London.

Geiger, W. (trans.) 1953. *Culavamsa, being the more recent part of the Mahavamsa.* Colombo

Gupta, H.P. 1976. 'Holocene Palynology from Meander Lake in the Ganga Valley, District Pratapgarh, U.P.' *Palaeobotanist,* **25**: 109–19.

Habib, I. 1963. *The Agrarian System of Mughal India (1556–1707).* Bombay: Asia Publishing House.

Habib, I. 1982. *An Atlas of the Mughal Empire.* Delhi: Oxford University Press.

Hamilton, F.B. 1819. *An Account of the Kingdom of Nepal,* Bibliotheca Himalayica, Series 1, Vol. 10. New Delhi: Manjusri, 1971.

Hill, C.V. 1997. *River of Sorrow.* Association for Asian Studies, The University of Michigan.

Hutton, J. 1795. *Theory of the Earth.* Edinburgh: Scottish Academic Press.

Islam, M.M. 1997. *Irrigation Agriculture and the Raj: Punjab, 1887–1947.* New Delhi: Manohar.

Ismail, M. 1995. *Early Settlements in Northern Sri Lanka.* New Delhi: Navrang.

Ives, J. D. and B. Messerli. 1989. *The Himalayan Dilemma: Reconciling Development and Conservation.* London: Routledge/United Nations University.

Kangle, R.P. 1972. *The Kautlīya Arthasastra: An English Translation with Critical and Explanatory Notes, Volume 2.* Bombay University Press.

Kanwar, S.S. 1997. 'A Soil Scientist Reminisces: Fifty Years of Soil Science Research to Serve Indian Agriculture'. *Asian Agri-History Journal,* **1**(2): 135–52.

Kenoyer, J.M. 2003. 'Uncovering the Keys to the Lost Indus Cities'. *Scientific American,* **289**(1): 58–67.

Lal, M. 1986. 'Iron Tools, Forest Clearance and Urbanisation in the Gangetic Plains'. *Man and Environment,* **10**: 83–90.

Leach, E. 1990. 'Aryan Invasion over Four Millennia'. In. E. Ohnuki-Tiernery (ed.), *Culture Through Time: Anthropological Approaches.* Stanford: Stanford University Press.

Moosvi, S. 1987. *The Economy of the Mughal Empire, c.1595: A Statistical Study.* Delhi: Oxford University Press.

Nene, Y.L. and N. Sadhale. 1997. 'Agriculture and Biology in *Rigveda*'. *Asian Agri-History*, **1**(3): 177–90.

Paton, T.R., G.S. Humphreys and P.B. Mitchell. 1995. *Soils: A New Global View*. London: UCL Press.

Possehl, G.L. 1997. 'The Transformation of the Indus Civilisation'. *Journal of World Prehistory*, **11**(4): 425–72.

Possehl, G.L. 2004. 'India, Ancient-Indus Valley'. In S. Krech, J.R. McNeill and C. Merchant (eds), *Encyclopedia of World Environmental History, Vol. 2 (F–N)* (New York: Routledge), pp. 672–5.

Prime, R. 2002. *Vedic Ecology*. Novato, CA: Mandala Publishing.

Raikes, R.L. 1965. 'The Mohenjo-daro Floods'. *Antiquity*, **39**: 196–203.

Raychaudhuri, T. and I. Habib. 1982. *The Cambridge Economic History of India, Vol. 1: c.1200–1750*. Cambridge: Cambridge University Press.

Rennell, J. 1781. 'An Account of the Ganges and Burrampooter Rivers'. *Roy. Soc. London Phil. Trans.*, 71: 87–114.

Richards, J. 1993. *The Mughal Empire*. Cambridge: Cambridge University Press.

Saberwal, V. 1999. *Pastoral Politics*. Delhi: Oxford University Press.

Sadhale, N (trans.) 1999. 'Krishi-Parashara (Agriculture by Parashara)'. *Agri-History Bulletin*, **2** (Secunderbad, India: Asian Agri-History Foundation).

Sarkar, J. N. 1975. *Studies in Economic Life in Mughal, India*. Delhi: Oriental Publishers.

Saxena, D. P. 1976. *Regional Geography of Vedic India*. Kanpur: Grantham.

Sharma, A. 1983. *Studies in 'Alberuni's India'*. Wiesbaden: Otto Harrassowitz.

Sharma, C., M.S. Chauhan, S. Sharma, M. Sharma and I.B. Singh. 2001. 'Proxy Records of Holocene Vegetation and Climate Change from Sanai Tal, Central Ganga Plain, Uttar Pradesh. Nat. Symp. Role of Earth Science Integrated Development and Related Societal Issues'. *Geol. Surv. India. Spec. Publ.*, *65*(111): 199–202.

Sharma, R.S. 1996. *The State and Varna Formation in the Mid-Ganga Plains*. Delhi: Manohar.

Shinde, V. 2002. 'The Emergence, Development and Spread of Agricultural Communities in South Asia'. In Y-Yasude (ed.), *The Origins of Pottery and Agriculture* (New Delhi: Roli Books), pp. 89–118.

Singh, G. 1971. 'The Indus Valley Culture Seen in the Context of Post-Glacial Climatic and Ecological Studies in Northwest India'. *Archaeology and Physical Anthropology in Oceania*. **6**(2): 177–89.

Singh, G., R.J. Wasson and D.P. Agrawal. 1990. 'Vegetation and Seasonal Climatic Changes since the Last Full Glacial in the Thar Desert, Northwestern India'. *Review of Palaeobotany and Palynology*, **64**: 351–88.

Srivastava, P., I.B. Singh, S. Sharma, V. Shukla and A.K. Singhvi. 2003. 'Late Pleistocene-Holocene Hydrologic Change in the Interfluve Areas of the Central Ganga Plains, India'. *Geomorphology*, **54**: 279–92.

Starkel, L., and S. Basu (eds). 2000. *Rains, Landslides and Floods in the Darjiling Himalayas*. New Delhi: Indian National Science Academy.

Thapar, R. 1984. *From Lineage to State: Social Formations of the Mid-First Millennium BC in the Ganges Valley.* Bombay: Oxford University Press.

Thompson, M., M. Warburton and T. Hately. 1986. *Uncertainty on a Himalayan Scale.* Milton Ash Edition.

Tripathi, T.M. 1990. *Hydrology in Ancient India.* Rookee: National Institution of Hydrology.

Trivedi, K.K. 1998. 'Estimating Forest, Wastes and Fields, c. 1600'. *Studies in History,* **14**(2): 301–11.

Vasu, S. C. (ed.) 1962. *Panini's Astadhyayi.* Delhi.

Velayutham, M., D.K. Pal, T. Bhattacharya and P. Srivastava. 2002. 'Soils of the Indo-Gangetic Plains, India: Historical Perspective'. In Y.P. Abrol, S. Sangwan and M.K. Tiwari (eds), *Land Use: Historical Perspectives. Focus on Indo-Gangetic Plains.* Delhi: Allied Publishers.

Vishnu-Mittre. 1972. 'Neolithic Plant Economy at Chirand, Bihar'. *Palaeobotanist,* **21**: 18–22.

Walker, A. 1997. 'Indian Agriculture' *Asian Agri-History,* **1**(1): 1–21.

Wasson, R.J. 1984. 'The Sedimentological Basis of the Mohenjo-Daro Flood Hypothesis'. *Man and Environment,* **8**: 88–90.

Wasson, R.J. 2003. 'A Sediment Budget for the Ganga-Brahmaputra Catchment'. *Current Science,* **84**(8): 1041–7.

Wasson, R.J., G.I. Smith and D.P. Agrawal. 1984. 'Late Quaternary Sediments, Minerals, and Inferred Geochemical History of Didwana Lake, Thar Desert, India'. *Palaeogeography, Palaeoclimatology, Palaeoecology,* **46**: 345–72.

Wasson, R.J., M.S. Chauhan, C. Sharma, A.K. Singhvi, P. Srivastava and I.B. Singh (in prep). 'Soil Erosion of a Durable Agroecosystem During the Last Two Millennia on the Central Gangetic Plan'.

Wasson, R.J., N. Juyal, M. Jaiswal, M. McCulloch, M.M. Sarin, V. Jain, P. Srivastava and A.K. Singhvi (in prep). 'The Mountain-Lowland Debate: Deforestation and Sediment Transport in the Upper Ganges Catchment'.

Wessels, C. 1924. *Early Jesuit Travellers in Central Asia 1603–1721.* The Hague: Martinus Nijhoff.

Wheeler, R.E.M. 1947. 'Harappa 1946: The Defences and Cemetery R37'. *Ancient India: Bulletin of the Archaeological Survey of India,* **3**: 58–130.

Whitecombe, E. 1972. *Agrarian Conditions in Northern India. Vol. 1. The United Provinces under British Rule, 1860–1900.* Los Angeles: University of California Press.

Willcocks, W. 1930. *Lectures on the Ancient Systems of Irrigation in Bengal and its Applications to Modern Problems.* University of Calcutta.

Wojtilla, G. (trans.) 1985. 'Kasyapiyakrsisukti: A Sanskrit Work on Agriculture'. *Acta Orientalia Academiae Scientiarum Hung.,* **39**: 85–136.

4

A Soils History of Mesoamerica and the Caribbean Islands

Tim Beach, Sheryl Luzzadder-Beach and Nicholas Dunning

A forty-year-old quote about the soils of the Maya realm remains accurate for much of Mesoamerica and the Caribbean: 'A far more thorough examination of the soil is needed before we can begin to interpret the ecological factors in Maya culture history' (Stevens 1964: 302). Indeed, we are still on the rising slope of an exponential growth curve of knowledge of the history of soils (Dunning and Beach 2004b). In this region we have archaeology to thank for much of what we know about the history of soils, because otherwise far less research would have been done in nations that have little science infrastructure. Much of our knowledge of soil history surrounds essential archaeological questions like subsistence and degradation at ancient sites, and thus archaeology provides a foundation for understanding soils at a time when they are changing faster than at any time in history. In this chapter we show that soils share a long, complicated and dramatic history with humanity, and that throughout history, especially over the last several decades, the highest rates of change have often occurred with pioneers.

Rapid recent changes of soil landscapes have two principal drivers: population change and land-use changes by groups that disregard ancient indigenous soil management features. Mountjoy and Gliessman (1988) gave an example of the misunderstanding of indigenous conservation technology in their description of modern mechanised ploughing running over *cajeta* terrace systems, discussed below. The ploughing destroyed what had been successful adaptation to farming on steep slopes in Tlaxcala, Mexico by expunging canals that diverted water and the *cajeta* basin that stored runoff, and this accelerated erosion on these steep slopes (See Figure 4.1).

Soils and history in Mesoamerica and the Caribbean fall around two major topics: human impacts over time and human soil practices or management over time. At the outset, the first topic includes too little information about the current

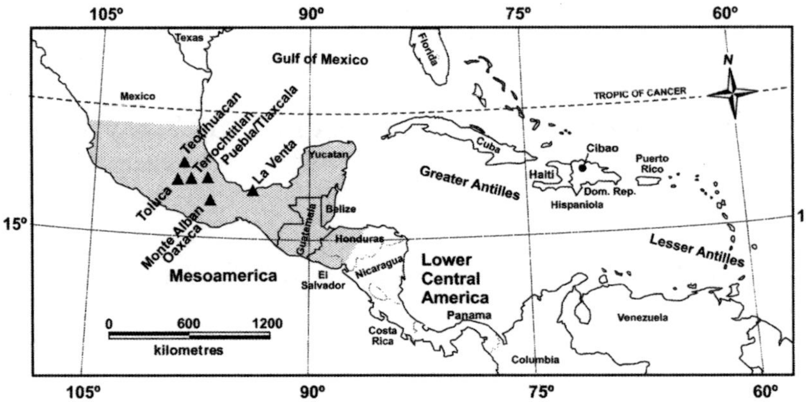

Figure 4.1. Central America and the Caribbean

region's soils diversity as a baseline for understanding trends in soil alteration. We also know too little about what are human-induced and what are natural processes. For example, in a recent, June 2003, excursion into the Maya mountains at the start of the rainy season, we confronted an intriguing dilemma. The Mountain Pine Ridge zone is eroding at a very high rate: rill and gully formation is rampant and runoff is saturated with sediment. The region has lumbering, but also an infestation of Southern Bark Beetle (*Dendroctonus frontalis*) that has killed a high percentage of the pine trees (e.g., *Pinus caribaea*). We could ascribe the erosion to lumbering, but the infestation has left more land bereft of canopy cover. We could also say that the infestation is probably human caused, in which case the erosion is mostly human induced. Today we can determine the key factors of natural and human induced soil erosion but we may never fully identify each factor from historical and archaeological records, and one factor alone may control the magnitude of erosion (Beach 1992).

CHRONOLOGICAL OVERVIEW

Recent soil history starts with the transition of the Pleistocene to the Holocene, because of the climatic, geomorphic, hydrologic and vegetation changes that dramatically changed soils and the advent of humans in the Western Hemisphere (though growing evidence supports a pre-Clovis and thus pre-Holocene presence in the Americas at several more and less controversial sites). Soil studies in the Teotihuacán Basin provide evidence for the changing climates of Late Pleistocene and Holocene, in that caliches formed during dry periods in the Late Pleistocene and argillic horizons formed during the warm and wet middle Holocene (McClung de Tapia et al. 2003). In the Basin of Mexico, butchered Mammoth remains occur with stone tools buried in peat soils that date to about 11,000 BP (Morett et al. 1998), but the first direct use and therefore alteration of soil comes much later with the advent of farming. In the Early Hunter Phase (11,000 to 7,000 years ago) hunters and gatherers certainly used products of the soil, collected plants and seeds, and managed these with their early domestication of squash and gourds and other species (Piperno and Pearsall 1998: 312); but the first large impacts came later with the Archaic Period (7,000–4,000 BP) associated with larger scale clearing by fire and the growing bundle of diverse crops like corn, beans and squash (Piperno and Pearsall 1998: 315). A remarkable group of cores and excavations near La Venta in Veracruz shows evidence for the first maize pollen, the first sunflower seeds and possibly the earliest Mesoamerican glyph (Pope et al. 2001; Pohl et al. 2002). Many still think of Central Mexico as the home of maize, though Piperno and Pearsall (1998, p. 314), even before the announcement of 7,100-year-old corn pollen near Veracruz, argued that the '... humid tropical lowlands were the major settings for the origin and development of agricultural systems ...' They see the development of the maize and manioc here as the driver for the diffusion of swidden agriculture from 7,000 to 5,000 years ago. Similarly, lake core studies in the heart of the Maya Lowlands show maize pollen and disturbance indicators around 4500 BP or earlier (Vaughan et al. 1985; Islebe et al. 1996; Wahl et al. in press). Such conclusions hearken back to Carl Sauer's more theoretical predictions of agricultural origins in tropical wet and dry riverine environments (Piperno and Pearsall 1998, pp. 18–21; Sauer 1952).

In the Caribbean Islands, (See Figure 4.1) the Archaic or Preceramic period (Caribbean Island cultural Period 1), begins with the arrival of humans on Cuba and Hispaniola between roughly 6000 and 5500 BP (Rouse 1992). There is mounting evidence that this first wave of colonisation originated on the Yucatán Peninsula (Wilson et al. 1998). A second wave of Preceramic colonists spread northward through the Lesser Antilles from northern South America, beginning around 4000 BP and reaching as far as Puerto Rico (Rouse 1992). Archaeological assemblages from this period consist chiefly of lithics and often large quantities of shells. Evidence for occupation on the islands typically occurs concentrated along coastal margins near marine resources (Petersen 1997). Human disturbance is also signalled by episodes of charcoal deposition in sediments in Lake Miragoane, Haiti and Laguna Tortuguero, Puerto Rico between 5500 and 5000 BP (Burney and Burney 1994; Higuera-Gundy et al. 1999).

The next period in Mesoamerica is the Formative or Preclassic Period (4000–1750 BP), during which the first environmental impacts on soils show up from pioneer farming and land clearance by 3400 BP as the start of the Maya Clays in the Lowlands (Brenner et al. 2003) and, after 3600 BP, as erosion-produced clays in the Highland Lake Patzcuaro (O'Hara et al. 1993). Likewise, wetland soil manipulation for agriculture appears by the Early Preclassic both in the Maya Lowlands and Central Mexico (Sluyter 1994; Jacob 1995; Pohl et al. 1996). This period coincides with the height of the Olmec at sites such as La Venta in the Gulf Lowlands of Mexico, but only the long beginning of Maya Civilisation and the Highland Zopotec site of Monte Alban (See Figure 4.1). Population growth occurs through this time in both zones and the first soil conservation features arise by the Middle Preclassic in the Central Petén (Hansen et al. 2002) and in Northern and Central Mexico (Hard and Roney 1998; Mountjoy and Gliessman 1988; Donkin 1979). Evidence also exists for some form of soil manipulation for wetland agriculture and water management in the Lowlands in the Preclassic (Pohl et al. 1996; Scarborough 1993). In the Valley of Guatemala, Late Preclassic urbanisation at the sprawling site of Kaminaljuyu was seemingly tied to the development of intensive, irrigation agriculture with the city and elsewhere in the valley (Popenoe de Hatch et al. 2002). Nearby in the Guatemalan Highlands, Preclassic Maya groups adapted themselves to a suite of fertile Andisols and a dynamic, changeable volcanic landscape (Robinson et al. 2002).

Archaeology (Hansen et al. 2002) and some lake cores provide evidence for abandonment in the Late Preclassic around Maya sites known for reaching their apex during this time such as El Mirador, Guatemala (See Figure 4.2). Pollen from disturbance indicators decline through this zone and lower magnetic susceptibility through the zone may indicate less input from erosion (Wahl et al. in press).

In the Caribbean Islands, the end of Period I and beginning of Period II (Early Ceramic Period) varies tremendously from island to island (Rouse 1992). Archaeologically, this transition is marked by the beginning of local ceramic traditions and often coincides with the arrival of agricultural groups from the Orinoco Basin known collectively as the Saladoid. The major exceptions to this transition were on Cuba and Hispaniola, where hunting-fishing-gathering economies persisted considerably longer and local ceramic traditions began prior to agriculture. Saladoid groups began moving into the Lesser Antilles around 2500 BP, with migration reaching Puerto Rico by at least 300 BP. Most Saladoid groups employed a mixed subsistence economy including the hunting of both native and introduced animals as well as farming (Petersen 1997). Agriculture was chiefly focused on the relatively fertile soils of the higher, wetter volcanic islands. While cassava was the dominant crop, a wide variety of fruits and palms were also exploited. Evidence for soil erosion during this period is so far lacking; however, the introduction of many plant and animal species from South America by Saladoid peoples and their subsequent subsistence activities had a profound, often systemic impact on many island ecosystems (Newsom and Wing 2004). Saladoid settlements also favoured more coastal locations and relied significantly on marine resources.

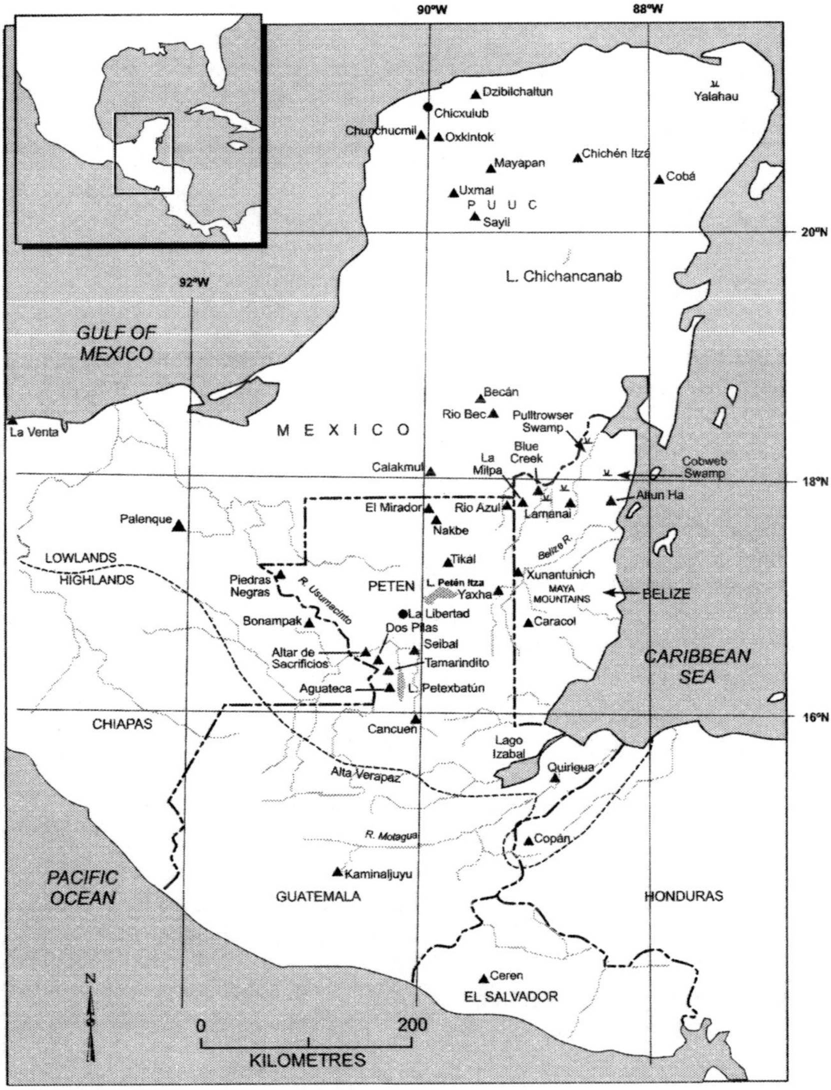

Figure 4.2. The Maya world

Manipulating soil for more intensive agriculture and erosion continues into the Classic Period (1750–1100 BP), when the Maya sites Monte Alban and Teotihuacán flourished and collapsed. The Classic Period had the highest populations, most intensive agriculture and, thus, soil manipulation and all the forms of intensive farming occur. Many writers have assumed that soil erosion was also most severe during this time and some even ascribed collapses to soil erosion, but soil conservation was also very widespread. With variable abruptness, Teotihuacán collapsed around 1400 BP, Monte Alban by 1300 BP, and the Maya cities of the Petén collapsed around 1100–1200 BP with widespread pollen-core evidence for rapid return of forest.

After 1500 BP in the Caribbean Islands (particularly the Greater Antilles), the transition from Period II to III (Late Ceramic or Ostionoid) saw significant increases in population as well as the regionalisation of cultures on the Caribbean islands and increasingly hierarchical societies, including the emergence of the first monumental architecture in the Greater Antilles (Curet 2003). Across much of the Caribbean Islands agriculture appears to have continued to be largely of slash-and-burn, digging stick variety and focused on cultivation of root crops (Peterson 1997). We see agricultural intensification evident in the form of terracing by at least 1300 BP in parts of Puerto Rico (Ortiz Aguilú et al. 1991; Oliver, et al. 1999) and evidence of soil erosion appears in the sediment record around the same time on Hispaniola (Brenner and Binford 1988).

Maya civilisation continued in the Northern Yucatán in the Post-Classic (1100–500 BP), especially at Mayapan, Yucatán, and Toltecs, Mixtecs and Aztecs flourish in Central Mexico. We can argue that this ushered in a golden era of soil use, manipulation and understanding in Mesoamerica because soil conservation flourished with many terraces, crop types, irrigation, fertilisation and even written/glyphic soil taxonomy in Aztec codices (Williams and Harvey 1997). This is a period of intense soil use in Mexico, but one of soil rejuvenation in most of the Maya lowlands as ninth-century Maya collapse, and then epidemics from European conquest, decimated Maya populations but allowed forest recovery. In the lakes and depressions of the region such as Lake Petén Itzá, organic sediments settled over the thick Maya clays deposited from eroding soils (See Figure 4.2).

In the Caribbean Islands, the years from about AD 1250 through the first years of Spanish contact (Period IV or Late Ancient History) saw the continued expansion and elaboration of chiefdoms in much of the Greater Antilles, although on Cuba non-agricultural groups continued to persist along with their horticultural neighbours (Rouse 1992). Intensive agriculture continued in parts of the Greater Antilles including terracing, ditching, and a type of field mounding known as ***conucos***. Farmers were also now exploiting an increasing variety of crops including maize, cotton, calabashes and tobacco (Petersen 1997). Most islands of the Lesser

Antilles were populated by the Island Caribs most of whom practised a mixed subsistence economy based on horticulture and marine resources.

After Conquest, populations declined precipitously throughout the Americas but a new form of environmental change came from livestock ranching and burning to maintain grazing lands. Centuries, indeed millennia, of soil use by successive peoples make soils here a complicated palimpsest, which has led scholars to divergent opinions about the causes of severe erosion in Central Mexico. In some cases, soil erosion has been so severe that farmers have actively eroded what is left of original soils, the cemented C-horizons that dominate surfaces, to tap softer sediments beneath. But different studies focus their blame for upland erosion and valley sedimentation on Pre-Columbian high populations using intensive agriculture or on Post-Columbian depopulation and abandonment, the nefarious impacts of livestock, and ham-fisted development schemes.

In recent times, rapid population growth and population displacements have led to pioneering mentalities again and severe erosion in some parts of the Maya regions and Central Mexico. Also, encouragement by state agencies and international development to spur production has led to agricultural expansion, especially since the 1950s. Much of the expansion was on to marginal lands, which are too steep, too arid, or have other features that make them prone to degradation. This led to deforestation: about half Mexico's forests and much more than half of its tropical forests have been removed (FAO 2005). Deforestation of course greatly affected hydrology, soil erosion and soil nutrient depletion. For example, gullying occurred on 65 to 85 per cent of dryland hillslopes in parts of Central Mexico (Bocco and Garcia-Olivia 1992). Thus the grand soil problems that this chapter will focus on are in many cases worse today than in most of the history of soils of Mesoamerica.

A PARADE OF PARADIGMS

For Central Mexico, many historical sources from the Conquest period document high populations and intensive cultivation in such forms as *chinampas* and terracing. Moreover, the large abandoned sites of Central Mexico, like Teotihuacán, seemed to show that ancient populations had also been high. Maya lowlands research before the 1960s, however, believed that ancient Maya populations were low and that soil manipulation was not intensive in nature. But broad, multidisciplinary regional surveys since the 1960s have found more and more evidence for high ancient populations and soil manipulations for intensive cultivation. This ushered in a new paradigm of large population centres and the litany of intensive soil and water management: dams, reservoirs, canals, terraces and wetland cultivation. There have been important critiques of the methods for estimating high populations (Bequelin and Michelet 1994, Rice and Culbert 1990), but settlement surveys have announced high population estimates for more and more sites. Hence, the current heterogeneous or mosaic model is that the Maya Classic had many large and smaller centres that used many possible kinds of intensive and extensive forms of soil management for subsistence (Dunning and Beach 2004b; Fedick 1996). Now many studies by archaeologists, geographers, biologists, geologists and soil scientists are testing the specifics of the Maya mosaic paradigm (Gomez-Pompa et al. 2003). Therefore, we have organised this chapter along the lines of many forms of evidence about the multiplicity of past soils and management.

Environments

Defining Middle America is an arbitrary process. We shall focus on the regions from the basin of Mexico to Central America, spanning from about latitude 22° to 6° North and longitude 78° to 100° West. We also compare this region with the islands of the Caribbean where the post-Columbian Old World/New World collision first occurred (See Figure 4.1). The environmental diversity of Middle America is impressive in geology, climatology, ecology, soils and human environmental history. Geologically, the region ranges from the Mexican Plateau, to the high and active volcanic peaks at the spine of most of this landmass, to metamorphic mountain ranges, to deeply incised canyons, to a fringing coastal zone, to a vast, flat carbonate plain, which is intruded by a small granite batholith in the Maya Mountains in Belize. Geological diversity is only matched by climatological and ecological diversity. Middle American ecosystems and climates range from near deserts to tropical forests and montane coniferous forests. The region lies mainly in the Trade Winds, with occasional intrusions of Westerlies' mid-latitude cyclones and seasonal dominance of the subsiding air and dry conditions of the Subtropical High or the precipitation of the Intertropical Convergence Zone. These large systems and complex topography lead to tropical wet and dry climates with high regional variations in precipitation, humidity and temperature, and the periodic occurrence of severe droughts (Hodell et al. 1995; 2000; 2001; Haug et al. 2003).

Soil Types: Modern and Folk Taxonomies

High ecological and geological diversity of course mean high soil diversity, and challenge any attempt at generalisation. Since the region ranges from sea level to nearly 6,000 m, and has vastly different bedrocks, climates and ecosystems, all the major soil orders occur. What stands out in terms of soils are the large extent of rendolls or rendzinas that are tropical mollisols, conditioned by a vast expanse of relatively young carbonate rocks of the Yucatán and an equally large extent of andisols (volcanic soils) and other soils with volcanic inputs, mainly in the Highlands.

The easiest region to generalise within Mesoamerica is the Maya Lowlands because this region is dominated by the Yucatán carbonate platform, a low mostly limestone terrain. This landscape arose differentially from the ocean since the late Cretaceous (Donnelly et al. 1990), which caused a chronosequence of soils to form based on how long they have been above sea level. One chronosequence lies in the northwest Yucatán, where concentrically older and older soils occur outward, and inland, from the Chicxulub impact crater, since the centre has been the most recent part to emerge from the sea (Pope et al.1996).

Five common soils that form on the limestones across Yucatán are entisols and inceptisols, thin and young soils that form in recently deposited material; rendzinas or rendolls (or ***pusluum*** or ***boxluum*** in Yucatec Mayan); alfisols or terra rossas (or ***Kankab*** in Mayan terms); vertisols; and histosols. Histosols or peats and mucks occur all over the world in low-lying areas with water tables near the surface (see Chapter 5). The inceptisols, entisols and rendolls are young, fertile, thin, clayey soils that range up to about 50 cm in depth and are differentiated by thickness and horizon formation. In the Maya Lowlands, the alfisols (Ustalfs or ***Kankabs***) are older, red, iron-rich, less fertile, generally thin, clayey soils. The rendolls form from organic matter, high amounts of carbonate and residual silicate minerals from the limestone parent material, and some wind and volcanic deposition. The alfisols form from the same materials but the organic matter generally decomposes more, little carbonate is left, and more silicate clay and iron minerals build up over greater periods of time (Beach 1998a). Vertisols are soils that form from expanding and contracting clays, which also form surface cracks and ridges called gilgai. Vertisols form in deeper profiles in karstic upland sinks or ***poljes*** such as the bajos discussed in Chapter 5 (Beach et al. 2003a). In all of these soils, silicate minerals, which dominate in many soils elsewhere, occur as impurities within limestones that are dominantly calcite ($CaCO_3$) (Isphording 1984; Beach 1998a). The $CaCO_3$ dissolves, leaving little mineral matter for soil formation. Thus these Yucatán soils are usually thin, relatively high in organic matter, and contain the residual silicate minerals, especially kaolinite, from the limestone, winds or volcanoes. Higher than expected organic matter in these tropical soils may be caused by fine-sized secondary carbonates impeding decomposition (Shang and Tiessen 2003).

Williams and Ortiz Solorio (1981) note that the view is widespread in the Mexican countryside that experts are outsiders who know more than peasants about everything except the soil, about which they are the experts. Pulido and Bocco (2003) show this expertise in a rural indigenous community in Central Mexico, in that the community values its soils and their management, which translates into sustainable soil management with relatively low gullying on very erosion-prone slopes. Thus ethnographic studies provide invaluable knowledge and contemporary insights into history, and we are fortunate to have ethnographies, histories and two extant codices for the Aztecs. These codices contain glyphs that provide soil data on 1100 agricultural fields in the Basin of Mexico. Williams and Harvey (1997, p. 30) stated that these complex, written Nahuatl soil classifications are unique in the Americas and they knew of only one other parallel in fifth-century BC China (see Gong et al. 2003). The classification in the ***Codice de Santa Maria de Asuncion*** shows 104 variants for soils in 200 hectares, to which peasant farmers today apply four classes, and modern taxonomy ascribes five soil phases. The 104 variants probably represent 18 taxa and three separate class levels. These descriptions include a number of aspects that relate to soil productivity such as texture (e.g., sandy or clayey); landscape positions (e.g., alluvium or hillslopes); irrigation; and colours (e.g., black or yellow). The most productive of the generic taxa was ***Atoctli***, which refers to alluvium of several kinds. Another interesting taxon is ***tepetlatlalli***, which refers to the well-known caliche soil C-horizons, known as ***tepetate***, that become exposed after about 70 cm of erosion. This Codex indicates a large area of ***tepetate***, and thus soil erosion, before the mid-sixteenth century AD (Williams and Harvey 1997, p. 34). Moreover, Sahugun's Florentine Codex from the same period even discusses the practice of purposely breaking up this eroded subsoil to make ***tepetlalli*** soils for planting.

What we know of the soil history of the Maya realm is less than what we know for Central Mexico, and it is based on ethnography and archaeology rather than texts. In most of the Petén region only light reoccupations by Maya groups disturbed the many centuries of abandonment that followed the Terminal Classic period (Palka 1999). These groups took with them their evolved soil knowledge (see Nations 1979). In the Maya-language areas of the Chiapan and Guatemalan Highlands complex indigenous soil knowledge persists (See Figure 4.2). One of the most distinctive features of these knowledge systems is their categorisation of various soils as either 'hot' or 'cold' (Maffi 1999). While these terms refer in part to the amount of insolation experienced by various soils, the terms also have significant overtones with regard to fertility and undertones of a political nature due to the historic displacement of many Maya groups on to poorer 'cold' land.

Many of today's Maya farmers of the Lowlands are shifted cultivators from other areas, often the Highlands, and they have either recently adapted their soil classifications or brought them from other places with very different soil types. Indeed, one study of the Kekchi Maya of the Petén, who had recently immigrated from the Highland Alta Verapaz, showed that they classified about 60 per cent of their new land as trash land (Dunning 1992; Dunning and Beach 2004a), but they began to recognise greater diversity over time. The Kekchi of the Petén have four basic levels of soil that differentiated warm and cold, texture and drainage, and color and depth (Dunning and Beach 2004a). Earlier Kekchi emigrants from Alta Verapaz to another tropical lowland near Lago Izabal, Guatemala also replaced their cold earth taxa with a drainage-based class (Carter 1969).

Yucatec farmers have been in place for much longer, and have direct connections to the historical and prehistorical Maya of their region. They also have a significant taxonomy which describes an impressive range of prospective harvest zones. Their classification defines seven main farmable landscape taxa and numerous amending taxa. The taxa also refer to drainage, colour, topography and texture; yet they are also replete with generations of accrued knowledge about what crop types grow best in particular soils (Dunning and Beach 2004a; Dunning 1992).

Maya farmers have also influenced soil formation over three millennia. Their presence and activities have consciously and unconsciously created middens, fertilised areas, depleted areas, and formed new soils. Their largest impacts have been on erosion of uplands and on the large areas of soils created by hundreds of abandoned Maya cities. Some literature showed that Maya sites had high soil quality and suggested this had been a pull factor for settlement. This may be correct, but we can say empirically that Maya activities at some sites have improved soils. For example, fertile mollisols around the colossal site of Chunchucmil in the northwest Yucatán are largely the creation of ancient limestone, plaster, floors and middens. Soils that formed with none of these are alfisols which are more leached, thinner and less fertile, and contemporary Yucatec farmers often ignored them, planting directly on and around ancient architecture in the anthropogenic mollisols (Beach 1998a; Dahlin et al. in press). This pattern occurs on many of the so called 'Garden Cities' of Yucatán, where the often black anthrosols lie next to and on top of older red soils. These Maya soils have many of the attributes of the Amazon's *terra preta*: dark colour, higher fertility, charcoal, many artefacts and active use, including mining, by farmers (Woods and McCann 1999; Lehmann et al. 2003; see also Graham in press).

Proxies for Past Soil Uses

The basis for interpreting past soil activities includes historical, physical, chemical and biological proxies in the context of dated sequences. Each of these provides a glimpse of human activities of the past but a major limitation has been quantifying these activities because of a legion of potential inaccuracies.

Historical and contemporary ethnographic methods provide some evidence of soil use in prehistorical times, but these earliest recorded or contemporary indigenous practices may still have nothing to do with antiquity. For example, the standard early source for the northern Maya area is by the Yucatán's second Bishop, Diego de Landa, a Franciscan who wrote of the language and culture of the Maya in 1568 (de Landa 1982). His book provides record of the soils and land uses of the sixteenth-century Maya in Yucatán, but this is still a century after Late Post Classic Mayapan, and many centuries after the Classic period when populations were much higher and land uses must certainly have been different (Dunning and Beach 2004a). Moreover, we have to consider that de Landa oversaw the burning of more than 27 scrolls, leaving only three for us today. From this source we get a notion of the dominance of land held in common, long-fallow swidden agriculture, and the riddle of high productivity in a region of depauperate soils.

Similarly, the many ethnographic studies of indigenous practices may or may not record any accrued knowledge from antiquity. For example, the Carnegie Institution of Washington studied Yucatec Maya farming practices in the first half of the twentieth century and concluded that the long-fallow swidden agriculture with low, dispersed populations was the only available practice for the ancient Maya (Steggerda 1941; Morley 1946, p. 141; Dunning and Beach 2004a). Yet, several intensification methods have been brought to the fore by archaeology and ethnobotany under a paradigm that recognised the imperative of high, concentrated populations.

Physical methods include physical comparisons of soils over time and geophysical methods to detect ancient soil landscapes. Datable buried paleosols, recognisable in excavation units by colour, texture, structure and other characteristics, provide a good source of information about environmental change. For example, a soil surface that must have been stable for millennia that is buried by deposition suggests high magnitude environmental change. Likewise, non-destructive geophysical methods such as magnetometer and ground-penetrating radar surveys can show buried ancient soil surfaces (Conyers and Spetzler 2002).

Chemical signatures in soils can also reveal past human activities. The oldest chemical prospecting tool has been phosphorous testing, used as an archaeological tool since the 1920s and recognised even earlier (Bethell and Máté 1989). Many human activities increase phosphorus in soils, where it tends to remain for long periods. As a whole, of course, human activities both remove and add elements (Eidt 1984). There are many other elemental additions to soil from pre-industrial human activities, including the major plant nutrients, such as nitrogen, magnesium, phosphorus and potassium, and elements and minerals such as mercury from cinnabar-based red paints, manganese from black paints, copper minerals from greens and blues (Cook et al. 2006), and iron from animal butchering and agave processing (Manzanilla 1996).

Many of these elements occur at high levels in human bone, plant and animal tissues, and in human and animal waste. Deposition of these materials in soil as burials, middens, production by-products or intentionally applied fertiliser may accumulate them significantly. But for some elements like nitrogen and potassium leaching and oxidation-reduction reactions are high and accumulations short-lived. Moreover, background signatures of calcium, and often magnesium, are already so high in most soils of the Maya Lowlands that they drown out human inputs. However, phosphorus is quickly 'fixed' by forming highly stable compounds with aluminium, iron and calcium. Thus, there is limited vertical or horizontal migration of the phosphate, though erosion and physical translocation will of course shift the soil particles. Hence, phosphorus is a good archaeological tool because it is relatively stable and reflects many human activities, and there are many approaches for phosphorus testing (Terry et al. 2000; Parnell et al. 2001).

Since the 1990s, more scholars have used a full range of analytical techniques such as soil micromorphology, carbon and oxygen isotopic studies, thin layer chromatography, gas chromatography/mass spectrometry, inductively coupled plasma-atomic emission spectrometry (ICP/AES),and inductively coupled plasma mass spectrometry (ICP/MS) with a variety of extractants. Several studies in Mesoamerica and elsewhere have found that archaeological features have distinct chemical signatures from natural soils and that floors, hearths, gardens and pigments can be distinguished based on this multi-elemental technique (Middleton and Price 1996; Wells et al. 2000; Parnell et al. 2002; Cook et al. 2006).

Biochemical techniques are also expanding, but again few studies have been published thus far, though biomarkers hold large potential for answering questions about ancient fertilisers and human activities on soils (Bull et al.1999). For example, current projects are attempting to work out ancient marketplaces and fertilisers to get at old questions about how ancient cities functioned and how enough food could have been produced in poor soils (Fernandez et al. 2002; Fedick 2003; Dahlin et al. in press). Barba and colleagues working in Mexico have analysed a variety of elements and biomarkers (e.g., albumin, carbohydrates, fatty acids) in short distance variation in archaeological and contemporary sites (Barba et al. 1987; Manzanilla and Barba 1990; Barba and Ortiz 1992).

Carbon isotope ratios through a vertical sequence of depositional soil profiles provide another approach to the question of what grew on paleosols because they reflect the amounts of C3 and C4 plants that formed different types of soil organic matter (Fernandez et al. 2005). Most of the tropical forest broadleaf plants have $\delta^{13}C$ ratios of -26 to -30 per mil, whereas the C4 plants such as many grass species have $\delta^{13}C$ ratios of -12. Many of the surface forest soils have $\delta^{13}C$ ratios similar to forest species, but buried soils, some dateable to the ancient Maya period have ratios that are much closer to C4 plants. Interestingly, one study even found that a contemporary ***milpa*** (swidden), which had grown maize for 20 years still had a ratio similar to the forest soil, which probably indicates that soils with ^{13}C ratios similar to C4 plants must have had maize and other C4 plants growing for long periods of time (Webb et al. 2004; Fernandez et al. 2005).

Fossils supply another traditional proxy for past soil uses and alterations. Here many studies have used the full range of fossils, including pollen, phytoliths, mollusk shells, diatoms and other skeletal remains (McClung de Tapia et al. 2003; Dunning et al. 2002; Hansen et al. 2002; Fedick 2003). As with biomarkers and chemical signatures, these preserve differently in different environments and they may not be very representative of past soil environments. As with all studies to reconstruct the past, multi-proxy approaches are the best, though the most expensive, because they provide multiple lines of evidence that may converge or diverge.

Soil Catenas

A common method for understanding soils is to study soils where all variables but one are constant, and to study how they differ by this one factor: climo-, bio-, topo-, litho- and chronosequences, discussed above (Birkeland 1999). A catena (similar to a toposequence) is a series of soils down a slope that vary due to drainage and topographic position (Jenny 1980; Birkeland 1998; Sommer and Schlicting 1997). Only a handful of projects have studied toposequences or other sequences across this region and they become important to history when they provide evidence for human induced change.

A comparison of enough soil profiles at each slope segment with natural versus altered land uses can provide a general picture of soil losses. For example, Furley (1987) compared the changes that occurred because of one milpa cycle between 1970 and 1981 on a catena near Belmopan, Belize. He concluded that soil losses were so high from one milpa cycle that soils could be eroded to bedrock in as few as 4 milpa cycles. Other studies of catenas in the Maya region using several methods echoed Furley's. Soil losses were very high in every method of evaluation and soils were truncated by nearly one-half on average over a decade and in some cases, erosion removed entire soil profiles in the same time period (Carlos Donado 1996; Coultas et al. 1997; Beach 1998b; Fernández et al. 2005; Beach et al. 2006). Some studies have also attempted to compare catena soil losses to estimates from the Revised Universal Soil Loss Equation (RUSLE) on these tropical, Mesoamerican slopes (Beach 1998b; Millward and Mersey 1999).

Soil Erosion in the Maya Lowlands

Several authors have proclaimed the thin soils of the Maya Lowlands as evidence of past erosion. Stevens (1964, p. 301), for example, speculated that Maya soils are still recuperating from Classic period soil losses based on the co-occurrence of Maya sites with the thin Yaxa soil series. This may have started with the eminent soil conservationist H.H. Bennett, who linked the Maya abandonment of the Petén with soil depletion in a 1926 article and later estimated that 50 per cent of Mexico and 30 per cent of Guatemala had been 'ruined for cultivation, nearly ruined and severely affected' (Bennett 1945, p. 106). The Conservation Foundation's ***Soil Erosion Survey of Latin America*** (1954) classified nearly all of the Petén into moderately eroded classes and a small fraction into slightly or severely eroded, even though the region was densely forested with low gradient slopes. This survey even maps the extremely flat northwest Yucatán as 10–25 per cent moderately or severely eroded, and others have likewise claimed that most of the original soil had eroded away (Robles Ramos 1950). As we have seen above, however, natural processes can also explain thin soils in the limestone-dominated Maya Lowlands. Indeed, all over the world, limestone barrens with no or thin soils occur, composed of high relative amounts of organic matter and sparse residual silicate matter from their parent materials. Because the soils are rocky, patchy and thin, modern, mechanised agriculture often fails, though traditional Maya digging sticks eke out yields.

A parallel contention of human-induced change by the ancient Maya is that the savannas and lateritic (oxisol) soils of the central Petén (around La Libertad, Poptun and elsewhere) were induced by repeated burning of forests and the soils subsequently laterised by intensified leaching and weathering (Stevens 1964, p. 299). Unfortunately, little modern research has taken up this interesting question (see Carlos Donado 1996).

Two giants of early twentieth-century Maya archaeology, Morley (1946) and Thompson (1954), took up the earlier suggestion of agricultural exhaustion as an explanation for the Terminal Classic collapse in the Petén and the subsequent rise of the Post-Classic centres of the Yucatán (Dunning and Beach 2004b). Stevens (1964, p. 302) also perpetuated the notion that soil depletion caused the Late Classic Maya collapse in an era when several studies linked soil depletion with collapses of ancient societies (Bennett 1945; Lowdermilk 1953; Morley 1956, p. 71). These ideas continue to the present in the assertion by Healy et al. (1983) that terrace soils degraded to toxic effects at Caracol and soil erosion and depletion contributed to the collapse of Copán (Abrams and Rue 1988; Wingard 1996), which Diamond (2005) takes up in more popular literature. The collapse scenario for Copán was based on pollen cores that showed high soil erosion rates and forest clearance in the Late Classic (Abrams and Rue 1988). Deforestation and accelerated erosion spread up into foothills and upper slopes as the population expanded, cut more wood, intensified swidden and decreased fallow times. Wingard (1996) built on the soil erosion/collapse scenario for Copán by using the Erosion Productivity-Impact Calculator (EPIC) computer model to simulate very high soil erosion rates and fertility declines with population expansion. The EPIC model estimated that fertility declines alone from the Late Classic would have required one hundred years for recovery.

Central Mexico

Many similar soil narratives exist for Central Mexico as for the Maya Lowlands, except that most soil practices and changes started at least as early and continued later in a more diverse environment. We have much more information from surviving indigenous soil use, taxonomy, agriculture and conservation in Central Mexico than in the Maya world since we have extant indigenous documents as well as many colonial documents.

The Central Highlands is a diverse landscape of cultures and ecosystems. The region includes central basins, mountain slopes, piedmonts and coastal plains. The Mesa Central consists of high, semi-arid, frost prone basins like Puebla, Touluca, Oaxaca and Mexico. Here the populations were dense, the cultures diverse and history long, with maize agriculture, pottery and sedentism stretching back farther than 4,500 years. These create a very complicated landscape for uncovering soil history, but the region is rich with studies that break down along the usual lines of soil history: soil erosion, folk taxonomies and soil management.

Studies of soils and soil erosion here range from process-based work on gully formation (Bocco 1993) to syntheses that attempt to make some broader sense of these complex interactions of soils and humans. For example, the special quincentennial issue of the *Annals of the Association of American Geographers*, edited by Karl Butzer (1992a and b), had several articles that reviewed the massive alterations present in these landscapes. These are often divided into studies that highlighted degradation induced by Pre-Columbian or European settlement. Denevan (1992) focused on severe Pre-Columbian Aztec erosion in Cook (1949, p. 86) and the *tepetate*, the Pre-Columbian exposed petrocalcic C-horizon that was such a widespread surface (Williams 1972; McAuliffe et al. 2001). Likewise, Butzer and Butzer (1993) working in the Central Mexican Bajio found no evidence for early livestock impacts being greater than those of Pre-Columbian times. These and many other works thus formed the basis for the Pristine Myth synthesis in the early 1990s, which showed that Pre-Columbian societies had severe environmental impacts and intensive land use. Nonetheless, many studies focus on European settlement as the degradation driver, especially in the last fifty years.

Another useful synthesis by C. Rincon Mauntner (1999, pp. 591–3) developed nine erosion themes: (1) humans caused vast soil erosion (Cook 1949, pp. 2–24; Heine 2003) and population density correlated with soil erosion (Cook 1949, pp. 14–86; Heine 2003); (2) climate and hydrologic change caused erosion; (3) soil erosion was too widespread to be caused by humans; (4) indigenous farmers are/were conservationists and most erosion is due to European land use changes; (5) soil erosion was linked to very recent socioeconomic processes; (6) severe soil erosion was linked to particular lithology (Kirkby 1972; Stevens 1964); (7) soil erosion was deliberate and induced since the La Natividad Phase (AD 1000–1530) (Sahagún 1577[1900]; Spores 1969, pp. 563–4); (8) importance of European livestock causing erosion (Melville 1994); and finally that (9) depopulation exacerbates erosion (Spores 1969; Kirkby 1972). Many of these themes run through soil erosion research around the world, and recall the complicated and heavily studied question of arroyo formation in the Southwestern United States (Bryan 1925; Tuan 1966; Cooke and Reeves 1976).

Often several of these nine themes run through an individual study and probably influenced later Maya studies. For example, Spores (1969) presented a soil erosion model that is similar to the model at Copán: low soil erosion in the Formative and Classic Periods because agriculture focused on valleys, now buried in many cases by 1.5–2 m of alluvium. Farmers started to move up slope during the Las Flores Phase (AD 150–1000) and after, and despite building cross-channel terraces, soils were truncated down to their calcium-cemented C-horizons (calcretes or caliches or tepetates). Soil erosion increased the higher farmers moved up hillslopes (Cook 1948, pp. 52–54). Farmers then turned to intentional erosion of the underlying softer Yanhuitlan Formation to produce farmable sediment. Similarly, Kirkby (1972, p. 30) also tied severe soil erosion to terrace building in the Natividad Phase (c. AD 1000). Soils still bare the effects of millennia of erosion because erosion is still much higher in the areas of bare *tepetates* (16 tons per ha per year) compared with areas dominated by intact soil profiles (3 t/ha/yr) (Figueroa-Sandoval 1975, cited in Cordova and Parsons 1997).

Soils in the Teotihuacán Valley show long and complex natural and anthropogenic influences from the Late Pleistocene to the present. McClung de Tapia et al. (2003) used multiple lines of soil evidence from soil chemistry, soil micromorphology, fossil pollen and phytoliths, and soil stratigraphy. They show erosion and sedimentation from the earliest human settlements about 1100 BC through intensive impacts at the height of Teotihuacan (AD 350–650), when expansion onto hillslopes led to the greatest landscape alteration. Likewise, in the Puebla-Tlaxcala region, Heine (2003) uses multiple lines of evidence to correlate severe soil erosion with intensive land use punctuated throughout history: high during intensive landuse periods like the present and low during abandonments like the sixteenth century. In contrast, Cordova and Parsons (1997) argue that abandonment and population declines drove two episodes of accelerated erosion. Heavy rilling during Toltec times (AD 750–1250) truncated soils and left relict soil pedestals. The Aztecs restored the region with terraces and check dams and built soils through the early 1500s. The post-Aztec, colonial period brought depopulation and insufficient maintenance of Aztec terraces and check dams that once again sped up soil losses.

A veterinarian and geographer, Carlos Rincon Maunter (1999), like Melville (1994), argued for the largest impacts over the last 500 years due to the landuse changes induced by human land management for sheep and goats. He, like Cordova and Parsons (1997), described curious fields with a few remaining soil pedestals that preserve small portions of the former 70 cm of topsoil above the present surface for most of the fields. Hence these soil pedestals stand testimony to the old surface and soil profile (Rincon Maunter 1999, p. 664). Cordova and Parsons (1997) show that these soils are already prone to erosion without livestock, and that livestock ranching varied greatly over time and place. It seems clear that in any given landscape, over time several of Rincon Mauntner's soil erosion themes could have applied.

OVERVIEW OF EROSION

These basic facts underscore three principles of land degradation through time in Mesoamerica: there was a diverse range of impacts; pioneer farming and land clearance produced more severe degradation than earlier work reflects; and conservation developed with growing societies and florescence in many, but not all, Classic sites. This is borne out in many of the Maya Lowland examples, where soil erosion starts in the Early Preclassic, conservation develops to its fullest extent in the Classic period, and massive alteration and soil erosion returns in the last few decades. This is also borne out in the Highland Mexico at Lake Patzcuaro, where the authors argued that most degradation occurred with site construction and again with abandonment in the Colonial period (Fisher et al. 2002). The story is somewhat different around the Basin of Mexico, where Cordova and Parsons (1997) argue that soil erosion accelerated with depopulation and abandonment of the rural landscape in both pre- and post- Aztec (Colonial) times. Thus half of the model also works for much of Mesoamerica in that erosion declined with active soil conservation during periods of high rural populations, but there seems to be a major difference between the moist, forested Maya Lowlands and the semi-arid steep lands of Central Mexico. Soil erosion essentially stops with abandonment and rapid reforestation in the Maya examples, whereas the highest erosion sometimes occurs with abandonment in Central Mexico. Significant alteration still comes with pioneer settlements in most studies of both regions. The debate, however, still rages about early post-settlement degradation from livestock, though recent rates of erosion from more intensive land uses on steep lands are the most severe yet.

Aggradation and Paleosols

Another source of empirical information on soil losses is aggradation or sediment accumulation over some base line. Such an empirical approach has several limitations because it only implies past erosion and it may be difficult to date when the erosion and sedimentation occurred. The most complete way to work out quantities of sediment movement and storage is with a full, long-term sediment budget study (Beach 1994; Reid and Dunne 1996), but there has been no such study in the entire Maya Lowlands. Thus far, we have only estimates of upland erosion and spotty measurements of the depths of sediment overlying ancient soil surfaces or Maya architecture. A number of studies have estimated sedimentation in valleys and depressions around the Maya Lowlands to range from 70 to over 200 cm deep (Olson 1981; Dunning and Beach 1994; Wingard 1996), and studies showed early slope destabilisation far northward into Veracruz somewhere between 400 BC and AD 550 (Sluyter 1997, p. 142). One well known example of aggradation comes from the coastal plain of northern Belize and the floodplain of the Belize River Valley are buried Ek Lu'um (Yucatec Mayan for black earth) paleosols from the Preclassic (1200 BC–AD 250) often buried by Preclassic through Classic fill (AD 250–850) (cf. Jacob 1995 and 1996; Jacob and Hallmark 1996; Pohl et al. 1996; Holley et al. 2000; Dunning et al. 2002; Beach et al. 2002 and 2003a; Gunn et al. 2002; Beach et al. 2006).

The first evidence for environmental change associated with sedimentation comes from lake cores with pollen that show declining forests and more savanna by as early as 5610 BP (Islebe et al.1996) and sediment derived from eroded soils by about 3000–3400 BP (Early Preclassic: 1500–900 BC) (Rosenmeier et al. 2002; Brenner et al. 2003). This changeover in the early pollen evidence could be from natural changes that were more conducive to savanna formation; nevertheless by 3000 BP, sedimentation increases and vegetation change indicated the widespread impacts of humans. Indeed in many lake cores from this region, the sediments associated with the advent of Maya agriculture stick out as dense mineral clays ('Maya clays') with increased charcoal and pollen of maize and disturbance taxa sandwiched by organic sediments. Hence erosion and sedimentation started soon after deforestation, even though the populations and tools of early Maya farmers must have been modest.

The evidence from most of the soil studies about 'Maya clays'only gives us a broad date for the early, buried soil surface and a *terminus post quem* date for the period of aggradation. In some sites, especially those with short, Late Classic occupations such as Cancuén, the impacts are datable to the Late Classic (Beach et al. 2006). But in many sites, as in the core studies, the first big impacts occur in the Early Preclassic. Indeed, in some wetlands the upper sediments have another paleosol and a sequence of artefacts that imply two episodes of sedimentation: from the Preclassic into the Early Classic and during the Late Classic or later (Beach et al. 2003b). These layers sandwich the later buried soil, which developed in the Early Classic through Middle Classic. The upper paleosol and overlying sediment in bajos and Late Classic terraces are littered with Late Classic ceramics, whereas most ceramics that extend down into the earlier paleosols buried under the terraces and on alluvial fans are too weathered to be identified.

Lake Patzcuaro

One of the most interesting fronts of the Pristine Myth debate that still goes on today is research by different teams at Lake Patzcuaro. Indeed, one of the death knells of this myth was a study of cores from this lake in Michoacan's Central Mexican Altiplano that found that erosion rates were at least as high during Late Preclassic and the Post-Classic as they were after the Spanish conquest (O'Hara et al. 1993; Butzer 1993). These findings made the authors question whether a return to indigenous methods as an approach to development was appropriate, given their findings that indigenous methods caused at least as much erosion as plough- and livestock-induced erosion. Fisher et al. (2003) in a study of the same lake using cores, trenches and exposures came to different conclusions. They argue that the Preclassic degradation was caused by initial clearance rather than agriculture, that high population is inversely related to soil erosion, and that the post-European soil erosion was tied to a loss in labour from the many contagions that accompanied the Spanish conquest.

In summary, since the late 1980s, many commentators have framed the discussion of human-induced soil change into debates over the Pristine Myth, which has pointed out once again how theory moulds perception of environmental change. What all of these erosion studies in Central Mexico underscore is our need for more refined soil histories using multiple lines of evidence and dating. Thus far, we have different chronologies of human impacts on soils, but one common sequence occurs as follows: (1) Holocene equilibrium soil development into the Early Preclassic; (2) soil erosion starting in the Preclassic, based on dates of the surfaces of buried soils and sediment cores; (3) soil development with less disturbance from sometime later in the Preclassic to the start of the Late Classic; (4) increased soil erosion in the Late Classic, but mitigated by widespread soil conservation; and (5) recovery and topsoil formation in the Post-Classic. The extensive terracing that diffused widely in the Late Classic did not expunge soil erosion but it may also have forestalled any widespread Malthusian menace that some scholars suggest for some sites. Yet, some sites such as Cancuén and Copán did have intensive soil erosion and sedimentation and little soil conservation.

Terracing

Agricultural terracing stood out to Maya archaeologists as early as the 1920s, but they did not recognise its implications for labour or as a form of intensive agriculture (Beach et al. 2002). Many studies have found that terracing across ancient Mesoamerica was a common and successful adaptation to the region's steep, tropical environments and eroded or skeletal soils. Donkin (1979, p. 131) writes that terracing occurs mainly in arid and semi-arid regions, though we know now that it also occurs in extremely wet areas. Indeed, in some areas with about 2,500 mm of precipitation, we have found terraces that continue to operate after 1200 years under the active geomorphic surfaces of tropical forests and milpas (Bocco 1991; Beach and Dunning 1995; Beach et al. 2002). In the early 1990s, there were still few real studies of terraces based on excavation and survey (e.g., Turner 1974 and Healy et al. 1983), but since then many studies have remedied this (e.g., Doolittle 1990; Dunning and Beach 1994; Fedick 1994; Treacy and Denevan 1994; Beach and Dunning 1995; Beach 1998a; Dunning et al. 1997; Beach et al. 2002). Excavation and survey of terrace systems are keys to understanding their construction, function, and soil evidence for past crops and cropping techniques.

The remains of ancient terraces include stone walls and support, soil sequences and artefacts; all remnants can provide dates for construction. Donkin (1979, p. 18), in his book on terracing in the Americas, wrote that terracing started by about 2500 BP in the Andes and Central Mexico. Since accelerated erosion started before 3000 BP, the need for terracing started earlier. Indeed, since Donkin's work, research in Northern Mexico, the Maya Lowlands and Central Mexico all points to terrace origins around 1000 BC in Northern Mexico. Hard and Roney (1998) and Hard et al. (1999) provide strong evidence for a sizable Archaic terrace group that dates to about 1050 BC, based on radiocarbon dates for maize and squash remnants and Late Archaic artefacts. In the Maya Lowlands, Hansen et al. (2002) reported late Middle Preclassic (c. 600–400 BC) terracing at Nakbe, Guatemala, and in the Central Highlands, Mountjoy and Gliessman (1988) stated that *cajeta* terracing starts by about 1000 BC. Despite the evidence for 3,000-year-old terracing, most terracing started in the Maya World only in the Classic Period (Beach et al. 2003a), whereas terracing is widespread geographically, culturally and chronologically across Central Mexico (Donkin 1979).

Worldwide, terraces function to provide a planting surface, maintain soil moisture favourable to crop growth by managing water infiltration, reduce soil erosion through slowing overland flow, catch or mound soils on eroded or thin-mantled slopes, and redirect water toward or away from lower lying fields or water sources (Doolittle 1985 and 1995; Beach et al. 2003). Many studies have interpreted evidence from excavation for each of these functions, and many others have developed taxonomies of terracing (Treacy and Denevan 1994; Beach and Dunning 1995; Beach et al. 2002). Such factors as landscape position, size, connection to drainage and irrigation, and materials form the bases for classifications. Poorly built or unmaintained terraces, however, are prone to mass wasting and gully erosion, because gullying by piping or waterfall action can undermine them (Beach and Dunning 1995).

What we know of construction materials reflects the limitations of methods: excavation and mapping on the active tropical slopes of Mesoamerica show well-made dams of heavy stones, but these may be the more persistent remnant anchors of a landscape that also included soil and vegetative berms. We know, for example, from the Basin of Mexico, vegetative walls such as maguey hedges (Sp. *bancales* and in Nahuatl *metepantlis*) served as semi-terraces that collected sediment around them over time (Palerm and Wolf 1957; Donkin 1979; Evans 1992). In these Mexica lands, aqueducts and canals irrigate the soil beds of rock-walled contour terraces (Cordova and Parsons 1997).

Interestingly, some terraces have unknown and perhaps unknowable functions in soils because they do not follow any usage rules we can ascertain. In northwestern Belize, for example, chert berms have erratic sizes and as often run parallel or diagonal to slopes as normal to them. The only clues are their Late Classic dates and largely chert construction. Perhaps these were field walls or rock piles removed for tilth and reapplied for mulch (Beach et al. 2002).

In the Maya Lowlands the greatest extent of terracing occurs around the Rio Bec in Mexico's south-central Yucatán Peninsula and around Caracol on the flanks of the Maya Mountains in Belize. Large scale terracing also occurs in Veracruz, Mexico. Sluyter and Siemens (1992) reported that more than 1000 km² of terracing are adjacent to the well-known wetland fields in Veracruz (see Chapter 5). These lie outside the Maya Lowlands, but the authors date them to a period that corresponds broadly to the Maya Classic. Turner (1974) estimated more than 10,000 km² of terraced lands in the Rio Bec region, and Healy et al. (1983) and Chase and Chase (1998) described Caracol's incredibly widespread terrace systems, which they argue must have been centrally controlled. Williams (1990) posits that many terraces evolve with little planning as only localised responses to landscapes, and many are built free-standing to dam eroding sediment and create a planting bed (Williams and Walter 1988). Beach et al. (2002) have also argued for more localised control near La Milpa and Dos Hombres in Northern Belize, though some terracing suggests broader watershed planning because the terraces connect to diversion channels and possibly intensively farmed bajos. Work in Tlaxcala in Central Mexico shows more integrated agroecosystems with sloping terraces that divert runoff into ***cajetas*** or basins, which farmers clean out and reapply to fields (Mountjoy and Gliessman 1988). Others studying the Aztec suggest that Late Post-Classic agricultural terracing was the result of household level decisions under growing populations (Smith and Price 1994). There is danger in reading central planning of soils and water management into an archaeological landscape because piecemeal building over time may appear the same as central planning, especially after many centuries (see Doolittle 2000, p.. 300–301).

Whether or not Mesoamericans centrally planned these soil conservation features, we are learning more and more about soil-terrace systems. Terracing occurred widely in Mesoamerica's lowlands and highlands, but did not occur in some sloping areas that warranted it. For example, many parts of Maya Mexico and Guatemala and seemingly everywhere in Belize had Classic Period terracing – Caracol, Xunantunich, La Milpa, Dos Hombres and the Petexbatún – but there was surprisingly little terracing at Copán, Honduras and around Tikal, Guatemala (Beach et al. 2002). Studies also suggest terracing was often a successful and sustainable agroecological system, though work at the same group of terraces at Caracol drew opposite conclusions on this point (Healy et al. 1983; Coultas et al. 1993). Research is also beginning to recognise the crops of ancient terracing, borne out in the latest chapter of Caracol terraces: carbon isotopes that suggest maize and other C4 grass species (Webb et al. 2004). Not surprisingly, studies have drawn more Malthusian conclusions about the sites without terracing such as Copán, with population-driven soil erosion and depletion (Wingard 1996), whereas high populations elsewhere such as in Central Mexico correlate with terrace construction to restore soils that had been truncated to their ***tepetate*** parent materials (Williams 1972: 626).

Field Walls

Albarradas or field walls occur across many sites in the Yucatán, especially Chunchucmil, Coba, Dzibilchaltun, Yalahau and Mayapan. Since these are such prominent parts of the sites, and even occur near the site centres, they must have some importance. In many cases they enclose a small area of a mound group and adjacent areas with no structures and often mediocre soils. The simple explanation is they were property boundaries of one kind or another, but chemical studies have not identified a proxy of soil use as yet.

At Yalahau, in the northeast Yucatán, however, Fedick and others have argued that low lying walled areas may have been for growing periphyton, microbial wetland mat communities dominated by cyanobacteria, as a superfertiliser that also has pesticide properties (Fedick et al. 2000). These research teams report evidence of possible periphyton use in the form of wetland sponge spicules and mollusc species in the forest soils near sites, and they are testing other evidence (Palacios-Mayorga et al. 2003). This should be a prime use of biomarkers as proxy evidence for wetland microbial inputs into upland soils.

Raised Beds and Ridges

Field ridges in upland, well-drained soils appear from prehistory to present indigenous techniques in many places in the Americas and in the world (Doolittle 2000, pp. 194–216 and 448–9; Whitmore and Turner 2001, pp. 123–7; Wilken 1987, p. 130). This is an early and comparatively low labour innovation that relies primarily on piling up planting areas either as mounds or ridges. Wilken (1987, pp. 130–44) described these techniques and debates their utility in the highlands of Mexico and Guatemala. Some indigenous farmers argue that they improve the planting bed, root penetration, rainfall interception, drainage, cold air drainage and aeration, but others are suspicious about the tradeoffs of labour for perceived gains. Whatever their utility, they appear so ubiquitously that they are a major part of the world and American soil landscapes both in active and fossil forms (Doolittle 2000).

Among the most intriguing example of upland field ridging are the fossil ridges at the ash-covered site of Ceren, El Salvador, which preserves a Classic Maya village from c. AD 650 (Sheets 2002: 8). Here with plentiful volcanic soil, low ridges formed the seed beds in all the milpas uncovered. Tell-tale for Classic Maya soil use, only one of eight milpas uncovered was in fallow, indicating intensive cultivation allowed by rich, volcanic soils; yet Parnell et al. (2002) found low phosphorus in two milpa samples. For the broader Middle America, Whitmore and Turner (2001, pp. 123–32) synthesised research that describes the techniques for building, cropping and maintaining these fields, which is based on soil 'pulverisation' that farmers with hand implements can only do with relatively soft soils. In many parts of Mexico, the ridges, called *camellones*, still echo the Nahautl word *cuemitl*. In contrast, the thin rocky soils of the northwest Yucatán cannot be manipulated thus, but farmers still focus on cropping the higher, naturally mounded karst uitzes (Beach 1998b).

Soil Management and Amendments

Until the 1960s, scholars guided by reports of Bishop de Landa from the sixteenth century assumed that Maya farmers had always used swidden or milpa techniques. These studies assumed milpa yields were sufficient to support low population densities that archaeologists had estimated (Cowgill 1962). But in the 1960s, researchers started to recognise that the great Maya cities were indeed habitation centres, rather than ritual centres, and that researchers turned up more and more mounds and some centres every year. Hence, population estimates exploded, and the search for means to feed the population led to evaluating new kinds of foods and new kinds of techniques to grow more food. Since ancient Maya civilisation lasted for so long and the population was so high, especially during the Late Classic (AD 600–900), the Maya must have used at least few intensive techniques to maintain food production.

Trying to explain ancient subsistence, researchers from several disciplines (including archaeology, geography, geology, soils and botany) evaluated many forms of evidence directly or indirectly related to soils. These included assessing basic soil productivity, assessing the use of lesser known crops like ramon, agaves and manioc (Bronson 1966, Evans 1992, Lentz 2000, Micksicek 1983, Wiseman 1983), and studying intensive and more efficient cropping techniques such as terracing, wetland raised and ditched fields (Dunning and Beach 2000), field ridging, fertilisation from animal and green manures, night soils, mulches and periphyton, recessional techniques, water management for irrigation, conservation and diversion, and the role of household or kitchen gardens.

Kitchen Gardens

Estimates vary for the importance of kitchen gardens or solares to Mesoamerican subsistence and their importance on soils. Kitchen gardens are a form of intensive or special soil management. As such, they usually require deeper soils and anthropogenic inputs, and may develop anthrosols that we can detect in ancient landscapes. These are probably partly the cause of the anthropic mollisols at sites like Chunchucmil and Mayapan (Beach 1998a).

There have been many studies of solares in Mesoamerica, and some show intensive production with high inputs and polycultural techniques that attempt to maximise land area by vertically arranging crops into multiple canopies (Atran 2003; Atran et al. 1996; Barrera et al. 1977; Killion 1992; Nations and Nigh

1980). For Central Mexico, Williams and Harvey (1997) suggests that kitchen gardens ranged from 0.3 to 0.9 ha and thus contributed greatly to subsistence. Yet for the Maya Lowlands, ethnohistorical analogy suggests more humble production. Caballero (1992) wrote that polycultural gardens produce mainly nutritional and dietary supplements, medicinal herbs and plants used as ornamentals and ritual objects, rather than bulk staples, and he concludes that kitchen-gardens produce only about 11 per cent of the average domestic calorie intake. Archaeological evidence, however, indicates that prehispanic gardens may have contributed more to subsistence, though as landscapes became more and more urbanised the distinction between former 'infields' and kitchen gardens likely became blurred (Folan et al. 1979; Smyth et al. 1995).

Assessing Soil Quality, Ancient Populations and Subsistence

Since numerous ancient Maya archaeological sites cover the Maya Lowlands and some of the sites are huge, spreading for many square kilometres, many have attempted to understand how the Maya fed themselves. Indeed, it is ironic that sprawling sites like Chunchucmil with thousands of inhabitants exist in areas with little to no soil, and water tables that lie perilously close to the surface and contamination from human waste. Today relatively small populations produce low yields with high labour inputs; for example, ancient Chunchucmil would have had to increase area or yield by about fourfold to feed the population (Beach 1998a). These puzzles have motivated a number of studies of soil productivity and potential productivity using different management techniques.

SOIL IMPACTS AND MANAGEMENT IN THE CARIBBEAN ISLANDS

Compared to Mesoamerica, the study of pre-Hispanic impacts on the soilscapes of the Caribbean Islands is in its infancy. Although the indigenous population of the Greater Antilles probablly numbered in the millions, by 1514 it had been reduced to a few thousands, chiefly as the result of the introduction of Old World diseases. This rapid obliteration of the Taino chiefdoms and other groups doomed them to relative obscurity and their record on the landscape has been only slowly recovered. A few early chroniclers documented many aspects of Taino culture, including some subsistence practices, most notably cultivation in *conucos*. Among these visitors was Bartolomé de Las Casas (who later became a famous protector of Native American rights in the Spanish American colonies), who also noted the apparent long record of Indian presence and environmental impact on Hispaniola:

> I have seen these mines of Cibao, one or two yards deep in the virgin earth, in the plains at the foot of some hills, burned wood and ashes as if a few days past a fire had been made there. And for the same reason we have to conclude that in other times the

> river came near there, and in such a place they made a fire, and afterwards the river moved farther away. Soil accumulated there, as the rains brought it down from the hills and covered the site. And because this could not happen except by the passage of many years and most ancient time, there is therefore a strong argument that the people of these islands and the mainland are very ancient (Las Casas 1990).

There has also been limited scientific study of ancient erosion and deposition, in part because of the existence of only a small handful of freshwater lakes suitable for paleoenvironmental investigation. Several cores have been analysed from Lake Miragoane, a cryptodepression in Haiti (Brenner and Binford 1988; Higuera-Gundy et al. 1999). These cores show evidence of deforestation and moderately increased sedimentation after about 1200 BP. However, the record of pre-Hispanic deforestation and sedimentation is comparatively minor in comparison to that in colonial times and especially the horrendous problems of the past 50 years.

By 1300 BP, farmers on Puerto Rico were sufficiently aware of soil erosion that they started building agricultural terraces. Although pre-Hispanic agricultural terracing has been reported in Puerto Rico since the 1930s, systematic investigation is more recent (Ortiz Aguilú et al. 1992; Oliver et al. 1999). The onset of terracing appears to be linked to the expansion of cultivation away from valley floors and coastal plains onto the lower slopes of mogotes in the island interior. The large majority of terraces investigated to date are stone-faced, dry slope, contouring terraces, but there are few examples of cross channel terraces and diversion weirs.

In drier areas of eastern Hispaniola ditch irrigation was employed in pre-Hispanic times, but the extent and dating of this practice remains poorly understood (Tabio 1989; Veloz Maggiolo 1976). By far the most widely practiced form of pre-Hispanic intensive agriculture was the construction of thousands of small raised beds or ***conucos*** (Sauer 1966). Typically ***conucos*** were constructed as a series of small circular earthen mounds on the order of 25 cm high and one metre in diameter, sometimes associated with shallow irrigation ditches. Often green mulching was used to help conserve the soils and moisture in the mounds. While cassava was the primary crop grown in ***conucos***, they were typically used for mixed-crop cultivation also including sweet potato, arrowroot, maize, bean, squash, peppers, peanuts, cotton and tobacco.

CONCLUSIONS

In our review of the soils history of Mesoamerica and the Caribbean Islands we have attempted to find patterns and trends for vastly different environments. One clear theme from many places is the advent of human impacts in the form of 'Maya clays' or eroded sediments over paleosols or organic lacustrine sediments. In the Maya Lowlands, the possible advent of human impacts comes first with change of more forest to more savanna species in the Petén of Guatemala about 5610 BP. This may just as well be caused by climate drying as human alteration, but by 3400 BP increased sedimentation of inorganic Maya clays and pollen of economic and disturbance taxa into depressions provides convincing evidence that deforestation and soil erosion had started (Brenner et al. 2003). Soil conservation and soil manipulation in the form of terracing also started in the Preclassic in the Petén and Belize (Hansen et al. 2002; Beach et al. 2002), but most conservation was slow in developing and did not become widespread till the Classic and, especially, the Late Classic. Thus the impacts of pioneer agriculture in the Preclassic aggraded many depressions and valleys (Beach et al. 2003a), but Late Classic soil conservation started a trend toward sustainable agriculture. Nonetheless, several important Late Classic Maya sites have little evidence of conservation and much evidence of erosion (Beach et al. 2003b). The loss of soil, however, stopped dramatically over much of the Maya Lowlands with collapses in the ninth and sixteenth centuries: forest returned, soil formed, and sedimentation slowed down to an organic trickle.

In Central Mexico, human induced soil alteration also started in the Preclassic, in the form of site construction and clearing for agriculture. Likewise, soil conservation started early, both in Central Mexico and Northern Mexico, by the Early Preclassic. Maize domestication must have started somewhere in this zone before 7000 BP (Pope et al. 2001), but evidence for major human-induced land use change that might have accompanied maize farming comes later, possibly as farmers moved up the slope. No single collapse quiets the environmental record, as occurred in the Maya Lowlands, and indeed the depopulation of before and after the Aztecs brought arguably more environmental impact because of the collapse of native populations that had maintained conservation features and the spread of livestock-induced erosion. The legacy of erosion on the upper piedmonts and hill slopes of Central Mexico are widespread, truncated soil profiles called ***tepetate***, and they still erode at nearly five times the rate of the areas with relict soils. In both Central Mexico and the Maya Lowlands, new pioneers have caused modern soil erosion from mechanised, modern techniques and ignorance about or indifference toward their newly deforested but ancient Mesoamerican fields. Some studies, though, have shown extremely unsustainable levels of erosion, such as the plot studies along the Pacific Coast of Mexico that show maize and grass soil loss as 70 and 49 t/ha/yr (Maas et al. 1988), soil erosion and GIS modeling studies to optimise the timing of land use actions (Millward and Mersey 1999), and a ^{137}Cs study under undisturbed deciduous forest that showed a soil loss of 13.2 t/ ha/yr (Garcia-Oliva et al. 1995). Soil erosion is particularly a problem in tropical soils because top soil layers hold a high percent of the major soil nutrients (Lal 1990). There can be little doubt that Mesoamerica has severe rates of soil erosion today, but we lack enough of these well designed and instrumented studies to give a full assessment.

In the Caribbean Islands, particularly the Greater Antilles, agriculture was slower to develop and most sedentary populations concentrated along coastal margins with relatively low erosion risks. However, as populations moved onto sloping inland terrain, the need to conserve soils became obvious and a tradition of terracing evolved. Unfortunately, this conservation knowledge was lost when indigenous populations were decimated in early Colonial times. The modern record of catastrophic soil loss in such places as Haiti stands in sharp contrast.

The rest of what we know about the history of soils in Mesoamerica has to do with soil manipulation to grow crops, ethnopedology studies, chemical assays, and archaeological excavation to discover what ancient people might have been doing on their lands to feed their hordes. We know that ancient urban and rural populations were extremely high, especially in Central Mexico (Williams 1989) and the ancient Maya Lowlands (Turner 1976), where agricultural productivity today is low and soils depleted. This riddle motivates our soil science today to try to answer questions of ancient subsistence, which might even provide us with some insights toward improving contemporary farming through agroecology (Gliessman 1998). We have started to gain insight into ancient soil knowledge, first from ethnographies of farmers about soils, combined in Central Mexico with codices that record remarkable Aztec soil knowledge. We have only the ethnopedologies from indigenous people and no such written records from the rest of Mesoamerica, and we can only ask in vain whether this Aztec knowledge has any precursors in the stream of literate cultures that came before.

Thus, we mainly get at ancient soil knowledge indirectly from archaeology – the different types and contexts of terraces and field walls, and from geoarchaeology – the biophysical, chemical and fossil evidence of remnant soils. These studies have shown impressively diverse agricultural architecture, especially since the Classic Periods, with many kinds of terracing and field walls that show extended control of the rural landscape: terraces across channels, on slope benches, on foot-slopes, on alluvial fans, on contours, and diversions toward and away from fields and away from freshwater sources. Thus, research does show the long stream of indigenous Mesoamerican soil knowledge, but we are just at the start of testing these terrace soils for the full suite of proxies that might give us ideas about how ancient people might have managed and altered soils for crop growth. Many papers have argued that these soil beds were heavily fertilised from night soil, organic matter and specifically periphyton to get at the riddle of ancient subsistence; whereas some have argued they were depleted by overuse to get at the riddle of collapse. Findings have been controversial as we have refined acceptable methods, but many new studies are coming on line that should begin to clarify our view of soil history.

Finally, the growing evidence for the impacts of pioneer farming parallel findings in many parts of the world, such as the buried soils and post-settlement alluvium of many parts of the United States (Beach 1994) and Mediterranean valleys (Wilkinson 2005). This is a fairly obvious idea because of its universality, but there is ample evidence that many development and natural resource planners – from

Mexico to China – still do not recognise the potential degradation of pioneering or the insight of groups with deep indigenous knowledge.

REFERENCES

Abrams, E. M. and D. J. Rue. 1988. 'The Causes and Consequences of Deforestation Among the Prehistoric Maya'. *Human Ecology* **16**: 377–95.

Atran, S. 2003. 'Anthropogenic Vegetation: A Garden Experiment in the Maya Lowlands'. In A. Gomez-Pompa, M. Allen, S. Fedick, J.Jiménez-Osornio (eds), *Lowland Maya Area: Three Millennia at the Human-Wildland Interface* (Binghamton, NY: Haworth Press), pp. 517–32.

Atran, S., D. Medin, N. Ross, E. Lynch, J. Coley, E. Ucan Ek' and V. Vapnarsky. 1996. 'Folkecology and Commons Management in the Maya Lowlands'. *Proceedings of the National Academy of Science*, **96**: 7598–603.

Barba, L., B. Ludlow, L. Manzanilla and R. Valadez. 1987. 'La Vida Doméstica De Teotihuacan: Un Estudio Interdisciplinario'. *Ciencia y Desarrollo* 77: 21–33.

Barba, L. and A. Ortiz. 1992. 'Análisis Químico de Pisos de Ocupación: un Caso Etnográfico en Tlaxcala, Mexico'. *Latin American Antiquity*, **3**: 63–82.

Barba, L., A. Ortiz, K. Link, L. López Luján and L. Lazos. 1996. 'Chemical Analysis of Residues in Floors and the Reconstruction of Ritual Activities at the Templo Mayor, Mexico'. In M. V. Orna (ed.), *Archaeolgical Chemistry: Organic, Inorganic, and Biochemical Analysis* (Washington, DC: American Chemical Society), pp. 139–56.

Barrera, M.A., A. Gómez-Pompa and C. Vázquez-Yáñez. 1977. 'El Manejo de las Selvas por los Mayas'. *Biótica*, **2**(2): 47–60.

Beach, T. 1992. 'Estimating Erosion in Medium-Size Drainage Basins'. *Physical Geography*,**13**: 206–24.

Beach, T. 1994. 'The Fate of Eroded Soil: Sediment Sinks and Sediment Budgets of Agrarian Landscapes in Southern Minnesota, 1851–1988'. *Annals of the Association of American Geographers*, **84**: 5–28.

Beach, T. 1998a. 'Soil Catenas, Tropical Deforestation, and Ancient and Contemporary Soil Erosion in the Petén, Guatemala'. *Physical Geography*, **19**(5): 378–405.

Beach, T. 1998b. 'Soil Constraints on Northwest Yucatán, Mexico: Pedoarchaeology and Maya Subsistence at Chunchucmil'. *Geoarchaeology*, **13**(8): 759–91.

Beach, T. and N. Dunning. 1995. 'Ancient Maya Terracing and Modern Conservation in the Petén Rain Forest of Guatemala'. *Journal of Soil and Water Conservation*, **50**: 138–45.

Beach, T. and N. Dunning. 1997. 'An Ancient Maya Reservoir and Dam at Tamarindito, Petén, Guatemala'. *Latin American Antiquity*, **8**(1): 20–29.

Beach, T., N. Dunning, S. Luzzadder-Beach and V. Scarborough. 2003a. 'Depression Soils in the Lowland Tropics of Northwestern Belize: Anthropogenic and Natural Origins'. In A. Gomez-Pompa, M. Allen, S. Fedick and J. Jiménez-Osornio (eds), *Lowland Maya Area: Three Millennia at the Human-Wildland Interface* (Binghamton, NY: Haworth Press), pp. 139–74.

Beach, T., S. Luzzadder-Beach, N. Dunning, J. Hageman and J. Lohse. 2002. 'Upland Agriculture in the Maya Lowlands: Ancient Conservation in Northwestern Belize'. *The Geographical Review*, **92**(3): 372–97.

Beach, T., S. Luzzadder-Beach, J. Lohse and D. Cook. 2003b. 'Pre-Columbian Aggradation and Ancient Maya Wetland Soils in Belize'. Second International Conference on Soils and Archaeology, Pisa, Italy, May 2003.

Beach, T., N. Dunning, S. Luzzadder-Beach, J. Lohse and D. Cook. 2006. 'Ancient Maya Impacts on Soils and Soil Erosion in the Central Maya Lowlands'. *Catena*, in press.

Bennett, H.H. 1926. 'Agriculture in Central America'. *Annals of the Association of American Geographers*, **16**: 63–84.

Bennett, H.H. 1945. 'Soil Conservation in Latin America'. In F. Verdoorn (ed.), *Plants and Plant Sciences in Latin America* (Waltham, MA: *Chronica Botanica*), pp. 165–9.

Bequelin, P. and D. Michelet. 1994. 'Demografía en la Zona Puuc: El Recurso del Metodo'. *Latin American Antiquity*, 5(4): 287–8.

Bethell, P. and I. Maté. 1988. 'The Use of Soil Phosphate Analysis in Archaeology: A Critique'. In J. Henderson (ed.), *Scientific Analysis in Archaeology*, Monograph No. 19. (Oxford: Oxford University Committee), pp. 1–29.

Birkeland, P. 1999. *Soils in Geomorphology.* Oxford: Oxford University Press.

Bocco, G. 1991. 'Traditional Knowledge for Soil Conservation in Central Mexico'. *Journal of Soil and Water Conservation*, **46**(5): 346–8.

Bocco, G. 1993. 'Gully Initiation in Quaternary Volcanic Environments under Temperate Sub-humid Seasonal Climates'. *Catena*, **20**(5): 495–513.

Bocco, G. and F. Garcia-Oliva. 1992. 'Researching Gully Erosion in Mexico'. *Journal of Soil and Water Conservation*, **47**: 365–7.

Bryan, K. 1925. 'Date of Channel Trenching (Arroyo Cutting) in the Arid Southwest'. *Science*, **62**(1607): 339.

Brenner, M. and M.W. Binford. 1988. 'A Sedimentary Record of Human Disturbance from Lake Miragoane, Haiti'. *Journal of Paleolimnology*, **1**: 85–97.

Brenner, M., D. Hodell, J. Curtis, M. Rosenmeier, F. Anselmetti and D. Ariztegui. 2003. 'Paleolimnological Approaches for Inferring Past Climate in the Maya Region: Recent Advances and Methodological Limitations'. In A. Gomez-Pompa, M. Allen, S. Fedick, and J.Jiménez-Osornio (eds), *Lowland Maya Area: Three Millennia at the Human-Wildland Interface*, (Binghamton, NY: Haworth Press), pp. 45–76.

Bronson, B. 1966. 'Roots and the Subsistence of the Ancient Maya'. *Southwestern Journal of Anthropology*, **22**: 251–79.

Burney, D.A. and L.P. Burney. 1994. 'Holocene Charcoal Stratigraphy from Laguna Tortuguero, Puerto Rico, and the Timing of Human Arrival on the Island'. *Journal of Archaeological Science*, **21**: 273–81.

Bull I.D., I. Simpson, P. van Bergen and R. Evershed. 1999. 'Muck 'n' Molecules: Organic Geochemical Methods for Detecting Ancient Manuring'. *Antiquity*, **73**: 86–96.

Butzer, K.W. 1992a. 'The Americas Before and After 1492: An Introduction to Current Geographical Research'. *Annals of the Association of American Geographers*, **82**: 345–68.

Butzer, K.W. 1992b. 'From Columbus to Acosta: Science, Geography, and the New World'. *Annals of the Association of American Geographers*, **82**: 543–65.

Butzer, K.W. 1993. 'No Eden in the New World'. *Nature*, **362**: 15–17.

Butzer, K.W. & E.K. Butzer. 1993. 'The Sixteenth-Century Environment of the Central Mexican Bajío: Archival Reconstruction from Colonial Land Grants and the Question of Spanish Ecological Impact'. In K. Mathewson (ed.), *Culture, Form, and Place: Essays in Cultural and Historical Geography*, Geoscience and Man, 32 (Baton Rouge: Louisiana State University), pp. 89–124.

Caballero, J. 1992. 'Maya Homegardens: Past, Present, and Future'. *Ethnoecologica* **1**: 35–53.

Carlos Donado, Luis. 1996. 'Characteristics and Potentials of Selected Soils in the Peten Region of Guatemala'. Unpublished Master's Thesis, Plant and Soil Science Department, Knoxville: University of Tennessee.

Carter, W.E. 1969. *New Lands and Old Traditions: Kekchi Cultivators in the Guatemalan Lowlands*. Gainsville: University of Florida Press.

Chase, A. and D. Chase. 1998. 'Scale and Intensity in Classic Period Maya Agriculture: Terracing and Settlement at the 'Garden City' of Caracol, Belize'. *Culture and Agriculture*, **20**(2/3): 60–77.

Conservation Foundation. 1954. 'Soil Erosion Survey of Latin America', reprinted in the *Journal of Soil and Water Conservation*, **9**.

Conyers, L. and H Spetzler. 2002. 'Geophysical Exploration at Ceren'. In P. Sheets (ed.), *Before the Volcano Erupted* (Austin: University of Texas Press), pp 24–32.

Cook, D., B. Kovacevich, T. Beach and R. Bishop. 2006. 'Deciphering the Inorganic Geochemical Record of Ancient Human Activity using ICP-MS: Two Examples from Late Classic Soil Floors at Cancuén, Guatemala'. *Journal of Archaeological Science*, 33(5): 628–40.

Cook, S.F. 1948. *The Historical Demography and Ecology of the Teotlalpan*, Ibero-Americana 33. Berkeley and Los Angeles: University of California Press.

Cook, S.F. 1949. *Soil Erosion and Population in Central México*, Ibero-Americana 34. Berkeley and Los Angeles: University of California Press.

Cooke, R.U. and R.W. Reeves. 1976. *Arroyos and Environmental Change in the American Southwest*. Oxford: Clarendon Press.

Cordova, C. and J. Parsons. 1997. 'Geoarchaeology of an Aztec Dispersed Village'. *Geoarchaeology*, **12**(3): 177–210.

Coultas, C.L., M.E. Collins and A.F. Chase. 1993. 'The Effects of Ancient Maya Agriculture on Terraced Soils of Caracol, Belize'. In J.E. Foss, M.E. Thompson and M.W. Morris (eds), *Proceedings of the First International Pedo-Archaeology Conference*, Special Publication No. 93–03 (Knoxville, TN: Agricultural Experiment Station, University of Tennessee), pp. 191–210.

Coultas, C.L., Y.P. Hsieh and T.J. Post. 1997. 'A Toposequence of Soils in Northern Belize, Central America'. *Soil and Crop Science of Florida*, **56**: 117–24.

Cowgill, U.M. 1962. 'An Agricultural Study of the Southern Maya Lowlands'. *American Anthroplogist*, **64**: 273–86.

Curet, L.A. 2003. 'Issues on the Diversity and Emergence of Middle-Range Societies of the Ancient Caribbean: A Critique'. *Journal of Archaeological Research*, **11**: 1–42.

Dahlin, B., T. Beach, S. Luzzadder-Beach, D. Hixson, S. Hutson, A. Magnoni, D. Mansell and D. Mazeau. In press. 'Reconstructing Agricultural Self-Sufficiency at Chunchucmil, Yucatan, Mexico'. *Ancient Mesoamerica.*

de Landa, Diego. 1982. *Relación de las Cosas de Yucatán.* Mexico City: Editorial Porrua.

Denevan, W. 1992. 'The Pristine Myth: The Landscape of the Americas in 1492'. *Annals of the Association of American Geographers*, **82**(3): 369–85.

Diamond, Jared. 2004. *Collapse: How Societies Choose to Fail or Succeed.* New York: Viking.

Donnelly, T.W., G. Horne, R. Finch and E. López-Ramos. 1990. 'Northern Central America; the Maya and Chortis Blocks'. In G. Dengo and J. Chase (eds), *The Caribbean Region: The Geology of North America* (Boulder, Colorado: Geological Society of America).

Donkin, R. 1979. *Agricultural Terracing in the Aboriginal New World*, Viking Fund Publications in Anthropology No. 56. Tucson, AZ:. University of Arizona Press.

Doolittle, W.E. 1985. 'The Use of Checkdams for Protecting Downstream Agricultural Lands in the Prehistoric Southwest: A Contextualist Analysis'. *Journal of Anthropological Research*, **41**: 279–305.

Doolittle, W.E. 1990. 'Terrace Origins: Hypotheses and Research Strategies'. *Yearbook of the Conference of Latin Americanists Geographers*, **16**: 94–7.

Doolittle, W.E. 1995. 'Indigenous Development of Mesoamerican Irrigation'. *Geographical Review*, **85**: 301–23.

Doolittle, W.E. 2000. *Cultivated Landscapes of Native North America.* Oxford: Oxford University Press.

Dunning, N.P. 1992. *Lords of the Hills: Ancient Maya Settlement in the Puuc Region, Yucatán, Mexico*, Monographs in World Prehistory 15. Madison, WI: Prehistory Press.

Dunning, N.P. and T. Beach. 1994. 'Soil Erosion, Slope Management, and Ancient Terracing in the Maya Lowlands'. *Latin American Antiquity*, **5**(1): 51–69.

Dunning, N.P. and T. Beach. 2000. 'Stability and Instability in Pre-Hispanic Maya Landscapes'. In D. Lentz (ed.), *Imperfect Balance: Landscape Transformations in the Precolumbian Americas* (New York: Columbia University Press), pp. 179–202.

Dunning, N.P. and T. Beach. 2004a. 'Fruit of the Luum: Lowland Maya Soil Knowledge and Agricultural Practices'. *Mono y Conejo: Journal of the Mesoamerican Archaeological Research Laboratory*, **2**: 3–15.

Dunning, N.P. and T. Beach. 2004b. 'Noxious or Nurturing Nature? Maya Civilization in Environmental Context'. In C. Golden and G. Borgstede (eds), *Continuities and Changes in Maya Archaeology: Perspectives at the Millennium* (New York: Routledge), pp. 125–41.

Dunning, N.P., T. Beach and D. Rue. 1997. 'The Paleoecology and Ancient Settlement of the Petexbatún Region, Guatemala'. *Ancient Mesoamerica*, **8**: 1–12.

Dunning, N.P., T. Beach, P. Farrell and S. Luzzadder-Beach. 1998. 'Prehispanic Agricultural Adaptive Regions in the Maya Lowlands'. *Culture and Agriculture*, **20**(4): 87–101.

Dunning, N.P., S. Luzzadder-Beach, T. Beach, John Jones, V. Scarborough and T.P. Culbert. 2002. 'Arising from the Bajos: The Evolution of a Neotropical Landscape and the Rise of Maya Civilization'. *Annals of the Association of American Geographers*, **92**(2): 267–83.

Eidt, R.C. 1984. 'Theoretical and Practical Considerations in the Analysis of Anthrosols'. In G. Rapp, Jr., G. Gifford and J.A. Gifford (eds), *Archaeological Geology* (New Haven: Yale University Press), pp. 155–90.

Evans, S.T. 1992. 'The Productivity of Maguey Terrace Agriculture in Central Mexico during the Aztec Period'. In T.W. Killion (ed.), *Gardens of Prehistory: The Archaeology of Settlement Agriculture in Greater Mesoamerica* (Tuscaloosa, Alabama: University of Alabama Press), pp. 92–115.

Food and Agriculture Organization of the United Nations (FAO). 2005. *The State of the World's Forests*. Rome: United Nations.

Fedick, S.L.1994. 'Ancient Maya Agricultural Terracing in the Upper Belize River Area'. *Ancient Mesoamerica*, 5: 107–27.

Fedick, S.L. 1996. 'An Interpretive Kaleidoscope: Alternative Perspectives on Ancient Agricultural Landscapes in the Maya Lowlands'. In S.L. Fedick (ed.), *The Managed Mosaic: Ancient Maya Agriculture and Resource Use* (Salt Lake City: University of Utah Press), pp. 107–31.

Fedick, S.L. 2003. 'Archaeological Evidence for Ancient and Historic Resource Use Associated with the El Edén Wetland, Northern Quintana Roo, Mexico'. In A. Gomez-Pompa, M. Allen, S. Fedick and J.Jiménez-Osornio (eds), *Lowland Maya Area: Three Millennia at the Human-Wildland Interface* (Binghamton, NY: Haworth Press), pp. 339–60.

Fedick, S.L., B.A. Morrison, B.J. Andersen, S. Boucher, J.C. Acosta and J.P. Mathews. 2000. 'Wetland Manipulation in the Yalahau Region of the Northern Maya Lowlands'. *Journal of Field Archaecology*, 27(2): 131–52.

Fernández, F.G., R. Terry, T. Inomata and M. Eberl. 2002. 'An Ethnoarchaeological Study of Chemical Residues in the Floors and Soils of Q'eqchi' Maya Houses at Las Pozas, Guatemala'. *Geoarchaeology*, **17**: 487–519.

Fernandez, F., K. Johnson, R. Terry, S. Nelson and D.Webster. 2005 'Soil Resources of the Ancient Maya at Piedras Negras, Guatemala'. *Soil Science Society of America Journal* **69**: 2020–32.

Fisher, C., H.P. Pollard, I. Israde, V.H. Garduno and S. Banerjee. 2003. 'Reexamination of Human Induced Environmental Change Within the Lake Pátzcuaro Basin, Michoacán, Mexico'. *Proceedings of the National Academy of Sciences*,**100**(8): 4957–62.

Folan, W.J., L.A. Fletcher and E.R. Kintz. 1979. 'Fruit, Fiber, Bark, and Resin: The Social Organization of a Maya Urban Center, Coba, Quintana Roo, Mexico'. *Science*, **204**: 697–701.

Furley, P.A. 1987. 'Impact of Forest Clearance on the Soils of Tropical Cone Karst'. *Earth Surface Processes and Landforms*, **12**: 523–9.

Garcia-Oliva, F., L. Martinez and J.M. Maass. 1995. 'Long-Term Net Soil Erosion as Determined by ^{137}Cs Redistribution in an Undisturbed and Perturbed Tropical Deciduous Forest Ecosystem'. *Geoderma*, 68: 135–47.

Gliessman, S.R. 1998. *Agroecology: Ecological Processes in Sustainable Agriculture*. Chelsea, MI: Ann Arbor Press.

Gomez-Pompa, A., M. Allen, S. Fedick and J. Jiménez-Osornio (eds). 2003. *Lowland Maya Area: Three Millennia at the Human-Wildland Interface*. Binghamton, NY: Haworth Press.

Gong, Z., X. Zhang, J. Chen and G. Zhang. 2003. 'Origin and Development of Soil Science in Ancient China'. *Geoderma* **115**(3): 3–13.

Graham, E. In Press. 'A Neotropical Framework for Terra Preta'. In W. Balée and C. Erickson (eds), *Time and Complexity in Historical Ecology: Studies in the Neotropics*, New York: Columbia University Press.

Gunn, J.D., J. Foss, W. Folan, M. del Rosario Domínguez and B. Faust. 2002. 'Bajo Sediments and the Hydraulic System of Calakmul, Campeche, Mexico'. *Ancient Mesoamerica*, **13**: 297–315.

Hansen, R.D., S. Bozarth, J.Jacob, D. Wahl and T. Schreiner. 2002. 'Climatic And Environmental Variability In The Rise of Maya Civilization: A Preliminary Perspective from Northern Petén'. *Ancient Mesoamerica*,**13**(2): 273–95.

Haug, G.H., D. Günther, L.C. Peterson, D.M. Sigman, K.A. Hughen and B. Aeschliman. 2003. 'Climate and the Collapse of Maya Civilization'. *Science*, **299**: 1731–5.

Hard, R. and J.R. Roney. 1998. 'A Massive Terraced Village Complex in Chihuahua, Mexico: 3000 Years Before Present'. *Science*, **279**: 1661–4.

Hard, R., J.E. Zapata, B.K. Moses and J.R. Roney. 1999. 'Terrace Construction in Northern Chihuahua, Mexico: 1150 B.C. and Modern Experiments'. *Journal of Field Archaeology*, **26**: 129–46.

Healy, P.F, J.D.H. Lambert, J.T. Arnason and R.J. Hebna. 1983. 'Caracol, Belize: Evidence of Ancient Maya Agricultural Terraces'. *Journal of Field Archaeology*, **10**: 397–410.

Heine, K. 2003. 'Paleopedological Evidence of Human-Induced Environmental Change in the Puebla-Tlaxcala Area (Mexico) During the Last 3,500 Years'. *Revista Méxicana de Ciencias Geológicas* **20**(3): 235–44.

Higuera-Gundy, A., M. Brenner, D.A. Hodell, J.H. Curtis, B.W. Leyden and M.W. Binford. 1999. 'A 10,300 ^{14}C yr Record of Climate and Vegetation Change from Haiti'. *Quaternary Research*, **52**: 159–70.

Hodell, D., J. Curtis and M. Brenner. 1995. 'Possible Role of Climate in the Collapse of Classic Maya Civilization'. *Nature*, **375**: 391–4.

Hodell, D., M. Brenner and J. Curtis. 2000. 'Climate Change in the North American Tropics and Subtropics since the Last Ice Age: Implications for Environment and Culture'. In D. Lentz (ed.), *Imperfect Balance: Landscape Transformations in the Precolumbian Americas* (New York: Columbia University Press), pp. 13–38.

Hodell, D., M. Brenner, J. Curtis and T. Guilderson. 2001. 'Solar Forcing of Drought Frequency in the Maya Lowlands'. *Science*, **292**: 1367–70.

Holley, G.R., R. Dalan, W.I. Woods and H. Watters. 2000. 'Implications of a Buried Preclassic Site in Western Belize'. In S.R. Ahler (ed.), *Mounds, Modoc, and Mesoamerica: Papers in Honor of Melvin L. Fowler*, Scientific Papers Vol. 28 (Springfield: Illinois State Museum), pp. 111–24.

Islebe, G., H. Hooghiemstra, M. Brenner, J. Curtis and D. Hodell. 1996. 'A Holocene Vegetation History from Lowland Guatemala'. *The Holocene* **6**(3): 265–71.

Isphording, W.C. 1984. 'The Clay of Yucatán, Mexico: A Contrast in Genesis'. In A. Singer and E. Galan (eds) *Palygorskite-sepiolite: Occurences, Genesis, and Uses* (Amsterdam: Elsevier Science Publishers), pp. 59–73.

Jacob, J. 1996. 'Archaeological Pedology in the Maya Lowlands'. In M. Collins (ed.), *Pedological Perspectives in Archaeological Research* (Madison, WI: Soil Science Society of America), pp. 51–80.

Jacob, J. 1995. 'Ancient Maya Wetland Agricultural Fields in Cobweb Swamp, Belize: Construction, Chronology, and Function'. *Journal of Field Archaeology*, **22**: 175–90.

Jacob, J. and C.T. Hallmark. 1996. 'Holocene Stratigraphy of Cobweb Swamp, a Maya Wetland in Northern Belize'. *GSA Bulletin*,**108**(7): 883–91.

Jenny, H. 1980. *Factors of Soil Formation: A System of Quantitative Pedology*. New York: McGraw Hill.

Killion, T.W. 1992. 'Residential Ethnoarchaeology and Ancient Site Structure: Contemporary Farming and Prehistoric Settlement Agriculture at Matacapan, Veracruz, Mexico'. In T.W. Killion (ed.), *Gardens of Prehistory: The Archaeology of Settlement Agriculture in Greater Mesoamerica* (Tuscaloosa, Alabama: University of Alabama Press), pp. 1–13.

Kirkby, M. 1972. *The Physical Environment of the Nochixtlán Valley, Oaxaca*, Vanderbilt University Publications in Anthropology 2. Nashville: Vanderbilt University Press.

Lal, R. 1990. *Soil Erosion in the Tropics: Principles and Management*. New York: McGraw-Hill.

Las Casas, B. de. 1990. *Bartolomé de Las Casas: History of the Indies*, edited and translated by A.M. Collard. New York: Harper and Row.

Lehmann, J., D.C. Kern, B. Glaser and W.I. Woods (eds). 2003. *Amazonian Dark Earths*. Dordecht: Kluwer Academic.

Lentz, D.L., (ed). 2000. *Imperfect Balance: Landscape Transformations in the Precolumbian Americas*. New York: Columbia University Press.

Lowdermilk, W.C. 1953. *Conquest of the Land through 7,000 Years*, Agriculture Information Bulletin 99. Washington, D.C.: Natural Resources Conservation Service.

Maas, J.M., C.F. Jordan and J. Sarukhán. 1988. 'Soil Erosion and Nutrient Losses in Seasonal Tropical Agroecosystems under different Management Techniques'. *Journal of Applied Ecology*, **25**: 595–607.

Maffi, L. 1999. 'Domesticated Land, Warm and Cold: Linguistic and Historical Evidence on Tenejapa Tzetzal Maya Ethnopedology'. In T.L. Gragson and B.G. Blount (eds), *Ethnoecology: Knowledge, Resources, and Rights* (Athens: University of Georgia Press), pp. 41–56.

Manzanilla, L. 1996. 'Corporate Groups and Domestic Activities at Teotihuacan'. *Latin American Antiquity*, 7: 228–56.

Manzanilla, L. and L. Barba. 1990. 'The Study of Activities in Classic Households: Two Case Studies from Coba and Teotihuacan'. *Ancient Mesoamerica*, **1**: 41–9.

McClung de Tapia, E., E. Solleiro-Robolledo, J. Gama-Castro, J. Villalpando and S. Sedov. 2003. 'Paleosols in the Teotihuacan Valley, Mexico: Evidence for Paleoenvironment and Human Impact'. *Revista Mexicana de Ciencias Geológicas*, **20**(3): 270–82.

McAuliffe, J.R., P.C. Sundt, A. Valiente-Banuet, A. Casas and J.L. Viveros. 2001. 'Pre-Columbian Soil Erosion, Persistent Ecological Changes, and Collapse of a Subsistence Agricultural Aconomy in the Semi-arid Tehuacán Valley, Mexico's "Cradle of Maize"'. *Journal of Arid Environment*, **47**: 47–75.

Melville, E. 1994. *A Plague of Sheep*. Cambridge: Cambridge University Press.

Middleton, W.D. and T.D.Price. 1996. 'Identification of Activity Areas by Multi-element Characterization of Sediments from Modern and Archaeological House Floors using Inductively Coupled Plasma-Atomic Emission Spectroscopy'. *Journal of Archaeological Science*, **23**: 673–87.

Miksicek, C.H.1983. 'Macrofloral Remains of the Pulltrouser Area: Settlement and Fields'. In B.L. Turner II and P.D. Harrison (eds), *Pulltrouser Swamp: Ancient Maya Habitat, Agriculture, and Settlement* (Austin: University of Texas Press), pp. 94–104.

Millward, A.A. and J. Mersey. 1999. 'Adapting the RUSLE to Model Soil Erosion Potential in a Mountainous Tropical Watershed'. *Catena*, **38**: 109–29.

Morett, L.A., J. Arroyo-Cabrales and O. Polaco. 1998. 'Tocuila, a Remarkable Mammoth Site in the Basin of Mexico'. *Current Research in the Pleistocene*, **15**: 118–20.

Morley, S.G. 1946. *The Ancient Maya*. Palo Alto, CA: Stanford University Press.

Morley, S.G. 1956. *The Ancient Maya*, 3rd edition, revised by G. Brainard. Palo Alto, CA: Stanford University Press.

Mountjoy, D.C. and S.R. Gliessman. 1988. 'Traditional Management of a Hillside Agroecosystem in Tlaxcala, Mexico: An Ecologically Based Maintenance System'. *American Journal of Alternative Agriculture*, **3**: 10.

Nations, J. 1979. 'Population Ecology of the Lacandon Maya'. PhD diss., Southern Methodist University, Dallas, Texas.

Nations, J.D. and R. Nigh. 1980. 'The Evolutionary Potential of Lacandon Maya Sustained-Yield Tropical Forest Agriculture'. *Journal of Anthropological Research*, **36**: 1–30.

Newsom, L.A. and E.S. Wing. 2004. *On Land and Sea: Native American Uses of Biological Resources in the West Indies*. Tuscaloosa: University of Alabama Press.

O'Hara, S.L., F. Street-Perrott and T. Burt. 1993. 'Accelerated Soil Erosion Around a Mexican Highland Lake Caused by Prehispanic Agriculture'. *Nature*, **362**: 48–51.

Oliver, J.A., J. Rivera Fontán and L.A. Newsom. 1999. 'Arqueología del Barrio Caguana, Puerto Rico: Resultados Preliminares de las Temporadas 1996–1997'. In *Trabajos de Investigación Arqueológica en Puerto Rico*, Tercer Encuentro de Investigadores. (San Juan: Occasional Publication of the División de Arqueología, Instituto de Cultura Puertorriqueña), pp. 7–26.

Olson, G.W. 1981. 'Archaeology: Lessons on Future Soil Use'. *Journal of Soil and Water Conservation*, **36**: 261–4.

Ortiz Aguilú, J.J., J. Rivera Meléndez, A. Príncipe Jácome, M. Meléndez Maiz and M. Lavergne Colberg. 1991. 'Intensive Agriculture in Pre-Columbian West Indies: The Case for Terraces'. In A. Cummins and P. King (eds), *Proceedings of the XIV International*

Congress for Caribbean Archaeology (Barbados: International Association for Caribbean Archaeology), pp. 278–85.

Palacios-Mayorga S., A.L.Anaya, E. González-Velazquez, L. Huerta-Arcos and A. Gómez-Pompa. 2003. 'Initial Evidence for Use of Periphyton As an Agricultural Fertilizer by the Ancient Maya Associated with the El Edén Wetland, Northern Quintana Roo, Mexico'. In A. Gomez-Pompa, M. Allen, S. Fedick and J.Jiménez-Osornio (eds), *Lowland Maya Area: Three Millennia at the Human-Wildland Interface* (Binghamton, NY: Haworth Press), pp 389–400.

Palerm, A. and E.R. Wolf. 1957. 'Ecological Potential and Cultural Development in Mesoamerica'. In L. Krader and A. Palerm (eds), *Studies in Human Ecology: A Series of Lectures Given at the Anthropological Society of Washington*, Pan American Union Social Science Monographs 3. (Washington: Pan American Union).

Palka, J.W. 1999. 'Lacandon Maya Culture Change and Survival in the Lowland Frontier of the Expanding Guatemalan and Mexican Republics'. In J.G. Cusick (ed.), *Studies in Culture Contact: Interaction, Culture Change, and Archaeology* (Carbondale, IL: Southern Illinois University Center for Archaeological Investigations), pp. 457–75.

Parnell, J.J., R. Terry and C. Golden. 2001. 'The Use of in-Field Phosphate Testing for the Rapid Identification of Middens At Piedras Negras, Guatemala'. *Geoarchaeology*, **16**: 855–73.

Parnell, J.J., R.Terry and P. Sheets. 2002. 'Soil Chemical Analysis of Ancient Activities in Ceren, El Salvador: A Case Study of a Rapidly Abandoned Site'. *Latin American Antiquity*, **13**(3): 331–42.

Petersen, J.B. 1997. 'Taino, Island Carib, and Prehistoric Amerindian Economies in the West Indies: Tropical Forest Adaptations to Island Environments'. In S.M. Wilson (ed.), *The Indigenous People of the Caribbean*,(Gainesville: University Press of Florida), pp. 118–30.

Piperno, D. and D.M. Pearsall. 1998. *The Origins of Agriculture in the Lowland Neotropics*. San Diego CA: Academic Press.

Pohl, M., K.O. Pope, J.G. Jones, J.S. Jacob, D.R. Piperno, S.D. deFrance, D.L. Lentz, J.A. Gifford, M.E. Danforth and J.K. Josserand. 1996. 'Early Agriculture in the Maya Lowlands'. *Latin American Antiquity*, 74(4): 355–72.

Pohl, M., K. Pope and C. von Nagy. 2002. 'Olmec Origins of Mesoamerican Writing'. *Science*, **298**: 1984–7.

Pope, K., A. Ocampo, G. Kinsland and R. Smith. 1996. 'Surface Expression of the Chicxlulub Crater'. *Geology*, **22**: 527–30.

Pope, K.O., M. Pohl, J.G. Jones, D. Lentz, C. von Nagy, F.J. Vega and I. Quitmyer. 2001. 'Origin and Environmental Setting of Ancient Agriculture in the Lowlands of Mesoamerica'. *Science*, **292**: 1370–73.

Popenoe de Hatch, M., E. Ponciano, Q. Tomás Barrientos, M. Brenner and C. Ortloff. 2002. 'Climate and Technological Innovation at Kaminaljuyu, Guatemala'. *Ancient Mesoamerica*, **13**: 103–14.

Pulido, J. and G. Bocco. 2003. 'The Traditional Farming System of a Mexican Indigenous Community'. *Geoderma*, **111**(3–4): 249–65.

Reid, L. and Dunne, T. 1996. *Rapid Evaluation of Sediment Budgets*. Cremlingen, Germany: Catena Verlag Press.

Rice, D.S. and T.P. Culbert. 1990. 'Historical Contexts for Population Reconstruction in the Maya Lowlands'. In T.P. Culbert and D.S. Rice (eds), *Precolumbian Population History in the Maya Lowlands* (Albuquerque: University of New Mexico Press), pp. 1–36.

Rincon Maunter, C. 1999. 'Man and the Environment in the Coixtlahuaca Basin of Northwestern Oaxaca, Mexico: Two Thousand Years of Historical Ecology'. Unpublished Ph. D. Dissertation, Austin: University of Texas.

Robinson, E.J., P. Farrell, K.F. Emery and G.E. Braswell. 2002. 'Preclassic Settlements and Geomorphology in the Highlands of Guatemala: Excavations at Urias, Valley of Antigua'. In M. Love, H. Escobedo and M. Popenoe de Hatch (eds), *Incidents of Archaeology in Central America and Yucatán* (Lanham, MD: University Press of the Americas), pp. 251–76.

Robles Ramos, R.1950. 'Apuntes Sobre la Morfologia de Yucatán'. *Boletín de la Sociedad Mexicana de Geografía y Estadística*, **69**: 27–106.

Rosenmeier, M., D. Hodell, M. Brenner, J. Curtis and T. Guilderson. 2002. 'A 4,000-Year Lacustrine Record of Environmental Change in the Southern Maya Lowlands, Petén, Guatemala'. *Quaternary Research* 57: 183–90.

Rouse, I. 1992. *The Tainos: Rise and Fall of the People Who Greeted Columbus*. New Haven, CT: Yale University Press.

Sahagún, Fray Bernardino de. 1900 [1577]. *General History of the Things of New Spain* [1577], 2nd revised edn, 4 vols. Santa Fe, New Mexico: School of American Research.

Sauer, C.O. 1952. *Agricultural Origins and Dispersals*. New York: American Geographical Society.

Sauer, C.O. 1966. *The Early Spanish Main*. Berkeley: University of California Press.

Scarborough, V. 1993. 'Water Management in the Southern Maya Lowlands: An Accretive Model for the Engineered Landscape'. In V. Scarborough and B. Isaac (eds), *Economic Aspects of Water Management in the Prehispanic New World*, Research in Economic Anthropology Supplement 7 (Greenwich, CT: JAI Press), pp. 17–68.

Shang, C. and H. Tiessen. 2003. 'Soil Organic Sequestration and Stabilization in Karstic soils of the Yucatán'. *Biogeochemistry*, **62**: 177–96.

Sheets, Payson (ed.). 2002. *Before the Volcano Erupted: The Ancient Ceren Village in Central America*. Austin: University of Texas Press.

Sluyter, A. 1994. 'Intensive Wetland Agriculture in Mesoamerica: Space, Time and Form'. *Annals of the Association of American Geographers*, **84**(4): 557–84.

Sluyter, A. 1997. 'Regional Holocene Records of the Human Dimension of Global Change: Sea-Level and Land-Use Change in Prehistoric Mexico'. *Global and Planetary Change*, **14**: 127–46.

Sluyter, A. and A.H. Siemens. 1992. 'Vestiges of Prehispanic, Sloping Field-Terraces on the Piedmont of Central Veracruz, Mexico'. *Latin American Antquity*, **3**: 148–60.

Smith, M.E. and T.J. Price. 1994. 'Aztec-Period Agricultural Terraces in Morelos, Mexico: Evidence for Household-level Agricultural Intensification'. *Journal of Field Archaeology*, **21**: 169–79.

Smyth, M.P., C.D. Dore and N. Dunning. 1995. 'Interpreting Prehistoric Settlement Patterns: Lessons from the Maya Center of Sayil, Yucatán'. *Journal of Field Archaeology*, **22**: 321–46.

Sommer, M. and E. Schlicting. 1997. 'Archetypes of Catenas in Respect to Matter – A Concept for Structuring and Grouping Catenas'. *Geoderma* **76**: 1–33.

Spores, R. 1969. 'Settlement, Farming Technology, and Environment in the Nochixtlan Valley'. *Science*, **166**: 557–69.

Steggerda, M. 1941. *Maya Indians of Yucatán*, Publication 531. Washington, D.C.: Carnegie Institution of Washington.

Stevens, R.L. 1964. 'The Soils of Middle America and their Relation to Indian Peoples and Cultures'. In R.C. West (ed.), *Natural Environment and Early Cultures, Handbook of Middle American Indians*, (Austin: University of Texas Press), vol. 1, pp. 265–315.

Tabío, E.E. 1989. *Arqueología: Agricultura Aborigin Antillana*. Havana: Editorial de Ciencias Sociales.

Terry, R.E., P. Hardin, S. Houston, M.W. Jackson, S.D. Nelson, J. Carr and J. Parnell. 2000. 'Quantitative Phosphorus Measurement: A Field Test Procedure for Archaeological Site Analysis At Piedras Negras, Guatemala'. *Geoarchaeology*, **15**: 151–66.

Thompson, J.E.S. 1954. *The Rise and Fall of Maya Civilisation*. Norman: University of Oklahoma Press.

Treacy, J.M. and W. Denevan. 1994. 'The Creation of Cultivable Land Through Terracing'. In N. Miller and K. Gleason (eds), *The Archaeology of Garden and Field* (Philadelphia: The University of Pennsylvania Press), pp. 91–110.

Turner, B.L. II. 1974. 'Prehistoric Intensive Agriculture in the Mayan Lowlands'. *Science*, **185**(4146): 118–24.

Turner, B.L. II. 1976. 'Prehistoric Population Density in the Maya Lowlands: New Evidence from Old Approaches'. *Geographical Review*, **66**: 73–82.

Turner, B.L. II. 1979. 'Prehispanic Maya Terracing in the Central Maya Lowlands: Problems of Agricultural Intensification'. In N. Hammonds and G.R. Willey (eds), *Maya Archaeology and Ethnohistory* (Austin: Univ. of Texas Press), pp. 103–15.

Turner, B.L. II. 1983. *Once Beneath the Forest: Prehistoric Terracing in the Rio Bec Region of the Maya Lowlands*. Boulder, CO.: Westview.

Tuan, Y., 1966. 'New Mexican Gullies: A Critical Review and Some Recent Observations'. *Annals of the Association of American Geographers*, **56**: 573–97.

Vaughan, H.H., E. Deevey and S. Garrett-Jones. 1985. 'Pollen Stratigraphy of Two Cores from the Petén Lake District, with an Appendix on Two Deep-water Cores'. In M. Pohl (ed.), *Prehistoric Lowland Maya Environment and Subsistence Economy* (Cambridge, MA: Harvard University Press), pp. 73–89.

Veloz Maggiolo, M. 1976. *Medioámbiente y Adaptación Humana en la Prehistoria de Santa Domingo*, Colección Historia y Sociedad 24. Universidad Autónoma de Santo Domingo.

Wahl, D., R. Byrne, T. Schreiner and R. Hansen. In Press. 'Holocene Vegetation Change in the Northern Petén and Implications for Maya Prehistory'. *Quaternary Research*.

Webb, E., H. Schwarz and P. Healy. 2004. 'Carbon Isotope Evidence for Ancient Maize Agriculture in the Maya Lowlands'. *Journal of Archaeological Science*, **31**: 1039–52.

Wells, E.C., R Terry, J. Parnell, P. Hardin, M. Jackson and S. Houston. 2000. 'Chemical Analyses of Ancient Anthrosols in Residential Areas at Piedras Negras, Guatemala'. *Journal of Archaeological Science*, **27**: 449–62.

Whitmore, T. and B. Turner II. 2001. *Cultivated Landscapes in Mesoamerica on the Eve of the Conquest*. Oxford: Oxford University Press.

Wilken, G. 1987. *Good Farmers: Traditional Agricultural Resource Management in Mexico and Central America*. Berkeley: University of California Press.

Wilkinson, T. 2005. 'Soil Erosion and Valley Fills in the Yemen Highlands and Southern Turkey: Integrating Settlement, Geoarchaeology, and Climate Change'. *Geoarchaeology*, **20**(2): 169–92.

Williams, B. 1972. 'Tepetate in the Valley of Mexico'. *Annals of the Association of American Geographers*, **62**: 618–26.

Williams, B. 1989. 'Contact Period Rural Overpopulation in the Basin of Mexico: Carrying Capacity Models Tested with Documentary Data'. *American Antiquity*, **54**(4): 715–32.

Williams, B. and H.R. Harvey. 1997. *The Códice de Santa Mariía Asunción: Facsimile and Commentary: Households and Lands in Sixteenth-Century Tepetlaoztoc*. Salt Lake City: University of Utah Press.

Williams, B. and C.A. Ortiz-Solorio. 1981. 'Middle American Folk Soil Taxonomy'. *Annals of the Association of American Geographers*, **71**(3): 335–58.

Williams, L. 1990. 'Agriculture Terrace Evolution in Latin America'. *Yearbook of the Conference of Latin Americanist Geographers*, **16**: 82–93.

Williams, L. and B. Walter. 1988. 'Controlled-erosion Terraces in Venezuela'. In W. Moldenhauer and N.Hudson (eds), *Conservation Farming on Steep Lands*. Ankeny, Iowa: Soil and Water Conservation Society of America.

Wilson, S.M., H. Iceland and T. Hester. 1998. 'Preceramic Connections Between Yucatán and the Caribbean'. *Latin American Antiquity*, **9**: 342–52.

Wiseman, F. 1983. 'Analysis of Pollen from the Fields at Pulltrouser Swamp'. In B.L. Turner and P.L. Harrison (eds), *Pulltrouser Swamp: Ancient Maya Habitat, Agriculture, and Settlement in Northern Belize* (Austin: University of Texas Press), pp. 105–19.

Wingard, J. 1996. 'Interactions Between Demographic Processes and Soil Resources in the Copán Valley, Honduras'. In S. Fedick (ed.), *The Managed Mosaic: Ancient Maya Agriculture and Resource Use* (Salt Lake City: The University of Utah Press).

Woods, W.I. and J. M.McCann. 1999. 'The Anthropogenic Origin and Persistence of Amazonian Dark Earths'. *The Yearbook of the Conference of Latin American Geographers*, **25**: 7–14.

5

Wetlands as the Intersection of Soils, Water and Indigenous Human Society in the Americas

Sheryl Luzzadder-Beach and Tim Beach

> Standing in stark contrast to drylands are wetlands. Whereas the former generally suffer from a paucity of moisture that inhibits agriculture, the latter have an abundance that also places extreme limits on cultivation. There are several types of wetlands ... they can all be considered areas transitional between terrestrial and aquatic ecosystems where the water table is usually at or near the surface, or land is covered by shallow water and where at least one of the three following conditions exists: the land periodically, if not always, supports predominantly hydrophytes; the substrate is undrained hydric soil; or the substrate is saturated non-soil (Doolittle 2000, p. 413).

INTRODUCTION

One of the most important chapters of Pre-Columbian soil history is Native American manipulation of wetland soils. Many wetlands in the Americas are so altered by humans that it is often difficult to draw a line between what is cultural and what is natural (Denevan 1992, Erickson 2000, p. 317). One reason why is that human interaction with wetlands in the Americas stretches far back to the early migrations from Beringia. Wetlands often preserved the earliest remains of culture in artefacts and human remains (Purdy 2001), including at one of the earliest sites at Monte Verde (Meltzer 1997). Wetlands sediments also preserve ecological records of past environments and human activity in the proxy records of paleosols, pollen, mollusc shells and other organic remains, and oxygen isotopes. But some wetlands themselves may be considered cultural artefacts, either unintentionally or meticulously planned. Two examples are deforestation, which through reduced transpiration can induce water tables to rise and wetlands to form, and wetlands that have been ditched and reformed into wetland fields by ancient civilisations all over the Americas. This latter interaction closely links the history of soils to hydrology and culture and represents one of the most widespread and intensive landscape changes by Pre-Columbian people.

Native Americans transformed wetlands of many types into a variety of patterns, including ridged fields, ditched fields, sunken gardens, and ***chinampas*** or so-called 'floating gardens'. To find a baseline or 'pristine' environment for comparison with wetland alternation in the Americas, we need to look as far back as the end of the Pleistocene (Denevan 1992; Erickson 2000). Since this history is so long and social and biophysical factors are so complex, research on wetland complexes has used an extensive tool-kit from a broad cross section of disciplines. Some disciplines have been predisposed to a biophysical perspective that natural processes are ubiquitous and dominant in soil formation, and other disciplines are predisposed to a cultural perspective; wherein:

> The dominant perspective … is that cultural systems encapsulate, mediate, and translate natural systems so thoroughly that the properties of the physical world are, in essence, irrelevant to our understanding of human societies ... Culture is … capable of complete intermediation of the natural world (Kolata 2000, p. 163).

The examples in this chapter show that physical factors are also important to culture and a relevant part of the human history of soils.

Many social scientists continue to be wary of their long history with 'nineteenth-century geographical and environmental determinism' (Kolata 2000, p.164), and focus only on culture within the 'theory of mutual embeddedness of the material and the cultural' (Kolata 2000, p.163). But over the last few decades, growing scientific research has repeatedly demonstrated that human action and the environment are complex, coupled systems. Scholars have used historical, geological, archaeological, ecological, and particularly geographical perspectives to interpret anthropogenic wetland environments. To these ends, many have reviewed progress in this field in indigenous Mesoamerica (see Chapter 4, also Sluyter 1994; Whitmore and Turner 2001); and reviewed progress and these approaches for ancient South America (e.g., Denevan 2002; Erickson 2000). Denevan (1992) and Doolittle (1992, 2000) summarised much of what is known about aboriginal agriculture in North America. Herein, we follow the threads of the major evidence about wetland soils manipulation in the Americas.

Pioneering work discovering and interpreting anthropogenic wetland landscapes in Pre-Columbian Latin America began largely with the efforts of geographers and archaeologists in the late 1960s and early 1970s (Parsons and Bowen 1966; Denevan 1970; Armillas 1971; Turner and Harrison 1983; Sluyter 1994). This work had it roots in expeditions like the Carnegie Institution's 'An Archaeological Reconnaissance by Air in Central America' in 1929, piloted by Col. Charles A.

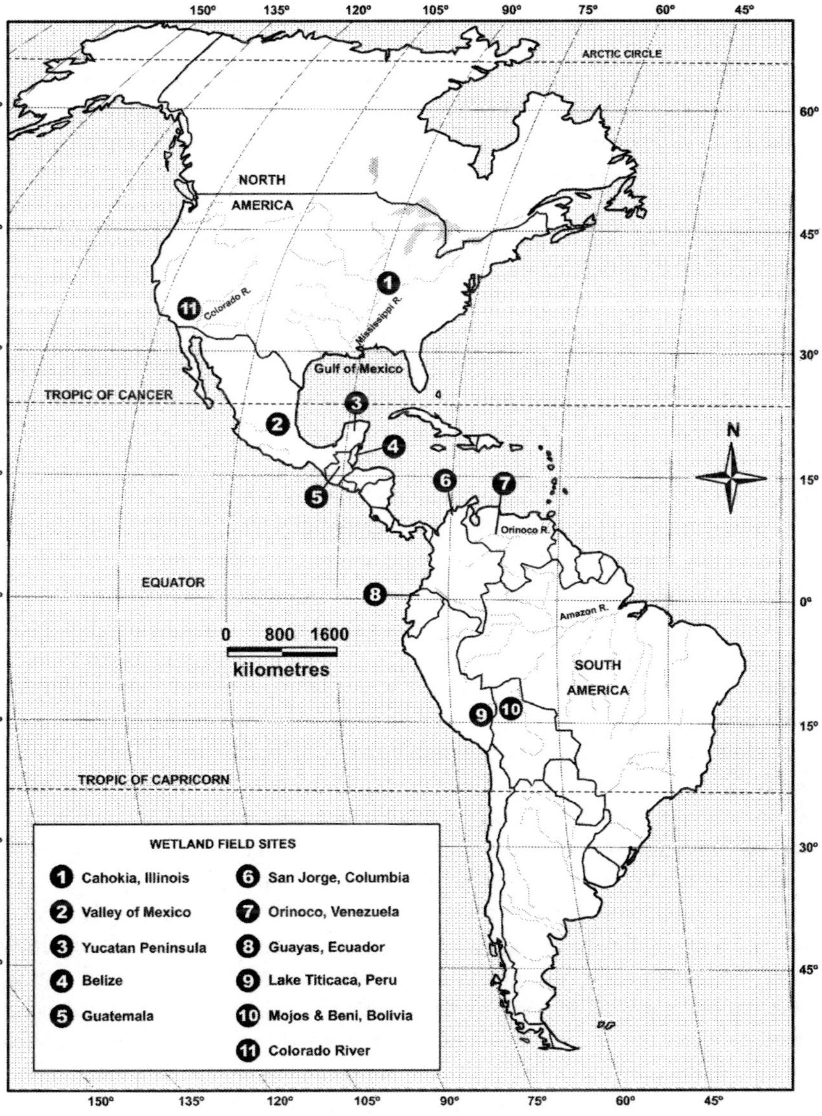

Figure 5.1. Wetlands in the Americas

Lindbergh and published in *The Geographical Review* (Ricketson and Kidder 1930, p.177). This research was an early venture in remote sensing and surveyed ancient Maya sites and environments in Guatemala, Belize and the Yucatán Peninsula (Figure 5.1). Other early twentieth-century research efforts, like the Carnegie's own work in the Yucatán (Steggerda 1941), studied the broad resource base of Maya sites, but many mid-century studies were focused on specific archaeology sites (Dunning and Beach 2004). As archaeological studies expanded to broader regional surveys in the 1960s and 1970s, they found more and more wetland patterns based on increasingly sophisticated aerial photography. For more than 40 years, since Palerm and Wolf (1957, p. 28) first hypothesised the ground patterns as ***chinampas***, scholars of several disciplines have worked to understand these patterns (Coe 1964; Parsons 1976; Nichols and Frederick 1993), but there is still much debate about how they formed in the changing environments of the Late Holocene.

Modern excavations of wetland fields include Armillas' (1971) groundbreaking research in the basin of Mexico, but because wetlands are difficult to excavate and their cultural artefacts unpromising, much early work on wetland fields in the Americas was on mapping patterns. Pioneering research by Turner and Harrison (1981, 1983) in Pulltrouser Swamp, Belize, began an era of more detailed studies of these remarkably patterned sites. The most frequent interpretation of the wetland sites is that humans created or modified them for agriculture. A variety of classifications of wetland fields' shapes and functions grew from this field work, but another interpretation was that they are patterned ground formed by a combination of natural and anthropogenic processes (Jacob 1995; Pope et al. 1996; Beach and Luzzadder-Beach 2002; Luzzadder-Beach et al. 2003). Turner and Harrison concluded that their particular wetland features in Pulltrouser Swamp were not natural features such as gilgai (patterned ground formed from vertisols under alternating wet and dry conditions). Continued fieldwork based on excavation and reconstruction has revealed more detail about the soil types, crops and ecology of these sites across the Americas (Turner and Harrison 1983; Denevan 2002; Erickson 2000; Kolata 2000; McClung de Tapia 2000). In this chapter we focus on significant Pre-Columbian sites in the Americas, including a typology of wetland soils and patterns, with regional discussions of the wetlands of Central Mexico; wetland and ***bajo*** agriculture in the Maya Lowlands of Mesoamerica; South America, regionally divided between the Andean Highlands and seasonally flooded lowlands in South America; and floodplain agriculture in North America (Figure 5.1).

WETLAND SOILS

Wetland soils in general are unique in that the climate component of soil formation is less important than in upland soils. Wetland soils are soils influenced by the seasonal or perennial influence of water. These soils vary according to how much of the time water inundates or floods them. Soils that have seasonal inundation tend to be mineral soils with mottled colours: greys from water-saturated reducing conditions, and yellows or oranges indicative of seasonally well-drained, oxidising conditions. Soil scientists give the horizons (e.g. B or C) of these mineral soils a 'g' designator for the gleyed-reduced minerals of the soil. Soils that have a constantly high water table are dominated by reduction and tend to be grey in colour, or sometimes pale blue and green. Soils that are inundated by water most of the time are dominated by higher fractions of organic matter, and thus are darker in colour, often dark browns and black. One soil order, histosol, is waterlogged and thus dominated by organic matter for more than half of the upper 80 cm (Soil Survey Staff 1998). Common terms for histosols are peats and mucks, depending on whether they have more or less organic matter. Mineral soils that are waterlogged fall into Aquic or other Great Groups of such soil orders as entisols, vertisols, inceptisols, mollisols and others. These soil characteristics have many implications for crops and natural vegetation communities that can grow in these conditions. Different crops have different requirements at different times in their growing cycles for moisture, but most grow best with plentiful micropore water held electrostatically to soil surfaces, and little macropore water that would reduce aeration of plant roots and aerobic microbial communities. This explains the imperative to drain inundated soils and build up (or raise) fields far enough above the water table for root growth to take advantage of micropore water and aeration in macropores.

TYPES AND FUNCTIONS OF WETLAND FIELD COMPLEXES

Wetland field research across the Americas shows that wetland fields range widely in pattern and dimensions (Turner and Harrison 1983; Parsons and Denevan 1989; Erickson 2000; Denevan 2002; Beach and Luzzadder-Beach 2003; Luzzadder-Beach et al. 2003). Their shapes range from herringbone patterns of ridged fields in South America to elongated ditched and raised fields in central Mexico, to rectilinear patterns resembling a lattice or spider's web pattern in the Maya Lowlands of Central America. Their heights range everywhere from less than 1 m to 3 m, and their spatial dimensions range from 4 m to 20 m wide, and from 10 m to 200 m long. This diversity in shape and size supports a variety of functions and necessitates different methods of manipulation.

From Pre-Columbian times people used wooden and stone implements such as stakes and spades to excavate the soils and form ditches between planting areas (Erickson 2000; Parsons and Denevan 1989, p. 220). They deposited the excavated soils on top to create raised planting platforms, to prevent crops from becoming waterlogged during part or all of the year. With minimum hydraulic manipulation afforded by perishable wooden or earthen dams, the canals excavated between the raised fields may also have served the dual purpose of draining away excess water in the rainy season, and delivering irrigation water during drier periods. William Denevan, one of the key scholars of Pre-Columbian landscape manipulation, has synthesised three main reasons for canalising wetlands in South America. The overall function of ditching fields was to expand soil resources (Denevan 2002, p. 264) by building anthrosols mounds. A secondary function was for irrigation or drainage, but further investigations of sites near Lake Titicaca, for example, revealed that ditch or depression patterns in many field systems did not seem to function to carry water away from the sites, but rather, may have been built to block flooding (Denevan 2002, p. 260). Third, the typical layout of ditches and fields may be for energy conservation, using water's high specific heat to help provide night-time frost protection in highland sites, especially depression areas prone to cold air drainage (Erickson 2000; Denevan 2002, p. 266).

Extensive excavation and mapping have discerned many other purposes of these raised fields beyond the basic three of reclamation, irrigation-drainage and warming. Raised field agriculture manages soil water by controlling water tables to maintain aeration, drainage and soil moisture infiltration. Soils manipulation also supplies and recycles nutrients through transferring organic muck and accumulated sediments from the adjacent canals to the planting platforms (Erickson 2000, p. 334). Wetland manipulations also provide an integrative agriculture and aquaculture system (akin to paddy rice ecosystems) through expanded and mobile wetland canal and dam management (Erickson 2000). Another function is to control water quality. Ancient farmers could have managed the canal and dam systems to improve water quality, separating fresh water from saline and alkaline waters (Denevan 2002, p. 269, citing Erickson 1992).

Several studies have gone beyond excavation and mapping to understand wetland fields by reconstructing field systems, employing modern scientific agricultural research, and poring through ethnohistorical literature. In Peru and Bolivia, Erickson used all of these approaches, testing these functions through experimental rehabilitation of ancient wetland fields (Erickson 2000; Denevan 2002) and finding precedence in ethnographic studies and validation in modern agricultural research (Erickson 2000). In Tabasco and Quintana Roo, Mexico, Gliessman (1991) studied both ancient and modern indigenous agriculture on raised field sites. He found modern fields farmed using indigenous methods to be an average of four times more productive than fields prepared and farmed mechanically. The abundant yields of wetland raised fields form one of the keys to understanding the population levels, urbanisation and civilisation of several regions of Pre-Columbian America. There have been other successful ***chinampas*** rehabilitation projects elsewhere in Mesoamerica, but they have been far from panaceas (Gomez-Pompa 1982). One was a state-run project by Tabasco Mexico in the 1970s, called the Camelons Chontales Project. Chapin (1988) explains how they tried to build ***chinampas*** but made many mistakes reminiscent of the mistakes made by early wetland restoration projects in the United States (National Research Council 1992). The failure of this project does not reduce the potential or past importance of wetland agriculture, but exemplifies typical errors of agricultural development – not including local people, taking account of ancient knowledge, or precisely planning the soils and hydrology. Indeed, after the government gave up on the Chontales Project, a local Maya group made the fields productive (Erickson 1998, p. 42).

CENTRAL MEXICO

Chinampas in the Valley of Mexico

The floating gardens of the Valley of Mexico, or ***chinampas***, are perhaps the best-known example of wetland agriculture in the Americas and the world. The Mexica-Aztec capital, Tenochtitlán, had about 9,000–12,000 ha of land in ***chinampas***, which Denevan called 'one of the most intensive systems of agriculture, past or present, in the Western Hemisphere' (Denevan 1970, p. 647; Sluyter 1994; Crossley 2004). But the ***chinampas*** long predate Aztec use, dating back to at least AD 800, and may have originated as early as 2000 years BP (Denevan 1970, p. 647; Denevan 2002, p. 235). They represent 'a highly specialised form of intensive agriculture' in the Basin of Mexico (McClung de Tapia 2000, p. 134). But, not all wetland fields are full-fledged ***chinampas***. Other excavations in the early 1990s found that some wetland fields near Teotihuacán were more likely just drained fields (McClung de Tapia 2000, pp. 140–1). McClung de Tapia (2000) placed the dates for ***chinampas*** as most likely to be middle Post-Classic (AD 1150– AD 1350), based upon her review of data from others' publications and excavations. Sluyter (1994) dated the tenure of intensive wetland agriculture in the Basin of Mexico fields at Xochimilco from 1400 BC to 400 BC, and Teotihuacán's from 600 BC to the present, also based on an extensive review of others' publications and excavation data (Sluyter 1994). Everyone who has studied these fields knows the difficulty of dating agricultural features and the variable range of robustness of the evidence. Some ***chinampas*** fields at Xochimilco, on the edge of Mexico City, are still in use today for intensive flower, herb, and vegetable market gardening and tourism, and field size has grown steadily over the last century.

These artificial island gardens are similar in function to the raised field complexes of South America, and also vary in scale and form, but are more consistently elongated or rectangular (Denevan 1970, p. 647). Studies of fossil remnants suggested that the earliest ***chinampas*** were narrower than their modern counterparts (Denevan 1970). ***Chinampas*** occur in a variety of hydrologic settings: in shallow lake waters, in swampy ground, and on seasonally inundated lake margins (Parsons and Denevan 1989[1967], p. 219). The so-called floating islands parallel other wetland fields with open canals surrounding built-up garden islands, well-edged with deliberately planted trees such alders (*Alnus*) and willows (*Salix*) (Parsons and Denevan 1989[1967], p. 219).

The native alders and willows stabilise the island margins, and ancient farmers layered strips of aquatic vegetation across the islands, building up a vegetative net to hold and raise the soil bed. Indigenous farmers cleaned the canals and maintained island fertility by scooping lake-bottom mud and organic matter onto these vegetative mats to build up the artificial islands (Coe 1964; Armillas 1971). These island fields ranged from about 3 m wide and 30 m long, up to squares of 100 m or more per side (Parsons and Denevan 1989[1967]; Carrasco and Sessions 1998; Smith 2003). Crop irrigation occurred by the high water table, evident in canal surfaces, capillary rise and water applied by hand from the canals. Since Lake Texcoco is saline, farmers must guard against saltwater intrusion with a system of dikes (Denevan 1970; Coe 1964; Armillas 1971).

The population of the Mexica capital of Tenochtitlan was extremely high, estimated from 200,000 to at 300,000 in Aztec times (Sluyter 1994; Carrasco and Sessions 1998), and required remarkably intensive food production systems such as upland terraces and the *chinampas*. The *chinampas* used a fine-tuned rotation and multi-cropping of agricultural commodities (Denevan 1970). Sluyter used an estimate of 9,000 ha of *chinampas* and their yield ranges to estimate food production enough to sustain a population of 171,000 annually (Sluyter 1994). The types of crops that could have been grown on these plots included amaranth, squash, maize, beans, chili and flowers among others, rotated in from seed beds or floating seed mats (West and Augelli 1966; Carrasco and Sessions 1998; Smith 2003).

CENTRAL AMERICA

Ancient Maya Wetland Agriculture

Polygonal features in the wetlands of the Maya Lowlands have fascinated scholars since the 1960s. These patterns appear in a variety of wetland environments, ranging from perennially to seasonally wet. The patterns have different shapes in these environments. In perennial wetlands, they appear to be human made canals and drained and/or raised fields. Researchers have interpreted these features as the product of Ancient Maya intensive wetland agriculture in either the Preclassic (1200 BC–AD 250) or the Classic Period (AD 250–850) (see preceding chapter by Beach et al.). Indeed, some hailed this interpretation as an answer to the riddle of ancient Maya subsistence in the Late Classic (AD 550–850), when Maya population and population growth were at their highest. But thus far the studies disagree about how and when the canals and fields formed. The debate centres on the degree to which these polygonal patterns are natural versus anthropogenic features; their dating; and on what was the range of ancient Maya activities in the perennial wetlands

(Turner and Harrison 1983; Siemens and Puleston 1972; Siemens 1983; Siemens et al. 1988; Jacob 1995; Jacob and Hallmark 1996; Pohl et al. 1996).

The lines of evidence that bear upon these questions are surface patterns stratigraphically uncovered from excavations, soil and water characteristics, paleoecological proxies, and artefacts. Conclusions from any one area may not have regional implications, because wetlands are a complex of several finely tuned parameters of an environment that might change dramatically with slight changes in any one parameter. As in other regions of the Americas, polygonal surface patterns occur all over the perennial wetlands of the Maya Lowlands, and many appear quite strikingly as anthropomorphic and others as natural. In fact, some of these features could be any number of possible natural features, and to identify canals reliably requires mapping showing regional connections and careful excavation through the full sequence of sediments. Adding further to the problem is modern disturbance. Farmers are draining, burning and ploughing these areas, which is greatly altering soil stratigraphy and organic preservation.

There exist several major views in the literature about the patterns. The first research on the topic was careful to examine natural explanations such as gilgai, an Australian aboriginal term for mounds formed from expanding and contracting clays (Siemens and Puleston 1972). Much of this region has such clays, and the gilgai topography occurs all over the bajos or seasonally wet environments, but most gilgai are low enough for ploughing to remove them (Eswaran et al. 1999), though they may still impart curious surface patterns when viewed from aerial photographs. Some rectilinear features may indeed be gilgai, but gilgai could not be present in truly perennially wet areas, because they required drying for clay contraction and wetting for clay expansion to form features. Paleogilgai, though, could be inherited in perennial wetlands from drier periods.

The earliest discovered fields were those on the coastal plain margins of the Yucatán Peninsula (Figure 5.1, and Figure 4.2 in Chapter 4) (Siemens and Puleston 1972) and next in coastal Veracruz, Mexico in the Olmec and Totonac areas (Schmidt 1977; Siemens 1982). Siemens et al. (1988) studied wetland agriculture in Yucatán, Mexico by excavating and using a series of cores to obtain samples for analysing phytoliths, pollen and ceramics from different depths. They found cross-shaped maize phytoliths comprised 13–16 per cent of the fine silt fraction at 90–110 cm depths and maize pollen reached 7 per cent of all pollen at 150 cm. The grey, silty-clay layer with most of the ceramics, economic pollen, and phytoliths also had plentiful charcoal and no layering, whereas the overlying sandy layers had layering and few economic proxies and artefacts (Hebda et al. 1991; Siemens 1998). The dating based on ceramics could distinguish only that below 60 cm the sediments had Preclassic or Early Classic (Totonac) elements but the radiocarbon dates were curiously modern (Siemens 1998).

The major area of debate has been between the interpretations of Turner and Harrison's projects (1981, 1983, and Harrison 1996) and of Bloom, Pohl and Pope's projects (1983, 1990 and 1996). Turner and Harrison used archaeological site analysis together with surface mapping of rectilinear features and excavations across the features to show the presence of wetland agriculture with both drained and raised fields during the Late Classic. The main difference in interpretation of stratigraphy is substantial. The Turner group interpreted two types of field: channelised fields at the dryland edge of wetlands and raised fields deeper in the wetlands, which they reasoned Maya farmers built up from canal excavations. They focused their interpretation on dominance of Late Classic artefacts in the upper soils of these rectilinear features and reasoned that the underlying grey, mottled clay was the Maya fill used to build fields. They used the following evidence to interpret these as Late Classic planting surfaces: the geometric regularity of fields; the mottled fill; the buried soil under the fill; artefacts occurring mainly above and below the fill layer; and the fill's pollen, indicative of a mixture of upland and wetland species (Turner and Harrison 1983). Unfortunately the pollen had suffered much degradation and the notion that this pollen indicated economic activity is 'less secure' (Turner and Harrison 1983, p. 114).

The Pohl, Pope, and Bloom (Bloom et al. 1983; Pohl 1990; Pope et al. 1996; Jacob 1995; Pohl et al. 1996) group interpreted their many excavations across similar features and a large excavation through one of Turner's excavations to show a different pattern (Pope et al. 1996, p. 167). They substantially deepened the excavation to obtain a longer chronology and clearer picture of the stratigraphy. Based on thin section analysis of minerals and chemical analyses, they interpreted the grey clay 'fill' as a combination of naturally precipitated gypsum and carbonate and anthropogenically eroded and then deposited clay (Pope et al. 1996, p. 170). Both groups found the following pattern: an upper organic clay soil that is 10–50 cm thick, a sequence of sediments dominated by gypsum and calcium carbonate that is about 100 cm thick, and a variable paleosol (Pope et al. 1996, p. 167; Turner 1983, p. 45). Turner's group found the paleosol only in auger cores, and interpreted it to be the old wetland soil upon which the Maya built the fields, whereas Pope et al.'s (1996, p. 170) deeper pit provides evidence that it formed before the Preclassic (based on ceramics and radiocarbon dates), had tremendous amounts of maize pollen and charcoal from Preclassic Maya agriculture, and is similar to the current swamp-forest soil of today. Farther into the wetland, the paleosol is capped by 124 cm thick peat that could only form near the surface, thus implying the water table had risen. The upper peat of the wetter areas corresponds to the gypsum-rich grey clay that started to form farther inland from about the Preclassic to Classic Period boundary (c. AD 200). Pohl's group argued that the gypsum-rich layers formed quickly because there is very little organic matter, or that the high levels of sulphide catalysed organic matter oxidation. They explained that this grey clay formed as sea level rose and the mottled appearance was due to a mixture of gypsum, carbonate and organic clay shown in the thin sections.

Evidence for Maya canal building may show different epochs responding to different regional environmental impacts. Jacob (1995) studied ancient wetland fields and canals at the site of Colha in Cobweb Swamp, Belize (Figure 4.1), reasoning that Maya canal building by the Classic or Preclassic was intended to manage rising water levels. Pohl et al. (1996) confirmed Bloom et al.'s findings, placing the intensification of ditching in the Maya Lowlands at 1000 BC, and abandonment of the ditched wetland fields in Belize largely by the Classic Period due to permanent flooding. Pohl et al. further reported that '[f]ield manipulations often involved minor modifications of natural hummocks' (Pohl et al. 1996, p. 355). These authors introduced the idea that the patterned or 'rectilinear' fields may be more natural in origin than initially thought, and that '[c]anal systems are not as extensive in Northern Belize as previously reported' (Pohl et al. 1996, pp. 355, 367). Pohl et al. (1996) limited these initial findings to low-lying areas subject to sea level rise, and separated them from findings for highland areas.

Turner's team provided foundational evidence for these complex human-disturbed wetlands. Pohl's team refined and refocused research on the paleosol and the grey clay above it. They described a cascade of natural processes that explained the soil sequences here: sea level rise drove up sulphate-rich water tables which precipitated gypsum crystals into and above the Preclassic soil surface. All of this buried this Preclassic landscape and its old soil surface. Both groups mentioned the deep paleosol, but Pohl's group recognised its plentiful charcoal, some artefacts and that it had to form during times of much lower water tables, otherwise it would have been a lacustrine (lake deposit) marl formed well below the current water table. Turner's team provided an appropriately broad chronology for the upper fields based on ceramics: Late Preclassic to Terminal Classic (Turner and Harrison 1983, p. 254). Pohl's group provided a longer chronology, showing how water inundation massively altered this environment before or at the start of the Classic Period.

The rectilinear surface features formed in dynamic ecosystems were influenced both by natural and Pre-Columbian human impacts. Human-induced change resulted from increasingly widespread population growth and deforestation beginning in the Preclassic (1200 BC–AD 250). This deforestation led to soil erosion that depleted upland soils and aggraded lowlands, thus altering depressions and wetlands from the Preclassic through the Late Classic (Beach 1998; Beach et al. 2002, 2003 and 2006 in press; Dunning et al. 2002; Hansen et al. 2002). Throughout this period, several lines of evidence indicate that the Maya Lowlands were becoming drier and that sea level was rising (Hodell et al. 1995 and 2000; Pohl et al. 1996; High 1975). Droughts here have been linked to solar cycles (Hodell et al. 2001), but deforestation, which could lead initially to increased runoff and higher water tables through lower transpiration, has also been linked to longer term drying (Ray et al. 2005). Similar findings about deforestation and water table rise have been reported in wetland sites following deforestation elsewhere in the world, including Ireland after elm declines (Moore 1975).

Also during this period of environmental change and population growth, the ancient Maya developed complex and diverse agricultural systems that conserved upland soil and water and reclaimed lowland fields. Yet we have scant evidence for intensive wetland agriculture in perennial wetlands of the core of the Maya Lowlands. Sluyter (1994) points out that evidence within the Maya core exists for only 15 ha near Rio Azul, 1.5 ha near Cerros and 375 ha near Edzna, whereas evidence for the periphery in Northern Belize, Rio Candelario and Bajo Morocoy is much more extensive and may have persisted through the Classic Period. Since the Maya had no beasts of burden, long distance food transport from these fields was not economical, except for high-value items like cacao. Thus, the distance from these fields to high population centres questions their role in broader Maya subsistence.

The Yalahau region of the Yucatán Peninsula (Figure 5.1) provides another area of study of wetland soil and Maya subsistence in its infancy. Like so many other study areas, here the first clues come from surface alignments adjacent to sites and wetlands. Fedick (2003) conducted a field survey of the El Edén wetland area in Quintana Roo, Mexico, consisting in part of an intensive surface survey of 78 rock alignment features in this Yalahau region wetland. The survey team found no prehistoric artefacts nor did they find Historic Period artefacts to date the rock alignments, and they noted that dating such structures is necessarily 'problematic' (Fedick 2003, p. 348). There was, however, a small Late Preclassic community nearby that provided associated dates of 100 BC–AD 350/450 (Fedick 2003, p. 349). Pollen studies to identify and reconstruct native and cultivar plant history have not yet been undertaken in the El Edén wetland. Fedick provided a description for the wetlands soils: 'thin, sandy silts, and silty clay, about 20 cm over bedrock … [L]ower areas … contain soils up to about 60 cm deep with peaty deposits over silty to sandy clay' (2003, p. 342). These coastal wetlands are of course prone to any changes in groundwater flow and sea level change. The rock alignments of El Edén appeared to not be functional as agricultural features under today's higher sea and water table stands. They would only be functional under lower water levels, which also helps us infer minimum ages for these features based on our patchy knowledge of past sea levels. Fedick noted that the water table in this region was likely 0.75–1 m lower in the Late Preclassic than today (Fedick 2003). Because of the difficulty of dating wetland features and reconstructing their hydrologic and botanical histories clearly, physical evidence for bajo and terrace manipulation and agriculture within the core of the Maya realm is far more extensive than for wetland agriculture. Fedick and others suggest that the rock alignments might have partly functioned to raise periphyton for fertilising (see Chapter 4).

BAJO FIELDS AND FLOOD RECESSIONAL AGRICULTURE IN MESOAMERICA

Bajo Agriculture

The difference between Bajo and wetland agriculture is that bajos – karst depressions – are seasonally wet for four to six months in a year, whereas perennial wetlands have water tables near the surface for most of the year. The soils of wetlands are dominated by peats and marls (or histosols), and deposited or precipitated sediments, whereas Bajos generally have vertisols or rendolls with thin patches of histosol (Beach 1998; Beach et al. 2003). Not all bajos fit into this classification; for example, the large and well studied Bajo Morocoy in Quintana Roo does not, because it has water table that is 1.5 m from the surface in the dry season that makes it virtually a perennial wetland through capillary water rise (Gliesmann et al 1983). This is a bajo in which perennial wetland agriculture could have been practised.

Research since the 1930s, no doubt spurred on by their close proximity to major sites, has hypothesised that bajos had been lakes or more extensive wetlands during antiquity especially in the northern Petén of Guatemala (Figure 5.1). Indeed in this region, Mirador, the largest Preclassic site, collapsed in the Early Classic coincidental with the hypothesised drying of bajos. Some recent evidence suggests these were more extensive wetlands and that parts, especially their margins, were intensively used (Dunning et al. 2002; Hansen et al. 2002).

As noted above, it is significant that most of the major Maya sites had no or few perennial wetlands nearby that could be likely wetland field sites: Tikal, Caracol, Calakmul, Copán, Dos Pilas and Palenque; not to mention the northern Yucatán sites like Chichen Itzá, Chunchucmil and Uxmal (Figure 4.2). Hence, any description of the ancient Maya as an intensive wetland agricultural society is problematic, and the other forms of agriculture must have been much more important. Several current studies have therefore focused on the bajos as areas of agricultural and silvicultural production. Studies have focused on excavation of field areas, diversions and terraces, and on soil testing for proxy evidence of past crops (see Chapter 4). In the central Petén around the Petexbatún, Seibal, and Cancuen, agriculture may have been in small bajos on upland ridges. The heavily eroded backslope soils and the thick build-up of sediments in the karst drains suggests heavy past use of these areas (Dunning and Beach 2000).

One type of bajo agriculture occurs along the edges of bajos where many footslope terraces and diversion chutes attempt to maintain soil and moisture conditions. The footslope terraces may have had multiple purposes: catching sediment eroded from slopes, retaining sediments carted in from the wetlands, and building up soil platforms above wet season inundation but within a favourably moist microclimate and slope position (Beach et al. 2002 and 2003; Dunning et al. 2002; Hansen et al. 2002). Terrace studies show evidence of agriculture and soil importation, possibly from bajos.

As has been done with wetland fields, studies are using more and more proxies to study evidence of ancient agriculture in bajo soils. Maize and manioc pollen or phytoliths at dateable soil layers indicates their propinquity to fields; carbon isotopic ratios can indicate dominance by C4 plants like maize; and large amounts of charcoal, phosphorus and artefacts probably indicate heavy human use. Additionally, many other species show up in these pollen and other proxy counts that are not food-related but nevertheless are still economic taxa such as timber trees and thatch palms (Dunning et al. 2002; Dunning and Beach 2000; Hansen et al. 2002).

The formation of some vertisols in bajos is another interesting aspect of soils history. In many bajos, vertisols soils are thick and black throughout their vertical profiles because self-ploughing of the expanding and contracting clays mixes the entire profile. But in many other bajos, 'tiger' vertisols develop vertical and diagonal stripes of A-, B- and C-horizons, and in some others these vertical and curved stripes are buried under typical horizontal soil layers (Beach et al. 2003). There remain considerable questions about these patterns. In the first case, the diagonal layers result from the long diagonal cracks that form in the dry season when A-horizons, black topsoils, slough off into the cracks and fill them with black soil in a matrix of grey C- and yellow-brown B-horizons. The descending soil may also force lower horizons upward, thus further distorting vertisols into melanges (Eswaran et al. 1999). In the bajos around La Milpa, Belize (Figure 4.2) Vertisols have complicated natural and anthropogenic origins (Beach et al. 2003; Dunning et al. 2002). Here vertisols that have 100 cm deep diagonally oriented horizons are topped by up to 70 cm of typically horizontal horizons. These so-called 'tiger' vertisols originally formed as horizontal layers since they have a full suite of horizons. Thus two major questions are, what causes them to change to diagonal horizons, and why are the surface horizons horizontal? One explanation involves the following historical sequences: (1) shallow rendoll soils formed by 3500 BP; (2) during the Maya Preclassic these soils aggraded from upland erosion by up to 100 cm, and deforestation led to greater extremes in soil moisture and contraction and expansion; (3) sediment loading on the clays provided a counterbalance mass for swelling and shearing and ripping apart horizons; and finally (4) the upper horizontal deposition occurred during and after the Classic Period, forming weakly developed topsoils above the lower, deformed sequence. This later episode of lower erosion and deposition corresponds to urbanisation, soil and water management and abandonment at La Milpa.

Flood Recessional Agriculture in Mesoamerica

Bajos and other seasonal wetlands all over the world have contained flood recessional farming for millennia. In flood recessional systems, farmers plant crops in the drying zones as floods recede because there may be sufficient soil moisture for crops to mature through a dry season. In Mesoamerica, Wilk (1981) studied this in Belize, and Gliessman (1991) in Tabasco, Mexico. The concepts were applied to the Maya Lowlands by Gliessman et al. (1983). Siemens (1978, 1982, 1983, and 1998) envisioned a more complex form of this in floodplains, wherein ancient farmers cut canals to accelerate drainage and planted progressively lower areas with the progression of the dry season. The difficulty with studying this ancient system is that it leaves little archaeological evidence, only canals that may fill in and leave no obvious surface expression. Many techniques, however, are possible, ranging from remote sensing (especially radar) to find canals or field patterns and excavation to find buried agricultural soils, to a range of proxy evidence suggesting crops and soil manipulation. For example, Terry (2003) showed carbon isotopic evidence in buried soils for past floodplain maize cultivation near Aguateca, Guatemala, which experiences a 10 m inundation in the wet season. Hence farming had to occur in the dry season, presumably as recessional agriculture.

SOUTH AMERICA

In South America, most raised field features were apparently abandoned by the time the Spanish arrived, because no written evidence was presented by the Spanish chroniclers of the time (and only a few observations later) to suggest these raised fields were under cultivation (Denevan 1970, 2002; Parsons 1989, p.205). There were likely a very small number of raised fields still being used in the late nineteenth century (Erickson 2000), though most of these agroecosystems in the Rio Catari basin near Lake Titicaca (Figure 5.1) seem to have been abandoned prior to AD 1150 (Kolata 2000, p.172).

Andean Highlands

The broader group of raised fields around the margins of Lake Titicaca and those of the Andean Highlands of Peru provide South America's best known examples of wetland agriculture (Figure 5.1) (Denevan 1970, 2002; Erickson 2000; Kolata 2000). Raised fields occupy these waterlogged edges in regional conjunction with agricultural terraces and a variety of other earth works, house mounds and causeways (Parsons and Denevan 1989; Parsons and Bowen 1966; Erickson 2000). Human influence in this landscape far precedes this, to at least 8000 BP (Erickson 2000), but specific dates for wetland fields are from about 1800 BC and later. In a study of the Lake Wiñyamarka arm of Lake Titicaca, (the Rio Catari basin), Kolata dated the building of most raised fields from AD 600 to 1100 (Kolata 2000, p. 172). Erickson (2000, p. 336) added that it was likely that the earliest farming on raised fields was between 1800 and 900 BC, and that it was developing along the lake edges between 900 BC and AD 200, and was found more extensively in the area between 200 BC and AD 600.

The raised fields of the Lake Titicaca Basin consist of alternating soil platforms and canals. The platforms are variable in size, and range in shape from strips to island-like structure surrounded by ditches (Erickson 2000). Their patterns include several kinds of raised field geometries: open checkerboard, irregular embanked, riverine, linear, ladder and combed field patterns (Denevan 2002, pp. 258–62; Parsons and Denevan 1989, p. 218). These field types all represent massive transformations of this landscape, since their sizes range 'from 4 to 10 m wide, 10 to 100 m long, and 0.5 to 3 m tall' (Erickson 2002, p. 333). The soils of these structures also show manipulation to 'a depth of 2 m or more on a massive regional scale' (Erickson 2000, p. 347). Excluding all other human earth works in this region, and there are many, these agricultural raised fields alone cover some 120,000 ha, an area about one-half the size of Belgium (Erickson 2000).

Constructed lowland wetland field structures of the Titicaca Basin can be classified in two forms: raised fields and sunken gardens, or *q'ochas* (Erickson 2000). Q'ochas coexist with raised fields, and are used even today (Erickson 2000, p.337). Q'ochas are shallow depressions, ranging in size from 0.1 to 4 hectares. They are found in poorly drained areas, and are thought to be largely natural formations, but are linked by artificial canal networks (Erickson 2000, p.338). These wetlands are still used today for a variety of crops and animal forage and they are also documented in early colonial writings, but their ancient and historical uses are uncertain (Erickson 2000, p.341).

Given the tremendous labour investments implied in these earth works, they must have been an integral part of the food supply for the region (Denevan 2002). In reviewing population estimates made in various studies from 1982 to present, Denevan (2002) estimated that Titicaca's raised fields supported between 215,000 and 2.29 million people. Estimating populations is notoriously error-prone but even the least of this order of magnitude range is substantial.

Excavating, building and periodically maintaining these fields and ditches has left a complicated and lasting human imprint on soils development, structure and stratigraphy. Wetland fields may provide the quintessential palimpsest because they may have been worked and abandoned at several different times.

Few published soils studies existed when Smith, Denevan and Hamilton (1966–1968 regional survey) explored the raised fields of Titicaca (Figure 5.1) (Smith et al. 1968; Denevan 2002). One study covered an area north of Puno, in the Paucarcolla-Juliaca plain, and classified the soils as 'gley, humic Andean Planosols' (see United Nations [FAO/UNESCO] 1974). Denevan reported the soils to be poorly drained lacustrine (lake deposit) soils of fine sediment size with moderate fertility, though in some places strongly alkaline (Denevan 2002, pp. 262–3). Raised field construction attempted to overcome the two main limitations of these soils: high alkalinity and a high water table. Since alkalinity decreased down these soil profiles, building drained fields essentially turns over the soil, exposing soils of lower pH and potentially higher fertility on the raised planting beds. Building up the fields and canalising drainage was also an effort to overcome poor drainage, obviously the major limitation for agriculture in swamps (Denevan 2002).

Seasonally Flooded Lowland Environments of South America

Parsons and Bowen, and also Denevan, conducted studies of extensive raised field sites in the lowland environments of South America. These sites were of course different from the Highlands, but there were also more similarities than expected. Studies of wetland raised field agriculture occurred in the San Jorge, Magdalena, Orinoco, and Amazon watersheds, and in coastal Ecuador and Peru, among other sites (Figure 5.1). Denevan estimated in 1970 that some 170,000 ha of raised or ridged fields were known in the Lowlands regions of South America (1970, p. 648). One major difference between highland and lowland sites is that lowland ones were seasonal wetlands, flooded for several months and drained the rest of the year (Denevan 1970, p. 648). Moreover, lowland sites are not threatened by frost, but limited by flooding and drought, and more akin to the bajos of Central America (Denevan 1970, p. 648). Indeed, like Central American Bajos, these South American 'wet savanna' ridged sites in the Beni (Bolivia) Basin (Figure 5.1), had clay-rich, low organic matter soils (Denevan 1970, p. 649). Though little was known about their antiquity or origin, the Beni sites, first discovered in 1961, inspired others to survey a broader extent of the wet savannas and Amazonia, revealing more sites in the 1960s and 1970s (Denevan 1970).

The morphology and dimensions of these lowland raised field patterns are similar to those described previously for the highland regions (Denevan 2002; Erickson 2000). For example, a lowland wetland field complex in San Jorge, Colombia (Figure 5.1) had three major types of ridged fields: a 'cano' pattern, where parallel ridges are built perpendicular to river levees; a 'checkerboard pattern' where blocks of ridges are arranged largely perpendicular; and a comb-like 'clustering of loosely parallel ridges' not aligned with levees or slopes (Parsons and Bowen 1966, p. 329). Soils in these ridged fields were again clay-rich and mottled yellow and red, indicative of alternating periods of flooding and drying (Parsons and Bowen 1966, p. 334). Besides the usual motivation for building wetland fields, Parsons and Bowen postulated that these were tied to the need for irrigation during a major dry period there between AD 700 and 800 (Parsons and Bowen 1966). They based this on a pollen record, which Van der Hammen and Hooghiemstra later published showing a cooling and drier climate and a general lowering of lake levels during the last 3,000 years (van der Hammen 1974, pp. 12–5; van der Hammen and Hooghiemstra 2003).

NORTH AMERICA

Pre-contact North American Wetland Agriculture

Despite scholarly and popular misconceptions, Pre-Columbian North American peoples engaged in intensive agriculture and soil manipulation. The preceding chapter (Beach et al.) describes, for example, America's earliest terracing network in Northern Chihuahua, and here we show parallel evidence for sizeable and early manipulation of wetland agriculture. William Doolittle offers the most comprehensive review yet, based on hundreds of studies of the evidence for ancient wetland agriculture and its implications for soils in North America (Doolittle 2000). Many other scholars have made notable contributions to the body of work on aboriginal agriculture in North America (e.g., Fowler 1969; Denevan 1992; Gartner 1999; Dalan et al. 2003). Wetland agriculture here took two general forms, flood recessional agriculture in seasonal wetlands and drained/ridged field agriculture in perennial wetlands.

Flood Recessional Agriculture in North America

As in most of the rest of the world, south-western North America's aboriginal agriculture occurred on uplands and lowlands – on slopes in shallow, eroding soils and on floodplains with deeper, aggrading soils (Doolittle 1992, p. 388). In addition to complex canal irrigation systems (Doolittle 1992), the agriculture that took place on the stream floodplains left a more subtle physical imprint, but rewarded pre-Hispanic farmers with crops that allowed large populations to subsist (Doolittle 2000, p. 427). Doolittle (2000, p. 414) wrote that Native Americans needed little or no soil manipulation to grow hydrophytic crops, like wild rice, cattails and onions.

As discussed earlier, flood recessional agriculture implies that cultivation here was more seasonal and depended on soil moisture left as flood waters receded (Doolittle 2000). Although such forms of agriculture leave only subtle archaeological clues (canals and levees, perhaps buried), ethnographic and documentary evidence are strong (Doolittle 2000). Seventeenth- and eighteenth-century Spanish and Jesuit travellers left 'some of the most thorough portrayals of any agricultural system anywhere' (Doolittle 2000, p. 426) when they commented on the recessional farming systems they observed in what is now northern Mexico and the south-western United States, including the lower Colorado River (Figure 5.1). They compared them with floodplain farming along the banks of the River Nile in Egypt (Doolittle

2000). Doolittle noted that this highly productive 'flood recessional agriculture dominated the lower courses of all the rivers from the Yaqui to the Colorado' in the south-west, and Spanish settlers quickly recognised their value and colonised these productive lands (Doolittle 2000, p. 427).

Raised/Ridged Field Agriculture of North America

Compared with recessional flood agriculture, physical evidence for Native American raised and wetland fields was more common before ploughing broke it up in nineteenth-century North America (Denevan 1970). Since so much of the aboriginal landscape has been drained and ploughed, differentiating perennial from seasonal wetland soil use is problematical, and much of the discussion does not clearly differentiate these. Parsons cited nineteenth-century documentary evidence for 'planting beds in valley bottoms' in Indiana and Michigan, and others in Wisconsin and Missouri (Parsons 1989[1967], p. 219). Denevan reported about 180 such fields in Wisconsin and Michigan alone, noting that a majority occurred in the Mississippi River floodplains (Figure 5.1); hence the structures probably functioned for drainage (Denevan 1970, p. 651; Denevan 1992). According to Gartner, 'Wisconsin has the most ridged fields' in North America (1999, p. 671). Labour inputs were more considerable to farm wetlands in 'lakeshores, freshwater marshes, and swamps' (Doolittle 2000, p. 428), therefore agricultural efforts in these zones in North America left more permanent evidence on the landscape in the form of ridged and drained fields. Yet, the abundance of evidence is still a relative term in North American pre-history. There is still controversy over ambiguous evidence for the function, origin and use of many of these ridged field features (Doolittle 2000). So, contrary to South and Central America, the physical evidence for North American drained or raised fields remains 'sparse' and there are no documentary or ethnographic sources (Doolittle 2000, p. 430). The best archaeological evidence for these fields occurs in three regions: the south-east, the Mississippi Valley and the Great Lakes area (Doolittle 2000, p. 431). Doolittle (2000, pp. 430–5) identified prominent wetland field sites in eastern North America such as: Fort Center near Lake Okeechobee in Florida; the Parker Earthwork in southern Ontario; the bottomlands where the Ohio, Illinois and Mississippi Rivers meet in Southern Illinois, most notably the Cahokia site; and the Sand and Senator Lake sites in Wisconsin.

Near Cahokia in Southern Illinois (Figure 5.1), Fowler (1969) studied agricultural fields presumed to be middle Mississippian in age (AD 1150–1500), based upon surveys and excavations. He proposed that subsistence here was based on drainage and ridge-furrow farming that allowed reclamation of wet floodplains (Fowler 1969, p. 365; Woods 1992). Cahokia is the site of Monk's Mound, the largest mound north of central Mexico (Fowler 1969). Cahokia was one of the 'most complex and populous cultural developments in all of United States prehistory'

(1969, p. 365), and he hypothesised that this region was influenced by 'expansion northward … of Mesoamerican civilisation … frontiers' (1969, p. 365). But there remains no hard evidence linking Mesoamerican culture and its various forms of intensive agriculture with Cahokia. The evidence for raised field agriculture itself at Cahokia is still considered sparse, compared with other North American sites such as the Parker Earthwork in Ontario (Doolittle 2000). Since Cahokia was so large, the spotty evidence for intensive raised fields here parallels the situation for the Maya Lowlands, where wetland fields are also distant from the main population centres. The possible functions of the raised or ridged field sites in the Mississippian culture might have included making weed removal easier, delivering irrigation, reclaiming bottom land, draining cold air, and raising plants above water-logged soils (Fowler 1969, p. 374; Doolittle 2000, p. 441). But some of the ridged fields were located higher than the 'saturated flood plain' zones (Fowler 1969, p. 374). Regardless of site, the fields represent an efficient strategy for increasing food production for 'the Mississippian temple-town type of cultural expansion' (Fowler 1969, p. 374).

Gartner (1999) has offered further insight into the function and meaning of Pre-Columbian raised fields in Wisconsin. As noted above, some but not all of the raised fields in the eastern North American landscapes were found in valley bottom and wetland contexts. The raised/ridged fields perhaps had a socio-political function in addition to a practical function (Gartner 1999), which may explain their more wide-ranging geomorphic settings. For example, the Hulbert Creek fields site in Wisconsin was located in a less than ideal choice for agriculture because of thinner soils (Gartner 1999, p. 675). These raised or ridged fields are highlighted in a study of the Oneota people, who practised agriculture near Lake Koshkonong between AD 1000 and AD 1150 (Gartner 1999, p. 672). They functioned physically in ways similar to raised fields elsewhere in the Americas, offering frost protection, improved soil fertility, disease protection, well-aerated and well-drained soil, soil moisture storage, minimised risk, and maximised solar exposure (by orienting the planting beds in a north-south direction creating a 'warmer soil') (Gartner 1999, pp. 674, 682). But these ridged fields may also have powerfully symbolic meaning. The raised fields can be viewed, along with effigy mounds and other earth works, as landscape elements of 'territoriality' (Gartner 1999, p. 672). Since ridged fields came into production as mound building ceased, they may 'constitute the symbolic successor for Mounds in Oneota societies', and represent 'communal labour' and a 'commitment to place' (Gartner 1999, p. 681). Ridged field motifs even made their way onto Pre-Columbian Oneota pottery of the time. Some motifs were very similar to 'plan view projections … [and] topological sequence[s] of the adjacent ridged fields' (Gartner 1999, p. 681).

SUMMARY

There can be no doubt that many wetland soils of the Americas have considerable anthropogenic inputs, and the evidence for human and environmental transformation of wetlands over the last millennia is overwhelming. In turn, human culture in this region has not developed independently from its environment, but in a coupled, complex system with the environment (Kolata 2000). The best evidence for intensive wetland soil manipulation comes from the Basin of Mexico and Lake Titicaca in South America. Here we have ethnographic, historical and archaeological evidence for intensive wetland agricultural systems and the anthropogenic soils within them.

The stories of North America and the Maya Lowlands are based more on archaeological and geomorphologic excavations, and are thus more difficult to ascertain and interpret. North America provides a surprisingly complex picture of human participation in the subsistence landscape in its Pre-Columbian ridged field and flood recessional cultivation systems. Here, wetland soil manipulation occurred on a smaller scale, though a high proportion of its wetlands have been altered and their evidence compromised.

For the Maya Lowlands, wetland manipulation for agriculture certainly occurred, but the evidence for intensive agriculture near the great population centres is lacking. Here, the last thirty years of Maya soil research has converged around a consensus that the Maya relied on a mosaic of soil-agroecological environments for subsistence. They exploited wetlands, but more so bajos, sinkholes, terraces, floodplains and uplands with varying intensity. The Maya soil environment also experienced much natural evolution from sea level, groundwater level, and droughts and other climate changes. The Maya soil environment was also altered by human-induced erosion and sedimentation, and possibly deforestation-induced increased runoff and water table rise. One verified impact on wetland soils in the Maya Lowlands was recurrent drought, especially in the Classic period. This is a major finding on the Maya environment and possibly on Maya civilisation, but we still need to model how these droughts, especially the Late Classic drought, affected wetlands, soils and subsistence (Hodell et al. 1995, 2000 and 2001; Haug et al. 2003). Another unverified but potential impact on bajos, floodplains, and wetlands in the Maya Lowlands was the impact of deforestation on runoff and water tables (Ray et al. 2005).

In all regions we need more focused fieldwork that uses all possible forms of evidence to understand the history and geography of these human manipulated wetland systems. As borne out in Maya lowlands wetland research, we especially need to appreciate the role of water chemistry in wetlands, and to consider all relevant environmental and human input factors. Patterned fields have been brought to our attention largely and serendipitously through early aerial photography. These fields promise to reveal much more about ancient subsistence, changing environments and potential development through further excavation, soil studies, restoration efforts and the many forms of proxy evidence.

But provisionally it is fair to conclude that over several millennia the peoples of the Americas, in hundreds of distinct locations, used their ingenuity and labour to alter soils and hydrology in their quests for food and fibre. Around and within a range of wetland environments, they ditched, drained, diverted and raised fields. Like agricultural peoples everywhere, they made their livings from soil and water, and as a result never left them just as they found them.

REFERENCES

Armillas, Pedro. 1971. 'Gardens on Swamps'. *Science,* **174**: 653–61.

Beach, Timothy. 1998. 'Soil Catenas, Tropical Deforestation, and Ancient and Contemporary Soil Erosion in the Petén, Guatemala'. *Physical Geography,* **19**(5): 378–405.

Beach, Timothy, and Sheryl Luzzadder-Beach. 2002. 'Pre-Columbian Aggradation and Wetland Agriculture in Belize'. Paper presented at 2002 AAG Annual Meetings, Los Angeles CA, March 2002.

Beach, Timothy, Sheryl Luzzadder-Beach, Nicholas Dunning, Jon Hageman and Jon Lohse. 2002. 'Upland Agriculture in the Maya Lowlands: Ancient Conservation in Northwestern Belize'. *The Geographical Review,* **92**(3): 372–97.

Beach, Timothy, Nicholas Dunning, Sheryl Luzzadder-Beach and Vernon Scarborough. 2003. 'Depression Soils in the Lowland Tropics of Northwestern Belize: Anthropogenic and Natural Origins'. In A. Gomez-Pompa, M. Allen, S. Fedick and J.J. Jimenez-Osornio (eds), *The Lowland Maya Area: Three Millennia at the Human-Wildland Interface* (Binghamton, NY: Haworth Press), pp. 139–74.

Beach, T., N. Dunning, S. Luzzadder-Beach, J. Lohse and D. Cook. 2006. 'Ancient Maya Impacts on Soils and Soil Erosion in the Central Maya Lowlands'. *Catena,* in press.

Bloom, Paul R., Mary Pohl, Cynthia Buttleman, Frederick Wiseman, Alan Covich, Charles Micksicek, Joseph Ball and Julie Stein. 1983. 'Prehistoric Maya Wetland Agriculture and the Alluvial Soils near San Antonio Rio Hondo, Belize'. *Nature,* **301**: 417–19, 3 February 1983.

Carrasco, David and Scott Sessions. 1998. *Daily Life of the Aztecs: People of the Sun and Earth.* Westport, CT: Greenwood Press.

Chapin, Mac. 1988. 'The Seduction of Models: Chinampa Agriculture in Mexico'. *Grassroots Development,* **12**(l): 8–17.

Coe, Michael D. 1964. 'The Chinampas of Mexico'. *Scientific American,* **211**(1): 90–8, July issue.

Crossley, Phil L. 2004. 'Just Beyond the Eye: Floating Gardens in Aztec Mexico'. *Historical Geography,* **32**: 111–35.

Dalan, Rinita A., George R. Holley, William I. Woods, Harold W. Watters, Jr. and John A. Koepke. 2003. *Envisioning Cahokia: A Landscape Perspective*. Dekalb, IL: Northern Illinois University Press.

Denevan, William M. 1970. 'Aboriginal Drained-Field Cultivation in the Americas'. *Science*, **169**(3946): 647–54, 14 August 1970.

Denevan, William M. 1992. 'The Pristine Myth: The Landscape of the Americas in 1492'. *Annals of the Association of American Geographers*, **82**(3): 369–85.

Denevan, William. 2002. *Cultivated Landscapes of Native Amazonia and the Andes*. Oxford: Oxford University Press.

Doolittle, William E. 1992. 'Agriculture in North America on the Eve of Contact: A Reassessment'. *Annals of the Association of American Geographers*, **82**(3): 386–401.

Doolittle, William E. 2000. *Cultivated Landscapes of North America*, Oxford Geographical and Environmental Studies. Oxford: Oxford University Press.

Dunning, N., and T. Beach. 2000. 'Stability and Instability in Pre-Hispanic Maya Landscapes'. In D. Lentz (ed.), *Imperfect Balance: Landscape Transformations in the Precolumbian Americas* (New York: Columbia University Press), pp. 179–202.

Dunning, N., and T. Beach. 2004. 'Noxious or Nurturing Nature? Maya Civilization in Environmental Context'. In C. Golden and G. Borgstede(eds), *Continuities and Changes in Maya Archaeology: Perspectives at the Millennium* (New York: Routledge), pp. 125–41.

Dunning, Nicholas P., Sheryl Luzzadder-Beach, Timothy Beach, John G. Jones, Vernon Scarborough and T. Patrick Culbert. 2002. 'Arising from the *Bajos*: The Evolution of a Neotropical Landscape and the Rise of Maya Civilization'. *Annals of the Association of American Geographers*, **92**(2): 267–83.

Erickson, Clark L. 1992. 'Prehistoric Landscape Management in the Andean Highlands: Raised Field Agriculture and its Environmental Impact'. *Population and Environment*, **13**: 285–300.

Erickson, Clark. 1998. 'Applied Archaeology to Rural Development: Archaeology's Potential Contribution to the Future'. In Michael B. Whiteford and Scott Whiteford (eds), *Crossing Currents: Continuity and Change in Latin America* (Upper Saddle River, NJ: Prentice Hall), pp. 34–45.

Erickson, Clark L. 2000. 'The Lake Titicaca Basin: A Precolumbian Built Landscape'. In David L. Lentz (ed.), *Imperfect Balance: Landscape Transformations in the Precolumbian Americas*, The Historical Ecology Series (New York: Columbia University Press), pp. 311–56.

Eswaran, H., F. Beinroth, P. Reich and L. Quandt. 1999. 'Vertisols: Their Properties, Classification, Distribution, and Management'. In *Monograph of World Soil Resources* (Washington, D.C: Natural Resources Conservation Service, USDA).

Fedick, Scott L. 2003. 'Archaeological Evidence for Ancient and Historic Resource Use Associated with the El Edén Wetland, Northern Quintana Roo, Mexico'. In A. Gomez-Pompa, M. Allen, S. Fedick and J. J. Jimenez-Osornio (eds), *The Lowland Maya Area: Three Millennia at the Human-Wildland Interface* (Binghamton, NY: Haworth Press), pp. 339–59.

Fowler, Melvin L. 1969. 'Middle Mississippian Agricultural Fields'. *American Antiquity*, **34**(4): 369–75.

Gartner, William Gustav. 1999. 'Late Woodland Landscapes of Wisconsin: Ridged Fields, Effigy Mounds and Territoriality'. *Antiquity*, 73: 671–83.

Gliessman, S.R. 1991. 'Ecological Basis of Traditional Management of Wetlands in Tropical Mexico: Learning from Agroecosystem Models'. In M.L. Oldfield and J.B. Alcorn (eds), *Biodiversity: Culture, Conservation and Ecodevelopment* (San Francisco: Eastview Press).

Gliessman, Stephen R., B.L. Turner II, F.J. Rosado May and M.F. Amador. 1983. 'Ancient Raised Field Agricuture in the Maya Lowlands of Southeastern Mexico'. In J.P. Darch (ed.), *Drained Field Agriculture in Central and South America*, BAR International Series 189 (Oxford UK: British Archaeologial Series).

Gomez-Pompa, Arturo, Hector Luis Morales, Epifanio Jimenez Avilla and Julio Jimenez Avilla. 1982. 'Experiences in Traditional Hydraulic Agriculture'. In Kent V. Flannery (ed.), *Maya Subsistence* (New York: Academic Press), pp. 327–42.

Hansen, Richard D., Steven Bozarth, John Jacob, David Wahl and Thomas Schreiner. 2002. 'Climatic and Environmental Variability in the Rise of Maya Civilization: A Preliminary Perspective from Northern Peten'. *Ancient Mesoamerica*, **13**(2): 273–95.

Harrison, Peter D. 1996. 'Settlement and Land Use in the Pulltrouser Swamp Archaeological Zone, Northern Belize'. In Scott L. Fedick (ed.), *The Managed Mosaic: Ancient Maya Agriculture and Resource Use* (Salt Lake City: University of Utah Press), pp. 177–90.

Haug, G.H., D. Günther, L.C. Peterson, D.M. Sigman, K.A. Hughen and B. Aeschliman. 2003. 'Climate and the Collapse of Maya Civilization'. *Science*, **299**: 1731–5.

Hebda, Richard. J., Alfred H. Siemens and Alistair Robertson. 1991. 'Stratigraphy, Depositional Environment, and Cultural Significance of Holocene Sediments in Patterned Wetlands of Central Veracruz, Mexico'. *Geoarchaeology: An International Journal*, **6**(1): 61–81.

High, L.R. 1975. 'Geomorphology and Sedimentology of Holocene Coastal Deposits, Belize'. In K.W. Wantland and W.C. Pussey III (eds), *Belize Shelf-Carbonate Sediments, Clastic Sediments, and Ecology* (Tulsa, OK: American Association of Petroleum Geologists) pp. 53–96.

Hodell, D.A, M. Brenner and J.H. Curtis. 1995. 'Possible Role of Climate in the Collapse of Classic Maya Civilization'. *Nature*, **375**: 391–4.

Hodell, D.A., M. Brenner and J.H. Curtis. 2000. 'Climate Change in the North American Tropics and Subtropics since the Last Ice Age: Implications for Environment and Culture'. In D.L. Lentz (ed.), *An Imperfect Balance: Landscape Transformations in the Precolumbian Americas* (New York: Columbia University Press), pp. 13–18.

Hodell, D., M. Brenner, J. Curtis and T. Guilderson. 2001. 'Solar Forcing of Drought Frequency in the Maya Lowlands'. *Science*, **292**: 1367–70.

Jacob, John S. 1995. 'Ancient Maya Wetland Agricultural Fields in Cobweb Swamp, Belize: Construction, Chronology, and Function'. *Journal of Field Archaeology*, 22: 175–90.

Jacob, J.S., and C.T. Hallmark. 1996. 'Holocene Stratigraphy of Cobweb Swamp, a Maya Wetland in Northern Belize'. *Geological Society of America Bulletin*, **108**(7): 883–91.

Kolata, Alan L. 2000. 'Environmental Thresholds and the "Natural History" of an Andean Civilization'. In Garth Bawden and Richard Martin Reycraft (eds), *Environmental Disaster and the Archaeology of Human Response,* Maxwell Museum of Anthropology Anthropological Papers No. 7 (Albuquerque NM: University of New Mexico), pp. 163–78.

Luzzadder-Beach, Sheryl, Timothy Beach and Jon Lohse. 2003. 'Pre-Columbian Wetland Soils Investigation, Blue Creek, Belize'. Paper presented at *Soils in Cultural Context: Mesoamerica Session,* Annual Meetings of the Association of American Geographers, 5 March 2003.

McClung de Tapia, Emily. 2000. 'Prehispanic Agricultural Systems in the Basin of Mexico'. In David L. Lentz (ed.), *Imperfect Balance: Landscape Transformations in the Precolumbian Americas,* The Historical Ecology Series (New York: Columbia University Press), pp. 121–46.

Meltzer, David J. 1997. 'Monte Verde and the Pleistocene Peopling of the Americas'. *Science,* **276**(5313): 754–5, 2 May 1997.

Moore, P.D. 1975. 'Origin of Blanket Mires'. *Nature,* **256**: 267–9.

National Research Council. 1992. *Restoration of Aquatic Ecosystems: Science, Technology, and Public Policy.* Water Science and Technology Board Commission on Geosciences, Environment and Resources. Washington DC: National Academy Press.

Nichols, Deborah L. and Charles D. Frederick. 1993. 'Irrigation Canals and Chinamapas: The Development of Hydraulic Agriculture in the Northern Basin of Mexico'. In B.L. Isaac and V. Scarborough (eds), *Water Management,* Research in Economic Anthropology Supplement 7 (Greenwich, CT: Jai Press), pp. 123–50.

Palerm, A. and E.R. Wolf. 1957. 'Ecological Potential and Cultural Development in Mesoamerica'. In A. Palerm (ed.), *Studies in Human Ecology: A Series of Lectures Given at the Anthropological Society of Washington,* Pan American Union Social Science Monographs 3 (Washington: Pan American Union, 1957).

Parsons, James J. 1989 [1969]. 'Ridged Fields in the Río Guyas Valley, Ecuador'. In William M. Denevan (ed.), *Hispanic Lands and Peoples: Selected Writings of James J. Parsons,* Dellplane Latin American Studies, No. 23 (Boulder CO: Westview Press), pp. 203–10.

Parsons, James J. and William A. Bowen. 1966. 'Ancient Ridged Fields of the San Jorge River Floodplain, Colombia'. *The Geographical Review,* **56**(3): 317–43.

Parsons, James J. and William M. Denevan. 1989 [1967]. 'Pre-Colombian Ridged Fields'. In William M. Denevan (ed.), *Hispanic Lands and Peoples: Selected Writings of James J. Parsons,* Dellplane Latin American Studies, No. 23 (Boulder CO: Westview Press) pp. 211–21.

Parsons, Jeffrey R. 1976. 'The Role of Chinampa Agriculture in the Food Supply of Aztec Tenochtitlan'. In C.E. Cleland (ed.), *Cultural Change and Continuity* (New York: Academic Press), pp. 233–57.

Pohl, Mary Deland (ed.). 1990. *Ancient Maya Wetland Agriculture: Excavations on Albion Island, Northern Belize,* Westview Special Studies in Archaeological Research. Boulder CO: Westview Press.

Pohl, Mary D., Kevin O. Pope, John G. Jones, John S. Jacob, Dolores R. Piperno, Susan D. deFrance, David. L. Lentz, John A. Gifford, Marie E. Danforth and J. Kathryn Josserand. 1996. 'Early Agriculture in the Maya Lowlands'. *Latin American Antiquity,* 74(4): 355–72.

Pope, Kevin O., Mary D. Pohl and John S. Jacob. 1996. 'Formation of Ancient Maya Wetland Fields: Natural and Anthropogenic Processes'. In Scott L. Fedick (ed.), *The Managed Mosaic: Ancient Maya Agriculture and Resource Use* (Salt Lake City: University of Utah Press), pp. 165–76.

Purdy, Barbara A. (ed.). 2001. *Enduring Records: The Environmental and Cultural Heritage of Wetlands*. Oxford: Oxbow Books.

Ray, D.K., U.S. Nair, R.M. Welch, R.O. Lawton, R.J. Oglesby, R.A. Pielke, T.A. Sever and D. Irwin. 2005. 'The Role of Deforestation in the Collapse of Classic Maya Civilization: Lessons for the Current Land Use Management in Northern Mesoamerica'. *EOS Transactions AGU* **86**(52), Fall Meeting Supplement, Abstract B33F-07.

Ricketson, Oliver Jr. and A.V. Kidder. 1930. 'An Archaeological Reconnaissance by Air in Central America'. *The Geographical Review*, **20**(2): 177–206.

Schmidt, Peter J. 1977. 'Un Sistema de Cultivo Intensivo en la Cuenca del Rio Nautla, Veracruz'. *Boletin del Instituto Nacional de Anthropologia e Historia*, Epoca **3**(20): 50–60.

Siemens, Alfred H. 1978. 'Karst and the Pre-Hispanic Maya in the Southern Lowlands'. In P. D. Harrison and B.L. Turner II (eds), *Pre-Hispanic Maya Agriculture* (Albuquerque NM: University of New Mexico Press), pp. 117–43.

Siemens, Alfred H. 1982. 'Pre-Hispanic Use of the Wetlands in the Tropical Lowlands of Mesoamerica'. In K. Flannery (ed.), *Maya Subsistence: Studies in Memory of Dennis E. Puleston* (NY: Academic Press), pp. 205–25.

Siemens, Alfred H. 1983. 'Wetland Agriculture in Pre-Hispanic America'. *Geographical Review*, **73**(2): 166–81.

Siemens, Alfred H. 1998. *A Favored Place*. Austin TX: University of Texas Press.

Siemens, A.H., and D.E. Puleston. 1972. 'Ridged Fields and Associated Features in Southern Campeche: New Perspectives on the Lowland Maya'. *American Antiquity*, 37: 228–39.

Siemens, Alfred H., Richard J. Hebda, Mario Navarette Hernandez, Dolores R. Piperno, Julie Stein and Manuel G. Zola Baez. 1988. 'Evidence for a Cultivar and a Chronology from Patterned Wetlands in Central Veracruz, Mexico'. *Science*, **242**(4875): 105–7.

Sluyter, Andrew. 1994. 'Intensive Wetland Agriculture in Mesoamerica: Space, Time and Form'. *Annals of the Association of American Geographers*, **84**(4): 557–84.

Smith, Clifford T., William M. Denevan, and Patrick Hamilton. 1968. 'Ancient Ridged Fields in the Region of Lake Titicaca'. *The Geographical Journal*, **134**: 353–67.

Smith, Michael E. 2003. *The Aztecs*, 2nd edn. Malden MA: Blackwell Publishing.

Soil Survey Staff. 1998. *Keys to Soil Taxonomy*, 8th edn. Washington DC: USDA Natural Resource Conservation Service/U.S. Government Printing Office.

Steggerda, M. 1941. *Maya Indians of Yucatán*, Publication No. 531. Washington DC: Carnegie Institution of Washington.

Terry, Richard E. 2003. 'Stable Carbon Isotopes of Soil Organic Matter Hold a Record of Ancient Maize Agriculture in Mesoamerica' (Presentation). In Giovanni Boschian (ed.), *Second International Conference on Soils and Archaeology Pisa, 12th-15th May, 2003, Extended Abstracts* (Pisa Italy: University of Pisa and International Union of Soil Sciences), pp. 125–6.

Turner, B.L. II. 1974. 'Prehistoric Intensive Agriculture in the Mayan Lowlands'. *Science,* **185**(4146): 118–24, 12 July 1974.

Turner, B.L. II. 1983. *Once Beneath the Forest: Prehistoric Terracing in the Rio Bec Region of the Maya Lowlands.* Boulder, CO: Westview.

Turner, B.L. II and Peter D. Harrison, 1981. 'Prehistoric Raised-field Agriculture in the Maya Lowlands'. *Science,* **213**(4506): 399–405.

Turner, B.L. II and Peter D. Harrison (eds). 1983. *Pulltrouser Swamp: Ancient Maya Habitat, Agriculture, and Settlement in Northern Belize.* Austin TX: University of Texas Press.

United Nations (FAO/UNESCO). 1974. FAO/UNESCO Soil Map of the World: Legend. Paris, France: UNESCO. Online at: http: //www.fao.org/ag/agl/key2soil.stm , FAO, 2004.

van der Hammen, Thomas. 1974. 'The Pleistocene Changes of Vegetation and Climate in Tropical South America'. *Journal of Biogeography,* **1**(1): 3–26, March 1974.

van der Hammen, Thomas and Henry Hooghiemstra. 2003. 'Interglacial-Glacial Fuquene-3 Pollen Record from Colombia: an Eemian to Holocene Climate Record'. *Global and Planetary Change,* **36**(2003): 181–99.

West, Robert C. and John P. Augelli. 1966. *Middle America: Its Land and Peoples.* Englewood Cliffs, NJ: Prentice-Hall.

Whitmore, T. and B. Turner II. 2001. *Cultivated Landscapes in Mesoamerica on the Eve of the Conquest.* Oxford: Oxford University Press.

Wilk, R.R. 1981. Agriculture, Ecology, and Domestic Organization among the Kekchi Maya. PhD Dissertation, University of Arizona.

Woods, William I. (ed.). 1992. *Late Prehistoric Agriculture: Observations from the Midwest,* Studies in Illinois Archaeology No. 8. Springfield IL: Illinois Historic Preservation Agency.

6

A History of African Soil: Perceptions, Use and Abuse

Kate B. Showers

INTRODUCTION

African soils cannot be separated from African history because they are, in themselves, historical bodies and they have interacted with human history since its beginning. The history of soil other than of its genesis (a concern of pedology) is a history of its perceptions and the consequences of its use. Such a history must necessarily be one of exploration and discovery, of experimentation and classification. The late nineteenth- and twentieth-century European experience is only the latest in a long series of human encounters with the continent. An essential feature of Africa's soils is their diversity. This is to be expected, when one considers the sheer size of the continent, and its many ecosystems. Africa's approximately 30,221,392 sq.km. is greater than the land areas of China, Europe and the United States combined, see Figure 6.1. It accounts for 20.4 per cent of the earth's land (National Geographic Society 1990, p. 130). This vast area contains snow-capped mountains, hot deserts, rivers and wetlands with diverse vegetation as shown in Figure 6.2.

The diversity of African soils was demonstrated in the twentieth century by the many complications encountered when 'modern' or 'improved' agricultural techniques were implemented throughout the continent. African soils simply did not fit preconceived notions, and did not respond in a manner similar to soils in other places (Lal and Sanchez 1992). Purveyors of modern and improved agriculture were, to use T. Beach's (1998, p. 400) concept, 'environmental pioneers', as had been Africans before them when they migrated and settled in different parts of the continent.

Europeans first encountering the tropical African high forest concluded that the red-coloured soils must be extremely fertile to support such luxuriant growth. In time, red soils became known as the curse of Africa, liable to harden irreversibly and become useless under cultivation. Eventually, all colours of soil were said to

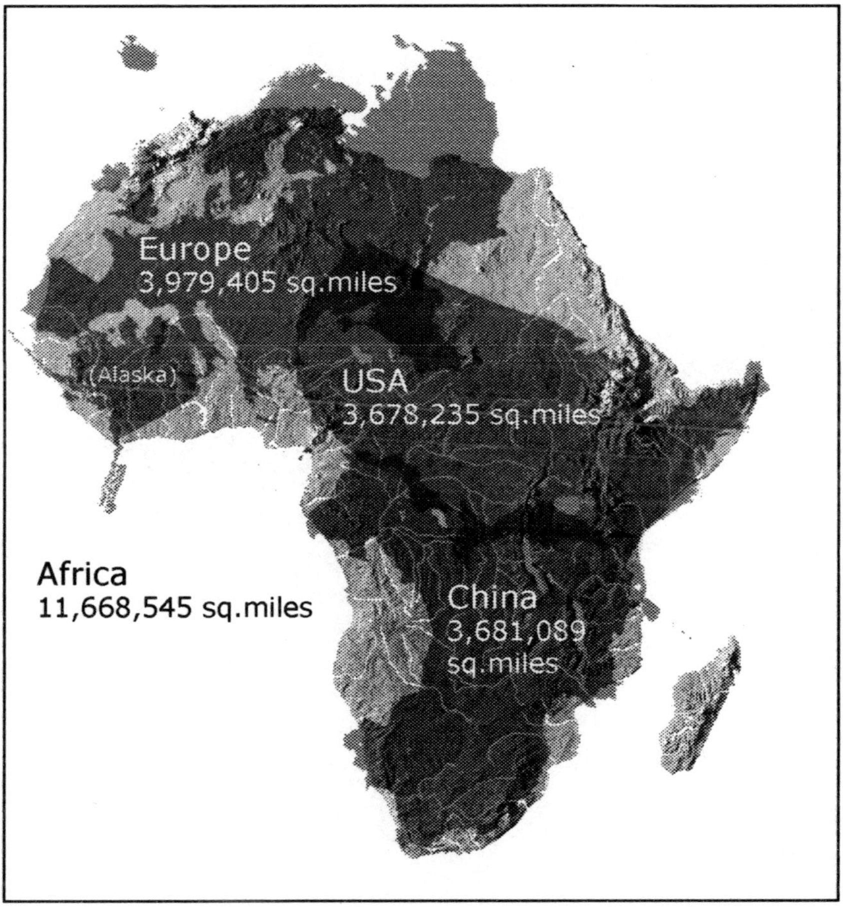

Figure 6.1. The land mass of Africa exceeds that of Europe, the USA and China combined. Based on 'How big is Africa?', Trustees of Boston University n.d..

erode with ferocity, when not blowing in desiccating winds. Finally, it was said that African soils were worn out by African smallholders' agricultural 'mining' of their fertility, and destroyed by trampling of their livestock. To gain some perspective, this chapter will attempt to view African soils as understood and used by Africans and Europeans for as much of the past as there is some kind of evidence.

Charting the ideas and activities of various waves of environmental pioneers on the African continent is a complicated process. Africa remains an understudied continent, and African knowledge systems – a profoundly endangered resource

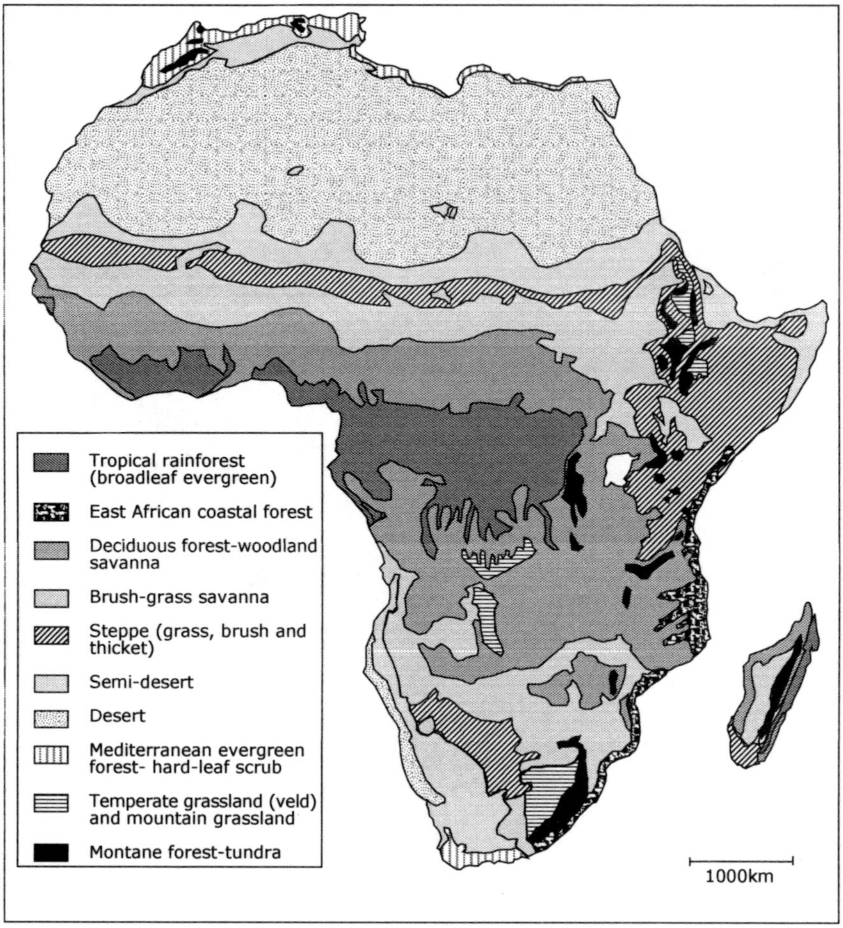

Figure 6.2. Principal vegetation zones of Africa. Based on map at http://library.berkeley.edu/EART/maps/africa-veg.gif

– are generally unknown to the larger world community. The historical record (in its broadest sense) is incomplete. Few documents have been analysed from an earth perspective, oral histories of societies and their landscapes have not been collected in most locations, and archaeological research has been very limited. However, materials from a wide range of sources do exist from which examples can be drawn for consideration. While the absence of reliable evidence is problematic, the confusion created by apparently authoritative documentation that lacks a firm factual basis must also be clarified. The twentieth century was a time when overlapping knowledge systems and practices existed, but not necessarily to the benefit of the newly arrived environmental pioneers – or the earth. Prejudice and ignorance often prevented information exchanges.

To provide a base for considering a history of African soil, general discussions of the perceptions of Africa and its soils, measurable properties of African soils, and soil knowledge systems will be presented. Attention will then turn to examples of African and European encounters with African soils. Four categories of activities that could have affected soil properties or entire soil bodies will be considered: crop production, livestock management, architecture/urbanisation, and mining. Discussion will centre around specific studies of particular places or systems in an attempt to identify mechanisms of change, rather than a continuous continent wide analysis in careful chronology. Generalisations from relevant studies will be made hesitantly, keeping in mind Richards' (1985, pp. 12, 13) cautions about working from a base of ecological particularism and seeking to avoid both 'glib generalisation' and '***ad hoc*** arguments' or 'special pleading', as well as concerns about the distortions inherent in extrapolation. Topics in need of further consideration or research will be noted, as this chapter is intended as a preliminary survey of the topic.

PERCEPTIONS OF AFRICAN SOILS

For outsiders (and not a few Africans), Africa – as landscape, vegetation or people – has been defined by European perceptions and experiences, particularly in the late nineteenth and twentieth centuries. Without understanding and acknowledging the nature of this definition, it is difficult to discuss Africa as perceived and experienced by others. The first complication is scale. Africa is a huge continent with extremely different climates, landscapes and cultures – not a country with a range of characteristics. Descriptive examples or scientific measurements from specific places or time periods – often extremes – have come to be cited as 'typical', and have been so generalised that a large, varied continent has been intellectually collapsed into a smaller, more homogeneous entity. Africa is often referred to as if it were a country defined by a few (largely negative) characteristics.

The creation of British Colonial Africa in the first two decades of the twentieth century nurtured this process at a bureaucratic level; a separate Tropical Africa Service was created within the Colonial Service to staff British Africa, just as the older, separate India Service staffed British India (for discussion of the creation of British Colonial Africa see Showers forthcoming). Officers in the Tropical Africa Service could be – and frequently were – moved from territory to territory. As they met and compared their experiences, and discussed them with other local non- Africans (traders, missionaries), consensus built. It was these generalised perceptions that informed London – insofar as Africa was discussed at all. Until the 1930s, few British Government officials had ever visited the continent (Killingray 2000, p. 42). A similar process of consensus based upon superficial observation informing a colonial bureaucracy took place in France.

With colonial administrations in Africa (British, French, Belgian, German, Portuguese) came expert opinion. An individual with (or without) subject matter specialty would arrive, tour, collect documents, meet with local authorities, and produce a report. In many of these reports opinion, rather than data, dominated. A.T. Grove (2000) reviews this literature of opinion, discussing some of the more notorious – and highly influential – reports responsible for codifying and spreading misinformation about Africa. It is these documents that established the framework for Roe's (1991, 1995) 'crisis narratives'. Soil has been central to many of these crises, from laterisation, degradation and desertification through fertility decline and erodibility. Modern expert reports and analysis continue the process. A late-twentieth-century example is the debate about 'nutrient mining' in the Sahel. A ten-year 'comprehensive data bank' was used to generate numbers that apparently demonstrated a dramatic depletion of soil fertility by traditional land users. The numbers were derived from combinations of soil erosion estimates with semi-quantitative nutrient loss data and 'best estimates' (some with coefficients of variation of 50 per cent), that were then 'upscaled' and projected across an entire region (Dreschel et al. 2001, pp. 412–23). 'Science' had once again validated preconceived belief.

In the late-twentieth century many of the 'truths' behind the crisis narratives, as well as the narratives themselves, were challenged (reviews of some of the crisis literature can be found in Scoones 1997; Batterbury et al. 1997; Mazzucato and Niemeijer 2000, pp. 16–21). Nevertheless, crisis narratives and the beliefs they support are so widely accepted that presentation of alternative perspectives is frequently derided as being romantic.[1]

For this reason, any attempt at writing a history of African soils must contend not only with the enormous size of the continent and its ancient human history, but also with both the lack of reliable information and an excess of generalisations and misrepresentations. To move beyond a debate about competing

[1] Discussions not based on crisis narratives are often accused of asserting the idea of 'Merrie Africa'. In defence against such a charge, Helge Kjekshus reminded readers in a footnote to the Introduction to the 1996 impression of his 1976 classic work, *Ecology Control and Economic Development in East African History*, 'The term "Merrie Africa" is first used by Hopkins (1973: 10) as a contrast to "Primitive Africa" of Alfred Marshall (1938) to set the stage for Hopkins' pioneering work on West African economic history. As the reader will recall, "Merrie Africa" denoted to Hopkins a mythical situation of populations living, without working, in abundance and plenty, but pursuing lives of ease and leisure of which major parts consisted of "interminable dancing and drumming"'. Kjekshus then states that in his book, 'The portrait of East Africans that emerges ... is that of hard workers, skillful planners, active learners and ultimate survivors. These are clearly not characteristics of the cast that made up A.G. Hopkins' myth of "Merrie Africa"' (Kjekshus 1996, p. xxix).

narratives, sources of information must be identified that can suggest the nature of interactions between humans and soils, as well as indicate whether they resulted in transformation, creation or destruction of soil bodies. A not inconsequential problem, as Scoones (1997) points out – and Chambers (1983) before him – is that even reliable studies are simply 'snapshots' of a particular place at a specific time. In Africa, 'some people in some places at some times' experienced events and conditions that didn't happen elsewhere, or at that place, but in a different time. Much of Africa has an extremely variable rainfall distribution, making it difficult for anyone to identify what is typical and what is extreme. In many cases, the extremes (drought and flood) are both typical and normal.

But it is the very perception of normal that makes historical documents difficult. As discussed by Showers (1989), normal to most people is that which is familiar, not a statistical concept of mean and deviation. Most documenters of Africa – travellers, residents, or experts – found whatever part of Africa they encountered to be quite different from their place of origin, and judged it to be abnormal. West Africa's climate and the presence of malaria inhibited European settlements, but not commerce or the establishment of plantations. The ubiquitous red soils were very distinct, even to short-term visitors. In some places, red soils cleared for crops hardened. These soils were described as 'laterite', the name coined by British geologist F. Buchanan-Hamilton in 1807 for red earth (it was not referred to as soil) that hardened and was used for bricks in the Kerala District of India (Russell 1973, p. 730; Schantz and Marbut 1923, pp. 118, 180–82). In 1820 P. Berthier analysed soil samples from the Fouta Djallon of West Africa, and found them to be similar to that of Kerala, and in 1911 J.D. Falconer described some Nigerian soils as 'iron clay' (*Encyclopaedia Britannica* 1929 vol. 13, p. 740). Soon all red West African soils were assumed to be laterite.

Semi-arid or arid regions of Africa were frequently perceived as being defective, or in a state of decline or degradation. These conditions were often blamed on local people and their land use system. In the Cape Colony (modern South Africa), R. Grove (1989) first argued, and Endfield and Nash (2002) documented in detail, how fundamentalist interpretations of the Bible led missionaries to believe that the dry landscape was a sign of God's punishment of the indigenous people; in neighbouring Basutoland (modern Lesotho), the relative lack of trees in the natural grassland was assumed by some to have been due to deforestation by the local residents (Showers 1989; Showers 2005); and in French North Africa, particularly Morocco, the lack of trees was said to be proof of the destructiveness of Arab land-use practices (Davis 2005). In both the semi-arid West African Sahel and southern Africa there is a literature attributing the low fertility status of the soils to traditional land use practices, particularly the burning of grass or dung.

Ideology, bias and perceptions of normal clearly have distorted observers' accounts of Africa. But these predispositions may well have been magnified by a confusing biophysical reality. Geomorphologist Martin Williams (2003) illustrates the difficulties involved with determining whether recent environmental change was due to earth processes or human activity. Examples from various savanna systems showed that major geomorphic events, such as tectonic activity, sand dune formation and historic floods and droughts, could account for recent environmental change in places where policy was directed at modifying human behaviour that had been identified as the cause of problems such as deforestation or soil erosion. Knowledge of the African biophysical world and its processes is fundamental to understanding human interactions with it.

In an attempt to dispel northern hemisphere biases and presumptions, and to begin to establish notions of African normalities, it is important to have an idea of the soil properties to be expected under the range of soil forming conditions on the African continent. Africa must be compared to itself, and not misconstrued as a deviation from other continents.[2]

AFRICAN SOIL IN THE ABSENCE OF HUMANITY

The African continent is ancient and the soils have been unaffected by glaciation (Mt Kenya's and Mt Kilimanjaro's glaciers' local effects being exceptions). Outside of volcanic regions, African soils did not have additions in the Quarternary. For this reason, along with Australia, the Amazon shield and southern India, Africa contains some of the oldest geomorphic surfaces on earth (Eswaran et al. 1992, p. 3). Soils formed on these surfaces are so ancient that they are no longer related to the underlying rock. Adjacent to these extremely old surfaces are newer ones, largely the result of uplift and peneplanation. Volcanic eruptions and stream deposition have also influenced local soil properties. Geomorphology has, therefore, been an important control on soil forming processes (Eswaran et al. 1992, p. 3). Although all of the rocks found in the temperate zones except glacial till are found in the tropics, some soil forming processes related to time are unique to that region,

[2] When considering environmental factors in an historical context, historians are confronted by two problems: the belief of some post-modernists that science is just a social construction, and a charge of environmental determinism. Those concerned with these ideas might consider Nancy Jacobs' discussion of the distinction between nature, a social construction, and 'an actually existing nonhuman biophysical world with its own integrity' (Jacobs 2003, p. 19). Jan Vansina (1990) argues that historians' 'underestimation of this diversity [within a rainforest] leads to a flawed understanding of the interactions between people and their habitats' (Vansina 1990, p. 41). Consideration of the biophysical world (of which soils are a fundamental part) is, therefore, not environmental determinism but, rather, an opportunity to look at the ways in which people engage with their surroundings, to see previously ignored expressions of creativity, and to more fully appreciate local environmental knowledge.

resulting in soil types unknown in the northern hemisphere (Eswaran et al. 1992, p. 3). Since the African continent lies in the tropics and subtropics, except in high mountain areas, the activities of flora and fauna are uninterrupted by cold, so that soil forming processes can occur year round. Soils in Africa more resemble those of Brazil, peninsular India and Australia than those of Europe or North America (Grove, A.T. 1970, p. 26).

The only statement that can be made with certainty about African soils is that they are enormously diverse. For every general statement made about African soil conditions, an example to the contrary can be found. With this in mind, let us consider some broad characterisations. According to the documentation accompanying the Africa sheet of the FAO/Unesco Soil Map of the World (1977), many African soils are highly weathered and have lower plant nutrient contents than northern hemisphere soils. Levels of phosphorus are low enough in much of the continent to limit plant growth. Elevated levels of aluminium and manganese, which can inhibit root growth of many plants, are also common in highly weathered soils. In some places the iron content approaches that of low-grade ore. Sub-Saharan West African soils have their fertility rejuvenated each year when dust, laden with basic cations (calcium, magnesium, potassium and sodium), blows in with the Harmattan winds. Arid and semi-arid regions of the north, east and south of the continent have sandy, stony and/or rocky soils that can be quite shallow (FAO/Unesco 1977, p. viii). Surface crusting and sealing, as well as salt deposition, are common in these soils. Areas of moving sand dunes exist in both the Sahara and Namib Deserts. Layers of dense or hard soil, called 'pans', can be found below the surface of soils in all rainfall regimes.

Many of Africa's highly weathered clay soils have a red colour, signifying high iron content (Eswaran 1988, p. 6). The fate of this iron depends upon soil conditions. It may simply accumulate as a stain on the surface of soil particles, or it can be concentrated. Where the water table fluctuates – either locally or on the scale of a floodplain – the soil solution can become enriched with iron, which is subsequently deposited as mottles, concretions, nodules or continuous sheets 2–5 metres – or even 10 metres – thick (Eswaran 1988, p. 7; Buckle 1978, p. 68). Referred to variously as laterite or plinthite, this material is characteristically soft and permeable when moist, but hardens upon drying. (Note: the word laterite has been used so broadly that its precise meaning is unclear. For this reason, some pedologists and soil scientists have refused to use the term. However, forms of the word are components of some international soil classification systems [see Faniran and Areola 1978, p. 161; Eswaran et al. 1992, p. 7]). The extent of hardness achieved, and its irreversibility, varies (Russell 1973, p. 731). When hard, it can be referred to as ferricrete, lateritic duricrust, or petroplinthite (Buckle 1978, p. 68; Eswaran 1988, p. 7). Soils with lateritic or plinthic layers have developed under both forest and savanna in West Africa, and are extensive in regions with annual rainfalls of

200–500 mm/year, or 10–16° N, such as Niger, Burkina Faso, northern Nigeria and Guinea's Fouta Djallon (Cassel and Lal 1992, p. 77; Buckle 1978, p. 68; Russell 1973, p. 729). Obeng (1978, cited in Cassel and Lal 1992, p. 77) reported that approximately 113 million hectares of forested soils and another 113 million hectares of savanna soils in West Africa had concretionary or iron hardpans. Although soil structure, texture and fertility establish potentials for plant growth, rainfall defines much of Africa's growing season(s).

African soils can be characterised in terms of their relations with water (FAO/Unesco 1977). Soils of the high rainfall region can have weak physical structures, making them susceptible to erosion. Soils lying between the semi-arid zones and the high rainfall regions (savanna and dry forest) can be sandy, and thus do not retain water, while others have physical and chemical properties which cause them to be poorly drained or prevent the downward movement of water (hardpans, ironpans, duripans). Those soils with high clay contents can hold water or block its movement, and some can be sticky when wet, and hard when dry. In the Niger and Congo River basins and in the Sudd region of the Sudan – as well as in many river valleys of the humid and temperate parts of Africa – there are large areas of poorly drained soils. The Mediterranean climate zones of northern and southern Africa have soils whose structures make them susceptible to erosion by water. Traditional soil use and management practices have been shaped by the combinations of soil physical and chemical properties and water relations at specific locations. For this reason, land use practices, like general statements, must be transferred from their landscape of origin with care.

SOIL KNOWLEDGE SYSTEMS

Soil comes to people's attention because of its properties – or lack of them. Soil knowledge systems begin when systematic observations are ordered. Soil properties are identified, named, and associated with particular soil bodies, allowing comparisons to be made. Distinctions among soils, and decisions about soil boundaries, are determined by the classifiers' perceptions, assumptions and needs. The resulting classification system provides a means of simplifying the complex and continuous soil reality into understandable, discrete classes (Krasilnikov and Tabor 2003, p. 203). Today, two major categories of soil knowledge systems are recognised: local and western scientific. Each type contains many classification systems, which reflect different perceptions of the world and have different purposes (Krasilnikov and Tabor 2003, p. 204). Local soil taxonomic systems are constructed to address particular concerns at a specific place. The relationships identified and names given may not have meaning in an adjacent region, but can identify particular soil-landscape relations extremely accurately, especially where there is great soil variability (Krasilnikov and Tabor 2003, pp. 209, 211; Niemeijer and Mazzucato 2003, p.

404). In contrast, western scientific soil classification systems seek to 'interweave' 'general principles and detailed observations' (Buol et al. 1973, p. 617) so that multiple purposes can be addressed over large regions, or even globally (Krasilnikov and Tabor 2003, p. 203).

In the late twentieth century, scholars in the emerging field of ethnopedology began to ask Africans about their understandings of soils (see WinklerPrins 1999 and the Ethnopedology Special Issue, *Geoderma* vol.11, 2003). Local soil knowledge was found to be widespread among rural people; to be somewhat specialised – people knew about soils they encountered; and to have strong historical dimensions. This dynamic knowledge base is the product of experience, observation and some systematic experimentation over time, and thus provides a long-term perspective to particular soil-plant systems. The length of these long-term perspectives can be centuries or millenia (WinklerPrins and Sandor 2003, p. 165).

Interpreting ethnopedological information is not simple, for each system of local soil knowledge reflects a unique way of perceiving the world (Krasilnikov and Tabor 2003, p. 204). To understand local soil knowledge, it must be considered within its own context, which means that outsiders must be open to different ways of knowing, or, in anthropological terms, avoid ethnocentrism (WinklerPrins and Sandor 2003, p. 165; Sillitoe 1998, pp. 190,192). An extreme example of this for most North Americans and Europeans would be the use of sound by the Bété people of Ivory Coast when classifying their soil. While these forest dwellers only distinguish three classes of colour (dark, light and bright/red), they listen to determine differences in soil texture (grittiness) (Birmingham 2003, pp. 486–7). Few residents of the northern hemisphere could imagine listening to soil as method of classification – but many would accept as scientific the shaking of particles in water and watching their different rates of settling. In addition, there is a tendency to treat knowledge as static or finite. Local knowledge systems have been shown to be dynamic, changing over time as a result of changes in its 'natural and social context' (Niemeijer and Mazzucato 2003, p. 409; WinklerPrins 1999, p. 155). As a result, categories, or members of categories, can change as circumstances change. This is particularly true for value – a soil that is considered to be good for one crop may not be good for another, and so on.

There is a direct link between local soil knowledge systems and those of western science. V.V. Dokuchaev's study of the chernozem soils of European Russia resulted in the first scientific classification of soils and a soil map (Dokuchaev Central Soil Museum, 2000–2002 undated; Buol et al. 2003, p. 9). Many of the terms he used to describe soil – and soil names – came from original folk names (Krasilnikov and Tabor 2003, p. 201). In his subsequent work Dokuchaev clearly distinguished soil from geological deposits, and identified it as an 'independent natural and historical body' whose origins were related to climate, vegetation, and time. This association between a soil and its factors of formation was the birth of

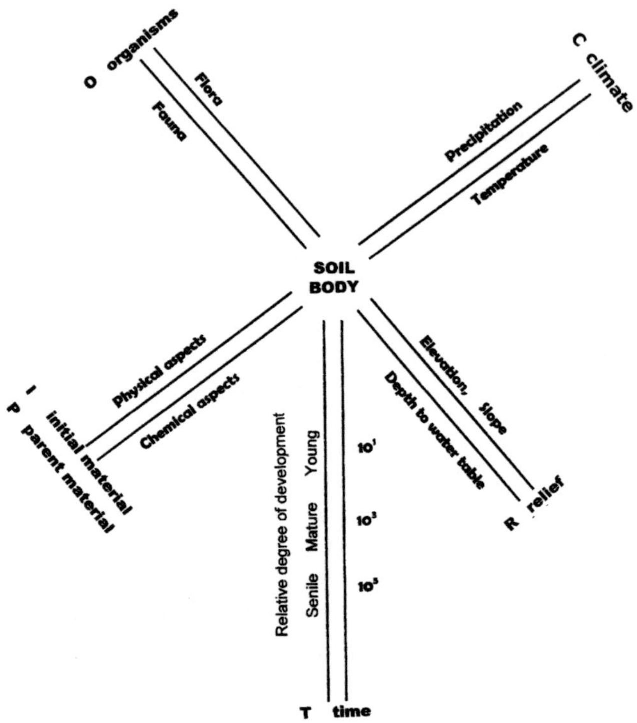

Figure 6.3. 'Flower' diagram showing factors involved in soil formation

genetic soil science. The idea of factors of soil formation was later expanded upon in the United States of America and expressed in the simplified 'flower diagram' shown in Figure 6.3.

Dokuchaev's ideas reached those who could only read English when *The Great Soil Groups of the World and Their Development*, written by his student, D.K. Glinka, was translated from German to English by C.F. Marbut and published in 1927 (Lipman 1933; Soil Survey Staff 1951, p. 3; Faniran and Areola 1978, p. 6). A variety of national soil classification systems were subsequently devised based upon physical and chemical properties of soils as revealed by soil profile analysis. Characteristics such as colour, texture (amounts of sand, silt and clay), pH (a measure of acidity), base status (amounts of calcium, magnesium, sodium and potassium), water relations, and position in the landscape were used as determining characteristics.

A History of African Soil

Soil classification and mapping has been an interest in Europe and North America since the turn of the twentieth century. According to the FAO/Unesco Soil Map of the World, two significant early soil maps of the African continent were those of C.F. Marbut in 1923 and Z.Y. Shokalskaya in 1944. Both were general maps of hypothesised soil distribution based on climatic, lithological or phytogeographical factors. Working in East Africa, chemist G. Milne used local soil names and knowledge in his early studies of soil catenas (sequences of soils down a slope) (Payton et al. 2003, p. 357). Until the 1950s, very few African soil maps existed that were based on actual surveys (FAO/Unesco, 1977, p. 1). The idea of a map of the African continent's soils based on regional soil surveys emerged at the Commission de coopération technique en Afrique au Sud du Sahara (CCTA)'s Inter-African Pedological Service (Service pédologique inter-africain – SPI)'s first Administrative Council meeting in Yangambi (then Belgian Congo) in 1953. At its second meeting in 1955, the SPI Administrative Council recommended that a 1: 5,000,000 map for soil conservation and use be drafted in close collaboration with the four regional committees – southern, eastern, central and western Africa. This decision stimulated preparation of new maps throughout the continent. Subsequent meetings produced compromises among Belgian, French, British, Portuguese and South African classification systems to reach agreement on a uniform map legend. When produced in 1963, it was the first map ever drawn as the result of international agreement, and 'made it possible to establish the relationship between pedogenesis [soil formation] and the development of major soil units' (FAO/Unesco 1977, p. 1). This map legend became the reference point for subsequent work. The idea of a World Soil Map arose at the seventh Congress of the International Soil Science Society in Madison, Wisconsin. The CCTA African soil map provided the base from which the first draft of the Soil Map of Africa was produced in 1968 (FAO/ Unesco 1977, p. 2). In the same year the Organization of African Unity (OAU) published its International Atlas of West Africa. Since the scales and much of the data were the same, the OAU map was incorporated into the second draft of the FAO/Unesco map. This 1971 draft became the final version of the Africa map of the FAO/Unesco ***Soil Map of the World*** (FAO/Unesco 1977, p. 2).

The years of consultative preparation described above demonstrate the extent to which the FAO/Unesco soil map embodied the international consensus of western scientific approaches to both soil identification and knowledge of African soils. Finally published in 1977, this 'paper map' is accepted as 'the appropriate source of soil information for studies at a continental, regional or global nature' (Nachtergaele, undated). It has been the basis for all subsequent African continental soil maps, including those involving 'simplifications and transformations', such as the USDA maps using their Soil Taxonomy classification system, and the Russian simplified 1: 15 million scale map of world landscapes, as well as electronic maps, including the recent digital maps (For historical description of the evolution of electronic soil maps of the world see Nachtergaele, undated, pp. 2–3). Changes in the maps have been largely due to data transformations, in which mathematical relationships between two or more soil characteristics are used to estimate or predict missing information (For discussions of pedotransfer functions and taxotransfer functions see Nachtergaele, undated, p. 4). Nachtergaele admits that 'some of the information contained in the [World Soils] map is of uneven quality, often outdated and completely wrong' (Nachtergaele, undated, p. 5). National and regional soil mapping exercises have been funded by a variety of institutions using a range of classification systems (USDA Soil Taxonomy, Orstrom, FAO and local systems), as listed in Nachtergaele's appendix, 'Country Soil Maps of Africa'.

However, Eswaran et al. (1992) argued that the FAO/Unesco map's lack of data perpetuates misinformation about tropical soils. As shown in Figure 6.4, when it was prepared, approximately 7 per cent of the African continent had been covered by large or medium scale survey maps with some ground-truthing for accuracy; 38 per cent had coverage by reconnaissance maps that showed soils in relation to other features such as climate, geomorphology or vegetation, and ***55 per cent of the continent was 'virtually unknown'*** (FAO/Unesco 1977, p. 7). Information from a few specific, detailed surveys was extrapolated to areas for which no information existed. These 'best estimates', once displayed in map form – on paper or with digital technology, have the appearance of authority (See Figure 6.5 for example). Boardman (1998) decries 'scientific myth making', in which specific and reliable data travel tortuous routes and undergo transformations, making them incredible.

While Boardman's discussion centres on the extrapolation of carefully collected measurements of soil erosion from a specific place to a regional, national or continental scale, the same could be said for maps based upon similar processes. For these reasons, people interested in specific places find local soil knowledge more useful than internationally constructed soil maps reflecting international consensus on the ways in which soils should be distinguished and classified (Niemeijer and Mazzucato 2003, pp. 404, 411). Similarly, those interested in making general statements about soil over large areas recoil from the regional variation and inconsistencies inherent in local soil knowledge.

A significant amount of effort has been put into comparing local soil classification systems with those of western science. Since the systems being compared have completely different origins and, often, purposes, the match is invariably imperfect. It is extremely important to avoid the common assumption that western scientific systems of knowledge are superior to other knowledge systems. Many studies by western scientists have validated distinctions identified by, and conclusions drawn from, local soil knowledge systems (WinklerPrins 1999, p. 153). Krasilnikov and Tabor (2003, p. 203) point out that soil classification systems provide a common means for talking about soils by simplifying a complex continuum into discrete classes. For this reason, they argue, all soil classification systems are artificial, based

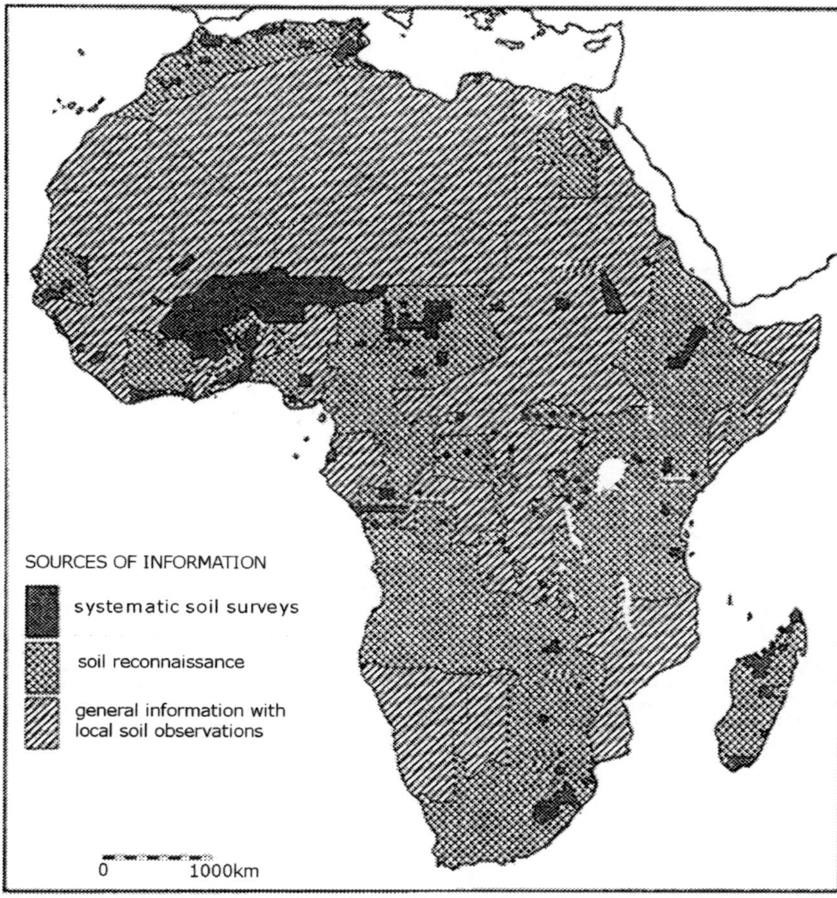

Figure 6.4. Sources of information on which soil maps are baseds

upon the information available and the beliefs and needs of the classifier. Each type of information has its purposes; care must be taken to ensure appropriate application. Although these systems should be understood and used as complimentary perspectives from which to learn about the soil and its processes, the apparent authority of western science with its generalised maps and specialised terminology threaten the very existence of complex and detailed local soil knowledge systems (Krasilnikov and Tabor 2003, p. 202).

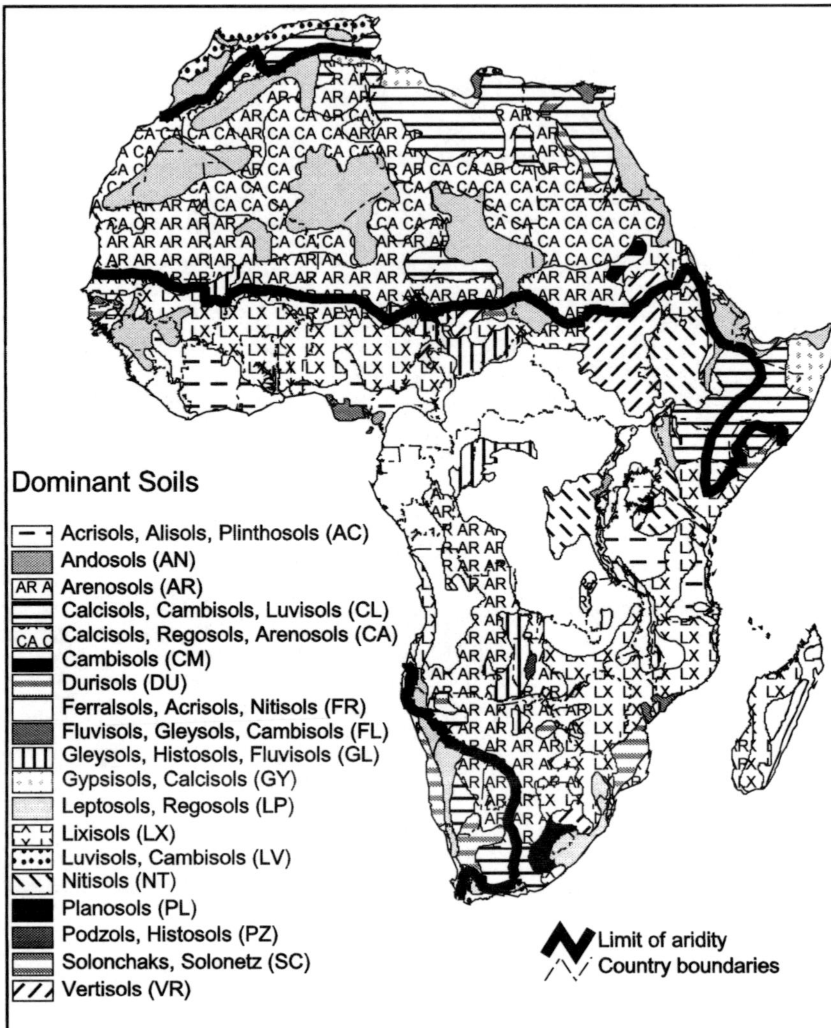

Figure 6.5. African soil types according to the FAO World Soil Resources map, 2003

AFRICAN SOIL IN THE PRESENCE OF AFRICANS

The African continental surface is ancient, as is human experience on it. Archaeological evidence shows the high grasslands of eastern and central Africa to have been the 'cradle of humanity' (Oliver 1991, p. 3), the region in which the human species emerged.[3] Oliver (1991) describes the pioneering aspects of humanity's move out from the abundance, or 'Eden', of the East African highlands 1.5 million years ago to regions of the continent that were both hotter and drier, or wetter and more densely vegetated, than the highland savannas.[4] They did so with simple tools, and encountered ecosystems with much less abundant game, and very different vegetation.

This pioneering did not take place in a static environment. Climate (and thus vegetation) change at the end of the Pleistocene and throughout the Holocene greatly affected the human experience. As precipitation regimes shifted, so did the extent of moist and dry forest, and of savanna (Casey 2005, p. 232). Learning from observation and experience, with subsequent adaptation of life styles, was essential for survival. With the advent of grinding stones and pottery (as far back as the sixteenth millennium BC) came the harvest of wild grasses, which, mixed with other kinds of foraging, hunting and fishing, sustained people for 6,000 years (Neumann 2005; Klieman 2003, p. 57). These earliest environmental pioneers were few in number (Casey 2005, p. 232) and they were not farmers. Presumably their impact upon soils was slight.

Whether these non-agriculturalists acquired soil knowledge by noting correlations between plant distribution or topography and soil types is not known, but African soil awareness is certainly old. The production of ceramics in West Africa in the early seventh millennium BC and in coastal Central Africa from the late sixth century BC, and their inclusion in Central African regional trade networks in the fourth millennium BC (Klieman 2003, pp. 52–4; Casey 2005, p. 234) suggests at least a developing knowledge of properties associated with different soil textures. The ancient name for Egypt was Kemet, which means fertile black alluvial

---

[3] African archaeology has been both under-funded and hindered by ideas generated by archaeological work in Western Asia and Europe, as well as by theories about stages of development. Vast areas of the continent have had little or no archaeological attention, and important questions within archaeology have not been addressed. Like African soil, the findings of African archaeology do not conform to experiences elsewhere, and have, therefore, been either ignored or accepted with great difficulty. For general discussion of these problems see Stahl 2005a, and for discussion related to specific topics, from hominims to urbanisation, see various authors within Stahl 2005b.

[4] Like all archaeological reconstructions, the detail of Oliver's model of humanity's spread is disputed. See Stahl 2005b.

soil. *Deshret* was the name for red desert land. Three thousand years ago Egypt's soils had established commercial values: *nemhura* soils cost three times more than *sheta-teni* soils (Krupenikov 1981 cited in Krasilnikov and Tabor 2003, p. 199). In other parts of Africa (for which written records are sparse or absent and historical linguistics methods have not yet been applied),[5] ancient soil names have not been documented and dates can be less clearly assigned to soil classification, value or management systems. However, Central African oral tradition makes it clear that the Bantu settlers arriving in the rainforest regarded knowledge as a key resource (Guyer and Belinga 1995; Klieman 2003). They eagerly acquired the environmental knowledge required for survival, as well as that for pottery production, from the Batwa, the autochthons living in the rainforest (today referred to as 'Pygmies'). Soil knowledge was certainly a component of this information transfer.

Archaeologists have shown that first herding (cattle, sheep and goats) and, much later, the cultivation of food plants, spread slowly across the continent, intermingling with existing lifeways based on foraging, hunting and fishing.[6] It is these newer forms of sustenance that had the potential to influence soil properties, and thus offer the opportunity to consider human beings as soil forming factors on the African continent. Unlike patterns of land use evolution in western Asia/the Near East and Europe (Stahl 2005a), the earliest food producers in Africa were mobile pastoralists who left limited archaeological traces (Shahack-Gross et al. 2003), not agriculturalists. As a stable and widespread way of life, pastoralism existed in the Sahara and Sahelian grasslands by around 4500 BC (Gifford-Gonzalez, 2005, p. 188; Neumann 2005, p. 257). By 2000 BC they produced ceramics, which were traded to hunter-gatherers (Shahack-Gross et al. 2003, p. 440).

While archaeological evidence clearly shows the development of African pastoralism, documentation for the rise of agriculture is less clear. Neumann (2005) usefully distinguishes between cultivation, 'any human activity that increases the yield of harvested or exploited plants', and domestication, the 'genetic, morphological, and physiological changes of the plants resulting from cultivation and conscious or unconscious human selection' (Neumann 2005, p. 250). Cultivation can be practised with wild plants as well as with domesticated plants, an important concept when trying to understand African land use systems, and one that chal-

---

[5] For discussion of the methods and applications of historical linguistics, see Klieman 2003.

[6] for excellent review articles with comprehensive bibliographies see Stahl 2005b. It is important to remember Kjekshus' (1996) restatement of a main theme in Johnson and Anderson (1988) that it is unhistorical to assert rigid classifications of African societies by ethnicity or economic roles such as pastoralist, agriculturalist or hunter-gatherer, since these societies were both socially dynamic and ecologically adaptive as they met survival challenges (Kjekshus 1996, p. xxii).

lenges common assumptions about the definition of agriculture. Like pastoralism, agriculture did not spread continuously throughout the continent; it occurred in different regions at different times. Cultivation was commonly added to – but did not replace – other means of food acquisition such as foraging, hunting and herding. This is why Neumann (2005) cautions that finding evidence of domesticated plants at an archaeological site does not indicate a reliance on agricultural production or warrant labelling the occupiers as farmers.

Although agricultural crops existed in Egypt from around the sixth millennium BC, a clear pattern of crop production is not found in archaeobotanical assemblages in the south-western and south-central Sahara until slightly before c. 1800 BC, and much later, from around the middle of the first millennium BC onwards, in the rest of the continent (Neumann 2005, pp. 263–4). There is no evidence to support notions of agricultural diffusion from Egypt to the rest of the continent, and no centres of plant domestication have been identified. Instead, there is evidence of a great diversity of agricultural production systems involving the cultivation of both wild and domesticated plants throughout the continent. For example, it was Near Eastern crops that were produced in about the sixth millennium BC Egypt, oasis agriculture based on 'emmer, barley, bread wheat, date palm and Mediterranean fruit trees' was practised in the northern Sahara from approximately the first millennium BC, and plants were domesticated in Ethiopia from around 500 BC. (Neumann 2005, p. 253). Although pearl millet (***Pennisetum glaucum***) was domesticated by about 1800 BC in the western Sahara and spread quickly through the dry regions of West Africa, rather than leading to an agricultural lifeway, its cultivation was simply added to existing practices of herding, hunting and foraging (Neumann 2005, p. 261). In humid African forests, yam cultivation and the management and cultivation of trees were practised (Vansina 1990). Many archaeological sites in the Central African rainforest provide evidence of a fully developed agricultural life without any direct evidence of domesticated plants or animals. These sites dating from the middle to late first millennium BC indicate a system of agroforestry based upon palm nuts, cola nuts, and/or the fruit of Aielé (***Canarium schweinfurthii***) (Klieman 2003, p. 58). Evidence of the cultivation of oil palm (***Elaeis guineensis***) in Central Ghana and the Niger Delta around 1000 BC, and of cultivated banana (***Musa sp***) in the Cameroonian rainforest from around 800/400 BC, supports notions of incipient arboriculture (Casey 2005; Neumann 2005). Very little work has been done to establish the development of agriculture in southern Africa (Neumann 2005).

Although a wide variety of plants were used for food, there was little possibility for soil disturbance. Crop production – whether clearing trees or planting seeds, tubers or cuttings – was initially based upon stone tools such as axes and digging sticks (Klieman 2003, p. 41). Associated with the widespread use of iron tools – axes and hoes – are more diversified agricultural systems, and larger field sizes, but not necessarily increased soil disturbance (Neumann 2005, p. 266). Hoed fields have rough surfaces that inhibit erosion, and hoed clods left in place serve as a mulch, protecting the soil surface (Showers 2005, pp. 12, 14).

Causality between the spread of iron tools and agricultural diversification is not clear. Where trees needed to be cut or other vegetation cleared or removed, the acquisition of iron tools made agricultural production easier. However, in Central Africa, it is argued, it was the introduction of a high-yielding, less labour-intensive crop – the banana – that created the possibility for the development of metallurgy. This more productive agricultural system could support non-agricultural specialisation (Klieman 2003, p. 99). Vansina (1990, p. 6) argues that it was the banana, not iron tools, that enabled Bantu immigrants settled along rainforest river systems to move deeper into the forest and 'colonise all of its habitats everywhere'.[7]

Oliver (1991, p. 49) states that modern African cultures reflect the fact that all of the life styles – hunting, fishing, herding and agricultural production – coexisted, and persist. What is difficult to show is the effect these diverse and complex land use systems have had on soil properties and soil bodies during the millennia in which they have been practised. It can be argued that life styles centred on fishing and hunting had little influence on soils. Although the East African trade in ivory dates to at least 2000 years ago, and hunting was an integral part of most African societies, hunting techniques other than the digging of pits in which to trap animals did not disturb the soil (MacKenzie, J.M. 1988, pp. 63, 77, 148). In contrast, livestock management and crop production have potentially significant consequences for African soils. The basic concepts of dispersed grazing with transhumance (seasonal movement of stock to a different region) and crop-fallow rotational cropping systems have been applied with variation over much of the continent for millennia. Even where there have been traditions of urbanisation in Africa, studies have shown that these settlements were always linked to larger landscapes; sub-Saharan iron-using urbanised people engaged in mixed farming date to around the first millennium AD (LaViolette and Fleisher 2005, pp. 332, 327; Oliver 1991, p. 93; Lewis and Berry 1988, p. 29).

Archaeologists have begun to study the prehistory of land use systems.[8] Part of the challenge of this work is to overcome such soil problems as acidity that cause

---

7 While discussion of the 'Bantu migration' is beyond the scope of this chapter, it should be kept in mind that, as Childs and Herbert (2005, p. 281) point out, 'Bantu-speaking peoples were borrowers rather than innovators; they acquired knowledge of ironworking, seed agriculture, and animal husbandry from their non-Bantu-speaking neighbors'. For critical discussion of the Bantu migration, see Vansina 1990, Klieman 2003 and Eggert 2005.

8 Neumann (2005, pp. 250–1) provides a useful history of the archaeology of African plant food production, noting the subject's very recent origins.

a rapid decomposition of organic materials and, thus, of evidence for subsistence practices (Klieman 2003, p. 24). Mobile populations (pastoralists as well as foragers and part-time cultivators) who carry their dwellings with them and conserve their tools left little for archaeologists to find (Gifford-Gonzalez 2005, p. 189). Shahack-Gross et al. (2003) point out that pastoral sites are often distinguished from those of foragers by the presence of dung deposits associated with night-time corralling of livestock. However, dung is only preserved if buried, decomposing at open-air sites. Sutton's (1989) inquiry into constructing histories of agricultural fields points out that existing techniques of 'archaeological recognition' could only identify ancient agricultural fields if they were 'mounded, ridged, bounded with banks or ditches or terraced in a pronounced way … Rarely can an excavator of a living site show just where ... fields were, let alone define their shapes and sizes' (Sutton 1989, p. 102). He concluded that the primary tools for reconstructing the 'typical' fields and farms that supported Africans since agriculture began were extrapolations back from documentation of existing and recent land use practices, aided by specialised soil and vegetation studies not then undertaken. Accepting Sutton's caution about the non-existence of a 'typical' field, it is, perhaps, useful to consider elements of the two most common African land use systems – extensive grazing systems and crop/fallow rotational agriculture – to see what possibilities for soil change might exist. As was indicated earlier, sharp separation of land use systems is not a reflection of reality. For the sake of analysis, each component will be considered separately, and then interactions will be noted. Finally, examples of traditional soil conservation practices will be discussed to suggest that for generations, centuries, or longer, awareness of soil erosion processes existed at least in particular places, and techniques were devised to control them either for field construction or stabilisation.

a. Livestock Management

Africa has experienced ancient traditions of dispersed or extensive grazing, since pastoralists move according to the needs of their stock for pasture and water. Cattle were tended on the continent before sheep and goats, and in Egypt before the Sahara, but by about the fifth millennium BC a successful cattle-based pastoral economy existed in the central Sahara. This ended during the arid years from approximately 3550–2550 BC as pastoralism shifted to the south and east. Cattle, sheep and goat pastoralism (the Saharan triad) first appeared in East Africa's Lake Turkana basin around 2050–2550 BC. Less archaeological evidence exists for southern Africa, but it seems that only sheep arrived in south-west South Africa in about the first millennium AD (Gifford-Gonzalez 2005, pp. 203, 204, 206). The existence of pastoralism in the absence of agriculture has been explained in terms of a system that could respond successfully to climate change (Gifford-Gonzalez 2005), and as a response to Africa's rich environmental resources, particularly in the savanna, that made labour-intensive agriculture unnecessary (Neumann 2005). Pastoralism was, however, limited to savanna regions by the presence of livestock disease, and it is these soils that might show the impact of millennia of pastoralism.

A striking characteristic of African savanna ecosystems is the seasonal occurrence of fire (Goldammer and de Rende 2004). Europeans were alarmed by it, while Africans used it (Laris 2004; Wardell et al. 2004). Set naturally by lightning strikes in storms at the end of the dry season, fire has been considered to be a beneficial disturbance, causing changes in vegetation structure and composition, as well as promoting nutrient cycling and distribution (Goldammer and de Rende 2004, p. 1). Recent studies of marine sediments taken off the coast of Sierra Leone show that during the past 400,000 years the greatest intensity of vegetation fires occurred in the West African region during periods when global climate was changing from interglacial to glacial mode (Bird and Cali 1998 cited in Wardell et al. 2004). Wardell et al. (2004) state this study also asserts that human activity has 'shaped the fire regime' for the last 10,000 years, making difficult historical distinctions between naturally occurring and human-set fires. At the turn of the twenty-first century the large number of fires set by African land users was causing economic damage, but little attention has been paid to the effects of fire on soils.

Burning has been widely used as a range management practice in savanna regions because it improves the nutritional value of vegetation. Van der Vijver et al. (1999) investigated the extent to which burning increased the concentration of nutrients in above-ground plant growth and contributed to the loss of nutrients from the system through volatilisation and ash convection. The study was carried out on the nutrient-rich savanna soils of Tarangire National Park, Great Rift Valley, northern Tanzania. Only a brief effect was found in vegetation immediately after burning, and burning did not affect root biomass. Measured plant changes were due more to changes in plant growth (increased leaf: stem ratios) than to added nutrient supplies, since the soils had inherently good plant nutrient status. This led to the conclusion that nutrient losses were not significant in comparison to the soil supplies, and that the repeated effects of burning were 'not as extreme' on these soils as compared to 'nutrient poor savannas', which were not included in the study (Van derVijver et al. 1999, p. 183). The extent to which wind-blown ash is 'lost' to the system depends upon how far it is blown, and how far 'the system' is considered to extend. The primary nutrient lost through volatilisation would be nitrogen. The amount of loss depends upon the nitrogen content of the material that was burned. At the end of the dry season, when most burning occurs, grasses have very low nitrogen contents. (The consequences of burning as a tool of crop production is discussed below under forest and savanna agricultural systems.)

Grazing stock can affect soil physical properties (Showers 2005, pp. 13, 14). The weight of animals can compress or compact soil, and hooves can dig down below the soil surface, destroying soil structure. The greatest damage occurs when soil is wet (Tanner and Mamaril 1959; Gradwell 1968; Twerdoff et al. 1999). Hooves dig deeper into wet soils, compressing and remoulding or puddling them (Gradwell 1968). Damage under these conditions could reduce pasture growth, and could be irreversible. The severity of compaction depends upon the intensity of grazing, soil type, and vegetation (Warren et al.1986; Krenzer et al. 1989; Twerdoff et al. 1999). Alternating periods of grazing with nongrazing allows some, if not complete, recovery (Gradwell 1968; McCarty and Maurak 1976; Haynes and Francis 1993). Studies have shown that compaction occurs only in surface soils (Krenzer et al. 1989), and dense vegetation can mitigate this consequence of grazing. Although soils in grazed pastures may be permanently denser or more compacted than those of nongrazed areas, this does not necessarily affect plant growth (Tanner and Mamaril 1959; Gradwell 1968). However, water enters a compacted soil with greater difficulty, and compacted soils can hold less water. Compacted soils can, therefore, be drier than non-compacted soils, so that plants growing on them are more likely to experience water stress (Gradwell 1968).

Livestock play a role in nutrient cycling as they eat and then deposit manure and urine. This is discussed in detail below, under savanna systems of agricultural production. Powell et al. (1996, p. 145), working in the semi-arid Sahel of West Africa, argued that cattle grazing on rangeland represents a closed cycling system, but as livestock producers increasingly settle, corralling of range-grazed livestock on crop land could create imbalances.

Because in many African pastoral systems livestock were herded to different locations in the landscape for food and water at different times of the year, what were recognised in the late twentieth century as conservation range management principles were being followed (Pendleton 1982; Coupland 1979). There is a possibility of some intensely grazed areas having permanently compacted soils, which affected soil-water relations. However, it is reasonable to assume that most African pastureland was unstressed by traditional pastoralists. This is not to say that there are no examples of soils disturbed by livestock management. River crossings and paths certainly had compacted soils, and where large herds of livestock were concentrated (such as the settlements of the Zimbabwean plateau in the eleventh–fifteenth centuries AD, particularly Great Zimbabwe), all the ills associated with over grazing and trampling undoubtedly occurred.

b. Agricultural Production

Nye and Greenland's (1960) classic study *The Soil under Shifting Cultivation* assesses the effects of crop-fallow rotation systems in both the evergreen and semi-deciduous high forests and the tall tussocky grass and tall and short bunch grass savannas of West Africa. Samples were taken and literature reviewed of soil and vegetation systems in the mature fallow state, immediately after clearing and burning, and in the succeeding three years of crop production. At the time this study was made, no generally recognised system existed for detailed soil classification, so discussion was of broadly grouped soils associated with distinct vegetation types. Although focusing on West Africa, this study establishes a base from which to consider the effects of rotational fallow systems on other African soils. Subsequent conceptually less comprehensive, but more detailed, investigations clarify points made and provide additional information. In particular, the 1970s droughts in the Sahel region prompted intensive study of grazing and cultivation systems in that semi-arid region. Similar studies began in the 1990s in arid Namibia.

Forest Systems. According to Nye and Greenland (1960), the moist evergreen forest common to coastal Liberia and Sierra Leone (see Figure 6.2) occurred on soils that were extremely weathered, had very low cation exchange capacity (CEC – the ability to retain exchangeable cations, such as the bases calcium, magnesium, potassium, sodium, as well as non-basic aluminium), and low pHs (a measure of acidity – low is acid). Soils under the drier semi-deciduous forest common to Ghana were richer – higher pHs (less acid) and had more bases to support plant growth. An analysis of the acid soils of the evergreen forest showed that the amounts of all exchangeable nutrients, except phosphorus, in the top 29 cm. of soil were considerably less than the amounts in the vegetation of mature forest fallows.[9] In contrast, the vegetation of drier Miombo woodlands (east, central and southern Africa) and thickets stored much less calcium and magnesium than the soil, and comparable amounts of potassium and phosphorus.

Detailed study showed that when plots in either the evergreen or semi-deciduous forests were cleared and burned, great changes took place in terms of soil chemistry, but unless wood was heaped, prolonging burning, the top 5 cm. of soil were little affected by heat (Nye and Greenland 1960, pp. 71–3). The resulting ash not only produced an influx of soil nutrients, but also sufficient basic cations to raise the soil pH (making it less acid). On acid soils, the change in pH was particularly significant. Burning initially reduced populations of soil microorganisms, including those that transform nitrogen, but was soon followed by a rapid increase in numbers, resulting in increased amounts of nitrogen in a plant-available form. Although carbon, nitrogen and sulphur from the fallow and forest litter were lost in the burning, the soil humus was not burned, so nutrients stored there were retained (Nye and Greenland 1960, p. 10). That the nitrogen lost in burning was not of great consequence had been shown by studies of green manure in the 1930s (reported by Sampson and Crowther 1943, cited in Richards 1985, p. 29). When incorporation of fresh green manure was compared to incorporating its ashes after burning, no difference was found.

[9] Vansina (1990) and Klieman (2003) note that recent research in the Central African rainforest points to greater diversity than previously imagined. See discussion of African soil maps.

On the acid, highly weathered soils of the West African moist evergreen forest, cropping could only be supported for one and a half to two years; a ten-year fallow was required to restore fertility. The relatively more fertile soils of the drier semi-deciduous forest could support crop production for three years, followed by an eight-year fallow. It was only early in the first year that the soil surface was affected by planting crops in either type of forest. During that time, in both systems, the soil of the cleared space was bare. With the first rains, seeds were planted in the surface soil (which was rich in organic matter and worm castings), with a digging stick – causing minimal soil disturbance. Annual grains and perennial crops were planted as an intercrop. Until the crops grew, the soil's surface was exposed to heavy rains that could destroy the structure of both exposed aggregated soil particles and worm castings, and ash could float away in runoff. Sheet or rill erosion could also occur. However, within four weeks the crop plants (and, likely, weeds) had produced a canopy that covered and protected the soil surface. In subsequent years the soil was never bare, as perennial crops were harvested and there was no replanting (Nye and Greenland 1960, pp. 3, 4).

Since animals were not kept by forest dwellers, the non-soil sources of soil fertility were, primarily, the decomposition of organic matter and additions from rainfall, dust, lightning, and the fixation of atmospheric nitrogen by microorganisms. The major losses from the system were considered to be due to leaching of nutrients below the rooting zone, removals in run-off water, the volatilisation of nitrogen by microorganisms, and of nitrogen and sulphur through oxidation when organic matter was burned. Nye and Greenland concluded that forest crop-fallow systems were nearly a closed cycle and, therefore, efficient and sustainable (Nye and Greenland 1960, pp. 43, 44). The system as described does not suggest modification of soil properties beyond temporary changes in the soil's chemistry (plant nutrients) and microbial populations. More recent international agroforestry research reviewed by Kass et al. (1999) supports these conclusions.

The Central African rain forests (modern Rwanda, Burundi, Uganda, Democratic Republic of Congo, Cameroon, Equatorial Guinea, Gabon and Congo Republic) have been occupied by human beings for millennia (Sayer et al. 1992, p. 43). The earliest evidence is for isolated groups of stone tool-using hunter-gatherers living on hilltops and near running water from 38,050–33,050 BC, during the Ndjilian humid phase, when Gabon's forest refuges were first formed. This was followed by 23,000 years of lower rainfall and temperatures associated with the Leopoldian era climate shift, when forest cover shrank and savannas expanded. During the subsequent humid Kibangian era (from 10,050 BC), the forest expanded to its current size, and residents produced small, less heavy stone tools and engaged in medium-distance trade (Klieman 2003, pp. 37, 40). The meagre evidence for subsistence suggests a reliance on wild food. However, at archaeological sites in the Ituri rain forest dating back 10,000 years, there are remains of the oil-producing plant ***Canarium schweinfurthii***, and later sites containing oil palm (***Elaeis guineensis***). The presence of these plants has been cited as evidence of possible early plant management (Casey 2005, p. 234). However, Casey points out, modern foragers use these plants, and lightning strikes and windfalls can open up spaces in forests providing the light required for oil palm survival (Casey 2005, p. 234, 235). Heeding Neumann's caution about assuming agricultural activities from evidence of cultivated plants, it is not reasonable to conclude that these were farming communities.

There is evidence of soil knowledge among ancient rainforest residents. Giles-Vernick's (2002) account of the knowledge system of the Mpiemu hunter-gather-farmers of the Central African Republic's Sangha River Basin, makes clear that local environmental knowledge, including soil knowledge, was part of their identity. Central to becoming a person in Mpiemu culture was the transmission of knowledge from one generation to the next. Giles-Vernick could only trace the Mpiemu's presence back 150 years, but her informants claimed an ancient origin for their knowledge, a time when lives lived in grasslands combined with those of the forest (Giles-Vernick 2002, pp. 49, 56, 57).

Vansina (1990) describes the process of multi-generational experimentation in West African forests that began 5000 years ago, leading to more sedentary lifeways and, ultimately, the agricultural systems of Bantu speakers based on yams, gourds, castor bean, black-eyed peas, and the ***Voandzeia*** groundnut (Klieman 2003, p. 41). These agriculturists began to move into the Central African rainforest along rivers, and by the middle of the second millennium BC had established many communities in the western equatorial forest. These settlements were involved with trade networks that carried ceramics by the fourth century BC , but it isn't until 500 BC that there is evidence of slash-and burn agriculture (Klieman 2003, pp. 40, 55). Sayer et al. (1992), concurring with Vansina (1990), make clear that all parts of Africa's tropical rainforest are now used by local residents, and that there is no area of forest that has not had a long history of human use.

According to Vansina (1990) the selective use of plants and animals by dispersed hunter-gatherers did not significantly alter rainforest dynamics, but the arrival of settlers who farmed did. Farming populations were kept low, he asserts, because large populations cut down so many trees that the supportive habitat was destroyed. Vansina's concern was for flora and fauna, and not soil. However, aside from the (unstudied) consequences of ceramics production, there is no evidence of soil disturbance – but none has been sought. Given Nye and Greenland's evidence of crop-fallow rotation systems in moist evergreen forests producing only temporary changes in soil properties, and noting the low density of human populations in the past, it could be argued that with respect to crop production, moist evergreen forest dwellers in the past were nothing more than participants in, and users of, an almost closed nutrient cycling system, and did not significantly alter soil properties or soil bodies.

This might not have been the case in the drier woodlands of central Africa. The Bemba, who inhabit northern Zambia's Miombo woodland, named their crop-fallow rotation system *chitemene*. *Chitemene* involves burning on a new field not only the trees cut in the process of clearing, but also branches and brush cut from adjacent land. The resulting intensive burning heated the top 2.5–5 cm of soil, with the result that soil structure improved (Nye and Greenland 1960, p. 71).

Soil properties were changed in the chitemene system for the short and medium term. Along with the expected initial transfer of nutrients from vegetation to soil, Stromgaard (1986) found that burning increased the sandy soils' ability to retain cations (cation exchange capacity, or CEC) for as long as 16 years after the burn, when the *chitemene* cycle could begin again. While the amount of basic cations (calcium, magnesium and potassium) retained in the soil might be considered to be low, they were essential for regeneration of the fallow vegetation. Stromgaard (1986) also documented the encroachment of grasses during a crop/fallow cycle, noting that the dense root systems prevented cultivation. Invasive grasses of the Hyparrhenia family secrete a toxin that suppresses the growth of nitrifying bacteria, thus reducing the amount of nitrogen available in the soil and retarding the regeneration of woody species (Stromgaard 1986, p. 104).

Stromgaard (1986) argued that the area cleared for a field and the surrounding land from which tree branches were cut must be viewed as components of a system consisting of an intensively farmed 'infield' and an 'outfield' exploited at low intensity (Stromgaard 1986, p. 97). The *chitemene* system concentrates nutrients from a given areas' vegetation in a smaller space, and improves that soil's ability to retain them. It also facilitates the growth of grass plants which affect the microbial environment. If these changes persisted over time, then *chitemene* could be considered to transform the properties of the soils on which it was practised. This form of cultivation is practised in most parts of the plateau in north-western and north-eastern Zambia (Moore and Vaughn 1994).

Savanna Systems. Although savanna vegetation is far more extensive than moist tropical forest, human relations with savanna soils are less clear. By definition, savanna soils occur where rainfall is too low to support a closed-canopy forest (See Figure 6.2). They are, therefore, less weathered and tend to have higher pHs and nutrient supplies. Perhaps surprisingly, soil organic matter levels of savanna topsoils were found to be lower than those under forest cover.[10] Rotational use of savanna soils for cropping has followed some of the principles and practices of forest systems,

[10] Plant roots and associated microbial communities, not aboveground growth, are the major source of soil organic matter (Bolinder et al. 1999).

but there are significant divergences. As in forest systems, low intensity burning did not affect soil temperatures below the surface 5 cm. of soil, but intense burns did (Nye and Greenland 1960, p. 11). The annual low intensity burning of the savanna was a mechanism for returning nutrients to the soil but, since grasses do not increase their store of nutrients year by year as forests do, when savanna vegetation was burned, fewer nutrients were deposited on the soil surface than when a field was cleared from forest (Nye and Greenland 1960, p. 36). The rapidity of annual burns and its positive effects on vegetation have been noted by recent researchers (Laris 2004; Wardell et al. 2004).

Parallels cannot be drawn between the values of mature forest and mature savanna. When cleared, old savanna not only had distinctly lower nitrogen levels than mature forests, but also less than savanna of two to six years. Nye and Greenland argued that because the long dry season meant that only annual crops could be planted in savanna regions, the soil was left bare each year after harvest. Additionally, surface soils had to be disturbed for planting because of the need to remove grass roots, and remained uncovered until crop canopies developed. Finally, plant cover was slow to establish when the fallow period began because of a scarcity of suckers and established seedlings (Nye and Greenland 1960, p. 124). This lack of cover, they argued, exposed savanna soils to erosion by water and wind. If eroded enough to expose subsoils containing plinthite, a very hard surface could form.

More recent research suggests that at the very least, dry-season soil was covered by weeds and unharvested plant parts (stover). In the moist high-grass savanna, as in the evergreen moist forest, livestock was limited by tsetse fly's trypanosomiasis, so manure was not part of the nutrient cycling system; soil fertility was maintained exclusively by fallow management. However, in the drier tall and short bunch grass savanna, where there were no tsetse flies, herds of cattle, sheep and goats were significant components of soil fertility management (Nye and Greenland 1960, p. 4). Hoffman et al. (2001, p. 268) described precise and careful management of post harvest stover and weeds for livestock grazing during the dry season in semi-arid northwest Nigeria. Grazing stock converted high carbon organic matter (stover) into more plant-available forms of nutrients and deposited them as urine and manure. The stubble mulch resulting from grazing protected the soil's surface. Fields in the study area (Kano close-settled zone) were only disturbed and bare at the onset of the rainy season, when roots had been dug and burned. Both resident livestock and those of pastoralists grazed the fields. So valued is manure for soil fertility maintenance that farmers in semi-arid West Africa have traditionally arranged with migratory pastoralists to corral their herds for a fixed number of nights on their fields (Williams 1999, pp. 17, 18). These arrangements were especially important for farmers without large numbers of livestock.

Detailed research presented in a series of papers by Powell et al. (1996, 1998) and Ikpe et al. (1999) investigated the role of livestock in maintaining soil fertility in the Sahel. Their experiments showed that while manure was important, it was the deposition of urine that had the most significant and long-term effects on subsequent crop yields. The research was carried out on a Labucheri soil, which is very low in organic matter content, has predominantly kaolinitic clays (very low ability to supply nutrients), and is poorly buffered (cannot easily resist acid-base changes). The addition of urine caused two major nutrient changes in this soil. As expected, levels of nitrogen in various forms increased. Surprisingly, levels of phosphorus also increased. Since there is virtually no phosphorus in urine, the researchers theorised that when the urine raised soil pH (the soils became less acid) during the first week, phosphorus dissolved out of the aluminium-iron complexes in the kaolinitic clays and became available to plants (Ikpe et al. 1999). Technically speaking, phosphorus was released when the structure of soil particles was ruptured. This process can be understood either as nutrient cycling – shifting phosphorus from unavailable to available forms in the soil – or destructive 'mining', since the release of phosphorus involves the destruction of individual soil particles. Powell et al. (1996) argue that cattle grazing a rangeland represent a closed nutrient-cycling system, but when grazed on rangeland by day and pastured on cropland by night, livestock are the 'principal vectors of nutrient transfer across the landscape' (Powell et al. 1996, p. 148; Powell et al. 1998, p. 260).

The Gourmantché people in the southern Sudano-Sahelian (900 mm mean annual rainfall) and more northern Sahelian (600 mm mean annual rainfall) regions of eastern Burkina Faso were found to have complex understandings of nutrient cycling systems, the role of organic matter, and the relative importance of water and soil fertility to crop production in a region with highly variable rainfall amounts and distribution (Niemeijer and Mazzucato 2003). The area consists primarily of flat to gently undulating peneplain with 'occasional lateritic hills'. Most of the soils have low fertility and low levels of organic matter; many are shallow, gravelly or have 'lateritic formations'. They can also crust, seal or become hard. The vegetation ranges from savanna woodlands in the south to tree and bush savannas in the north. The Gourmantché believe that the landscape is shaped by erosion, sedimentation and the transformation of organic matter. They think that erosion could be a source of soil fertility, as well as create gullies that could, over time, become a valley. They also think that few cultivable soils are inherently good; water relations – not soil chemical or physical properties – determine a soil's utility. A soil in a water collecting spot would be good in a drier year, and bad in a wetter one. No soil would be good in a drought. Soil fertility is understood to be a function of organic matter, and distinctions are made based upon its source, location (on top of, or in, the soil) and degree of decomposition. Darker soil colours are associated with organic matter so decomposed that it is no longer visible. Gourmantché farmers believe that plants consume organic matter from the soil, and that repeated cropping reduces soil fertility, making soils lighter in colour. They also realise that too much organic matter or manure limits crop production. While livestock are appreciated for the manure they produce, their trampling is associated with decreased ground cover and subsequent erosion, as well as with soil crusting and compaction. Because the Gourmantché farmers see a relationship between organic matter additions and soil fertility, they believe that there cannot be an irreversible loss of soil fertility. In dry years, when vegetation is sparse, soil organic matter levels decline, but in wetter years with more luxuriant growth, the levels increase again. Farmers, they believe, can work within this continuum of vegetation supply (and manure) to maintain soil fertility.

c. Soil Conservation Practices

Where topography and rainfall created a need for water control or the prevention of soil movement, locally evolved, specialist techniques developed, including the construction of mounds, ridges, pits and terraces (Sutton 1984, pp. 30–33; Grove and Sutton 1989, p. 115). Mounding and ridging, usually prepared by hoe, were components of agricultural systems in regions of medium rainfall, especially on lighter soils (Sutton 1989, p. 101). These techniques emerged from crop production needs. One function was the prevention of soil erosion in savanna soils. However, Nye and Greenland found mounding and ridging to accelerate erosion in forest areas of West Africa (Sutton 1989, p. 101; Nye and Greenland 1960, pp. 89–90).

A variation of this system is pit construction on steep slopes in the Matengo Hills of Tanzania. A series of pits constructed across and down the slope collect water for slow infiltration, preventing erosive overland flow (for description and photograph see Temu and Bisanda 1996). This pitting system was a variant on shifting cultivation. First observed as a coherent system by European visitors in the 1890s, pitted fields in the late twentieth century were cultivated for 4–6 years, and then left to fallow for 2–10 years.

Agricultural terracing was not widespread in Africa, Sutton (1984, pp. 30–33; 1989, p. 102) argued, because there was no need for it in most places. Soil movement could often be controlled with vegetation and timing of agricultural operations, or with non-permanent structures. In the west Cameroon Highlands, the traditional distribution of cultivated land into fallow, trees, and pasture promoted stability. If rills were discovered in cultivated fields, the land was covered with brush or returned to pasture. Trees maintained in the fallow area broke the slope length, thus inhibiting erosion. When this vegetative management system changed under late twentieth century pressures, rapid erosion resulted (Temato and Olson 1997). Further north in Morocco's Rif Mountains, shifting cultivation of grain together with arboriculture of raisins, figs and plums, olives or almonds (depending upon soil type), goat raising and charcoal production supported people for centuries in a dry Mediterranean climate. When the system was disrupted by twentieth-century land re-allocation and forest protection, serious erosion began (Heusch 1981, pp.

419–420). For generations stone bunds or lines were constructed in Ethiopia's Harerge Highlands as needed on the contour, their spacing determined by the slope. In addition, soil bunds were constructed to control erosion at planting time. They were easily removed when crop cover protected the soil, structures were no longer needed, and weed control was essential. Most of these structures were temporary, but some persisted for as long as 20 years (Asrat et al. 1996, p. 160). Elsewhere in the Ethiopian Highlands, water control and erosion prevention were achieved by ditching. Farmers varied the position and depth of ditches in their fields each year to avoid gradual widening and deepening over time (Alamayehu 1996, p. 167).

Permanent stone terraces were constructed for use in conjunction with irrigation systems in hilly or mountainous locations. A terrace system at Engaruka, Tanzania, was abandoned approximately 300 years ago after 200 years of use (Sutton 1984, pp. 36–39; 1989, pp. 103, 107). At Konso, Ethiopia one has been in use for at least 400 years, and in the Atlas Mountains, individual dry stone terrace walls strengthened by tree roots or interwoven with tree trunks have been used for more than 100 years (Amborn 1989, p. 73; Hamza 1996, p. 46). Stone terraces have been constructed in the Mandara Mountains of northern Cameroon for generations as a component of water control management, along with ridging and drainage canals. However, they were not built as fixed, permanent structures, but rather are continually modified by adjustment of height of wall or even location on the slope (Hiol Hiol et al. 1996, pp. 196, 198).

In other locations, irrigation was not obviously the major purpose of terrace construction. The terraces abandoned at Nyanga, northern Zimbabwe, 200–300 years ago were not clearly connected to irrigation (Sutton 1984, pp. 33–36; 1989, p. 103; Grove, A.T. and Sutton 1989, p. 115), while those from 800–900 years ago in the Great Zimbabwe region of southern Africa appear to have been related to erosion control (Oliver 1991, p. 110). Stone-walled terraces have been used on steep slopes in Maku, Nigeria to create agricultural fields. Oral tradition states that terrace construction began as fortification of villages in the era of slave raids. Soil accumulated behind the walls was used as small fields, and did not form a continuous terrace system on the hillsides. Maintenance was, and is, crucial, as a break in the terrace wall could result in washed out crops (Igbokwe 1996, p. 223). Evidence of terracing constructed with stone walls also exists in Nigeria and Cameroon (West Africa); in Darfur and Kordofan provinces of Sudan; in Tanzania and Kenya Highlands; and in Nyanga, Zimbabwe 200–300 years ago – that were not for irrigation (Grove and Sutton 1989, pp. 115, 122; Sutton 1989, p. 103).

Whether or not these interventions changed soil properties can be debated. In some locations normal soil movement downslope was arrested, and the accumulated soil was used for crop production. Such an intervention is an example of the creation of man-made soil bodies. In other locations, interventions counteracted accelerated erosion associated with farming. Soil surfaces were disturbed by both agricultural production and the construction of conservation structures, but there are no accounts of a landscape destroyed by agriculturally induced soil erosion before the arrival of Europeans.

d. Architecture/Urbanisation

Whether for building one-roomed house components, large mosques, markets or walls, soil as brick or plaster has been used for centuries in many parts of Africa (for photographs and descriptions of regional architectural styles see Denyer 1978). Not all soils are suitable for construction. It is only those with high concentrations of certain types of clays that make good bricks and plaster. This suggests that for as long as people have used soil for building, there has been a basic knowledge of soil textural properties, and an ability to distinguish one soil type from another. The excavation of these soils in sufficient quantity for construction undoubtedly resulted in local soil disturbance. The construction of one household's buildings, however, was not as consequential as the creation of urban areas that not only disturbed soils they covered, but also stimulated extraction from – and use – of larger landscapes.

Although it has been described as the least urbanised continent,[11] Africa actually has a long tradition of cities and towns. Since the architecture in many African settlements consisted of structures made from easily degraded materials such as straw, earth, or small trees, they were not easily preserved or recognised. With or without the monumental architecture associated with classical definitions of urbanity, concentrated settlements of human populations did exist in Africa, and did affect the larger landscape, including soils.

The Nile valley has a history of urbanisation dating from 3500 BC, and urban areas have existed along the Mediterranean Coast of Africa from 1200 BC (Lewis and Berry 1988, p. 97). Large-scale settlements existed in West Africa's Inland Niger Delta from the first millennium AD. One such place, Jenné-jeno, began as a seasonal settlement in about 250 BC, covered 12 hectares by 100 AD, and grew to a permanent village of daubed pole-and-mat houses measuring 25 hectares by AD 400. From AD 400 – 900 Jenné-jeno occupied 33 hectares, with a cautiously estimated population of 3,200–6,400 in town, and a further 6,700–13,400 in nearby settlements. A massive wall of sun-dried cylindrical bricks enclosed the 33-hectare site (LaViolette and Fleisher 2005, pp. 328, 334). Continuous urban settlement in the Chad Basin dates back to the ninth century AD. Ngazagarmo, the capital city from the fifteenth to early nineteenth century, was at its height in the seventeenth century. It covered about 1,500 ha and contained a quarter of a million people

---

[11] Until the 1980s, Northern Hemisphere conventions about – and definitions of – cities and urban areas largely excluded Africa's urban reality and past from consideration. For discussion, see LaViolette and Fleisher 2005.

(Denyer 1978, p. 35). In 1658 Ngazagarmo reportedly had four mosques and 666 roads cleared and widened (Denyer 1978, p. 35).

In the Sahara desert a chain of cities stretched from Tekrur in the west through Audoghast, Timbuktu and Gao to Agadez in the east from the tenth to the fifteenth centuries (Denyer 1978, p. 31). Further south, along the northern limits of the tsetse fly, a second line of cities developed that included Segou, Djenne, Ouagadougou, Katsina and Kano. From this line of Hausa urban areas south, many towns had extensive walls enclosing not only buildings but also open land (Denyer 1978, p. 33). The Yoruba cities of Western Nigeria, which probably date from the fifteenth century, had town centres consisting of a palace, a market and a temple with its sacred grove of trees (Denyer 1978, pp. 35, 36). Two main roads divided the city into quarters. Houses were built close together in family units, surrounding a well, rather than along a defined street (Lewis and Berry 1988, p. 350). According to Lloyd (1967, p. 220), Ibadan and Abeokuta were surrounded by hamlets from which the farming community commuted, while Ijebu Ode was surrounded by more than 100 permanent villages. Lloyd noted the 'antiquity of the villages' in the mid-1960s.

On the East African coast, the town of Rhapta (in modern Tanzania) was reported by Ptolemy to have 'grown from an emporium to a metropolis' by AD 300 (Chami 1996, p. 16). A series of towns and villages based on local, regional and long-distance trade dating from the eighth to the eighteenth centuries dotted the coast from modern Somalia to Mozambique (LaViolette and Fleisher 2005, p. 339). Building materials ranged from stone to earth-and-thatch, and populations from more than 5,000 to 15,000 people (LaViolette and Fleisher 2005, p. 340). The towns tended to be arranged around a well or market, not along streets, and household units included garden areas for domestic food production (Lewis and Berry 1988, p. 351). The merchants in these towns controlled local, regional and international trade, and thus the towns were connected to the larger landscape and inland settlements by paths and roads.

In southern Africa, cattle-keeping and farming populations on the Zimbabwean plateau built settlements from the first millennium AD. Buildings were made from stone and from coursed earth, called *daga*. In time, political states developed, based upon cattle-keeping and the gold trade, with major settlements at places such as Mapungubwe, Khami, Danangombe, and Great Zimbabwe (Beach, D.N. 1984). These thirteenth–fifteenth-century cities are renowned for their monumental stone structures, but *daga* housing also existed (LaViolette and Fleisher 2005, pp. 236–7). North of the Orange River in modern South Africa, eighteenth-century Tswana people lived in concentrated settlements while operating an extensive land use system. European travellers in the early nineteenth century reported that the Tlhaping (a Tswana speaking group) capital was the same size as the colonial settlement of Cape Town (Jacobs 2003, p. 45).

No matter what their construction materials, African urban areas affected the soils they occupied as well as the surrounding landscape. Soil surfaces were covered, roads and paths were created, water was drawn, used and disposed of, and waste was discarded. The result was soil compaction, disruption of soil water relations, and pollution. Increased demands were placed on food-producing areas to supplement urban gardens. If soil was used for construction, quarries had to be dug, and when a building was no longer used, the soil had to be either reused or disposed.

e. Mining

More obviously disruptive of soil than agriculture, livestock production or urban areas is mining. Holes or tunnels in the ground or in stream banks are dug to extract desired minerals or soil types. Not only does the removal of the desired material cause disruption of the soil, but purification can result in large quantities of discarded waste. For millennia Africans have mined iron and copper for tools, ornaments, currencies and ritual objects, and salt and clays for nutritional, medicinal or ritual use. Small- and large-scale mining and metal working centres existed throughout Africa, producing ore and finished products for local use as well as for regional and long-distance, international and intercontinental trade networks. Mining was a major source of wealth for African empires and kingdoms (Tráore 1994 cited in Hilson 2002, p. 154).

Metallurgy. Minerals can be collected from the soil's surface, retrieved from stream beds, or excavated from beneath the ground. Once mined, mineral ores must be separated (smelted) from non-ore materials by heating. Smelting results in some pure metal and much larger quantities of waste, called slag, that is usually deposited on the soil surface. To produce finished objects, the metal must be forged – heated and worked. Since each of these processes has potential to affect soils directly or indirectly, it is important to have an idea of their nature and extent.

Iron, copper and gold were the primary metals mined and processed in Africa – lead and tin to a lesser extent (Childs and Herbert 2005, pp. 281, 282). The centres of mining and skilled craftsmanship developed in areas rich in ores, from which regional and global trade networks spread. Among Africans iron was the most commonly used metal (both for domestic and ritual use), and in many societies copper was more valued than gold. Copper-poor West Africa imported it across the Sahara in about the twelfth century AD (Childs and Herbert 2005, p. 290). At approximately the same time, gold, whose primary demand was non-African, was being exported to the Arab and Indian Ocean worlds. The substantial trade with Europe came later (Childs and Herbert 2005, p. 287). Hilson (2002, p. 153) suggests that gold was first mined intensively from Archaen greenstone belts in modern Zimbabwe, Botswana, Tanzania, Mozambique and South Africa.

African metal workers knew a great deal about the properties of metal. They produced not only wrought iron and cast iron, but also low- to medium-carbon steel. Steel production requires the manipulation of conditions inside smelting furnaces (for descriptions of the great array of furnaces and unique African furnace design that eliminated the need for a bellows see Childs and Herbert 2005, p. 284). In southeast Nigeria, the lost-wax casting process was highly perfected.[12] Central and southern African smiths made copper and iron wire by both hammering and drawing (Childs and Herbert 2005, p. 286).

The development of Africa's sophisticated metallurgy traditions, like those of pastoralism and agriculture, does not conform to patterns identified by European archaeologists. There was no Bronze Age in Africa, and evidence points to the smelting of iron before copper (Childs and Herbert 2005, pp. 227, 228). Radiocarbon dates of wood (the primary fuel in smelting) from scattered sites around the continent going back to the second millennium BC, confirm metalworking to be an ancient African activity. Rather than finding evidence of iron working as an imported skill, calibrated radiocarbon dates from about 800–400 BC at archaeological sites in Niger, Nigeria, Gabon, Cameroon, the Central Africa Republic, and the Great Lakes supports the possibility of independent invention. However, there are not enough data to chart the spread of ironworking across the Africa (Childs and Herbert 2005, p. 280). Everything to do with metallurgy had soil consequences.

Mines. Most obviously, metals must be removed from the soils containing them. Knowledge of soil properties – both the staining of surface soils and the kinds of vegetation supported – was used by ancient prospectors to locate ore deposits (Childs and Herbert 2005, p. 282). Since iron is so common in Africa – from deposits on individual soil particles (as mentioned earlier) to concentrations of ore – there should be no surprise that it was the most widely and continuously mined mineral. Ironworkers were able to distinguish between lower and higher grade ores, and only used oxides such as haematite, magnetite and limonite (Childs and Herbert 2005, p. 282). The greatest number of less frequently occurring copper deposits are in central and southern Africa. With the exception of Zambia's Copperbelt that was identified with twentieth century prospecting technology, almost all of the known sub-Saharan copper deposits were worked by miners in the pre-colonial era (Childs and Herbert 2005, p. 282). Although gold is widespread throughout the continent in both alluvial and reef forms, the earliest archaeological evidence for its use is not until the seventh and eighth century AD at Jenné-jeno (West Africa) and in elite burial sites at Mapungubwe, near the Limpopo River in southern Africa, dated to

---

[12] Lost-wax or *cire perdue* casting 'involves making a model of the object to be cast in wax, then enveloping it in clay. When the wax is melted in an oven, the metal is run in to take the exact shape of the "lost" wax' (Childs and Herbert 2005, p. 286)

approximately the twelfth century AD (Childs and Herbert 2005, p. 282). A lot of southern African gold came from alluvial (river) deposits and was panned (Tlou and Campbell 1984, p. 42).

Establishing the history of mining has been complicated by the fact that modern mines have been dug on top of ancient mining sites. The most common type of mine was a shallow pit or shaft dug with hoes or digging sticks. In some locations, mines consisted of holes dug into hills, creating caves, which were used until they collapsed (Tlou and Campbell 1984, pp. 42, 43). In central Africa, some open-cast copper mines were very large. A pit in Katanga, Democratic Republic of Congo was ¾ of a mile long and 600–1000 feet wide, which compares favourably to modern pits made with steam shovels. In nearby Kanshani, Zambia, an area 7,000 yards in diameter had been excavated. Surveyors have calculated that over the centuries of mining, thousands of tons of malachite were removed (Childs and Herbert 2005, p. 283). A lot of gold mined came from alluvial (river) deposits and was panned (Tlou and Campbell 1984, p. 42).

Deep mines were required for both copper and gold in central and southern Africa's hard rock areas, particularly in Zimbabwe and South Africa. By AD 750, specialised mining communities had developed in southern Africa and underground mines were dug. Gold mining is thought to have begun in Zimbabwe around AD 900, and in northeast Botswana soon after (Tlou and Campbell 1984, p. 42).

In preparation for digging underground mines, fires were set to fracture surface materials, followed by digging shafts with iron picks and hammer stones (Childs and Herbert 2005, p. 283).

The deepest shaft found in Botswana measured 26 metres, a seventeenth-century-AD Zimbabwean copper mine had shafts that extended down 60 feet, and a gold mine's descended 180 feet (Tlou and Campbell 1984, p. 42, Childs and Herbert 2005, p. 283). Shaft depths were limited by an area's water table – there was no ancient technology for pumping. Once shafts had been dug, miners descended by rope to dig ore, which was sent back to the surface in baskets (Childs and Herbert 2005, p. 283).

Smelting and smithing. Along with mining, large-scale iron production centres existed throughout the continent. In West Africa, the best known are the Middle Senegal Valley, Futa Jalon in Guinea, Mema in Mali, Yatenga in Burkina Faso, the Bassar region of western Togo, Hausa-speaking regions in Niger and northern Nigeria, and the Ndop Plain of Cameroon. There were also major centres in Central Africa (the Batèkè plateaux of Gabon and Congo), the Great Lakes region of Rwanda/Burundi, Tanzania, and Malawi; the Shona-speaking region of Zimbabwe; and Phalaborwa in South Africa's eastern Transvaal (Childs and Herbert 2005, pp. 283–4). The technology of iron smelting reached Botswana approximately 2000 years ago (Tlou and Campbell 1984, p. 42), and has existed in Kondoa Irangi, central Tanzania, for a thousand years or more (Mapunda 2003). Evidence has been found of cultural affiliations between ironworkers of central and coastal Tanzania from the first century BC to AD 200 (Chami 1996, p. 16). Iron and copper mining and smelting in northern Zimbabwe date from about AD 200 (Tlou and Campbell 1984, p. 42). Where processing or refining of the minerals occurs near the mining site, waste products, or slag, are left behind. The assumption has been that slag was left in place, and could be used to determine the extent of metal working at a given location. However, at least some Africans practised recycling. Slag was used to temper Late Iron and pottery in East Africa, and was also used for house foundations (Childs and Herbert 2005, p. 286).

The environmental history of ancient African metallurgy has yet to be thoroughly researched and written. Mining is a 'migratory' industry, because it moves on when a given mineral has been removed, leaving a disturbed landscape behind (Hilson 2002). From a soil perspective, mining can disrupt soil topography and water movement, leave concentrations of non-soil materials which may be toxic to life processes, and stimulate erosion processes. Sedimentation of nearby waterways is a frequent consequence of small-scale mining, because soil washes away from piles left by digging, or is rinsed out to purify mineral ores. Holes and trenches resulting from quarrying can stimulate drainage of water from surrounding soil, as well as retain water – and toxins – that can percolate into (recharge) drier surrounding soils. Mining soils from river sediments not only disturbs the river's bed, but also can result in depositions of sand and other soil particles on river banks and in flood plains.

On-going research in Kondola Irangi, in the centre of Tanzania, is addressing the question of whether ironworking was a causative factor in the area's massive gully erosion (Mapunda 2003). Quarrying was not involved in mining operations – sand ore washed down by the Intela and Baura Rivers was collected and smelted. It is the smelting and smithing processes that have been posited as inducing erosion because the required high heat would have come from burning large amounts of wood. This fuel need, it has been argued, caused deforestation that exposed soil, resulting in catastrophic erosion. But Mapunda (2003) argues that local inhabitants knew that all trees did not burn equally, so only specific species capable of producing sufficient heat were harvested. Accordingly, although certain species could have been endangered or eliminated, the area would not have been denuded of vegetation, so soils would not have been exposed to erosive forces because of smelting. Childs and Herbert (2005, p. 294) support this line of logic by suggesting that the disappearance of slow-burning hardwoods was a major factor in the decline of African iron-smelting.

f. Medicine

Geophagy, the eating of earth, has been common in Africa for centuries for medicinal and ritual purposes. Clays from Uzalla, Nigeria that are sold widely in markets across West Africa, have a mineral composition similar to clays used in the commercial pharmaceutical Kaopectate, the commonly used anti-diarrhoeal remedy (Aufreiter et al. 1997, p. 293). Geophagy has been widely documented in pregnant women, and theorised as being a source of iron (Abrahams 1997). As well as providing a nutritional supplement, clays can also help with detoxification. Eight samples of edible clays from Gabon, Kenya, Nigeria, Togo, Zambia and the Democratic Republic of Congo (former Zaire) were analysed for their contents of calcium, copper, iron, magnesium, manganese and zinc and detoxification capacities. The samples were found to be able to provide nutritionally significant amounts, and could, therefore, serve as a mineral supplement (Johns and Duquette 1991, pp. 450,454). In addition, because they were all kaolin clays, they could serve the same function as Kaopectate. A clay recovered from an archaeological site occupied by ancestors of Homo sapiens was found to have nutritional and detoxification capacities that were identical to those of edible clays used in Africa in the 1990s (Johns and Duquette 1991, p. 455).

Very specific medicinal mining of clays has, therefore, been carried out since the beginnings of human history. The extent and significance of this not been assessed, perhaps due to the taboo which industrial societies attach to the worldwide historical (and modern) practice of eating earth.

g. Art and Ritual

In the Middle and Upper Pleistocene there was regular symbolic use of pigments. Archaeologists have found examples of pigments shaped into pencils, and of symbols inscribed onto pieces of pigment (Marean and Assefa 2005, p. 120). The earliest mines in Botswana date to 33,000 years ago when haematite (red ochre) and specularite (an iron ore) were dug, ground to a powder, and mixed with fat, blood, egg-white or honey to make paint for use in rock art or body decoration (Tlou and Campbell 1984, p. 42). Cave paintings using soil as pigments are also commonly found in Lesotho. Elsewhere, soil, particularly clay, was excavated to make washes with a range of colours to be used as a kind of makeup – covering all or parts of the body – or for building decoration. In the eighth–fifteenth-century Kongo kingdom, clay was associated with the land of the dead, which was called the white clay world. Clays were used in funerary art to decorate wooden sculptures and to make terra cotta stelas. Because earth from burial sites was considered to relate to the spirits of those buried there, traditional healers combined burial site earths with botanical and other materials to make medicines. White clay also symbolised innocence and purity, so it was used to anoint the winner of a lawsuit (Thompson and Cornet 1981, pp. 30, 31, 37, 40).

Documentation of the use of soil for art and ritual practices is scattered in anthropological literature, and has not been the subject of inquiry from an environmental perspective. But in Africa, as elsewhere, soil helped make art possible.

AFRICAN SOIL IN THE PRESENCE OF EUROPEANS

Unbeknownst to them, when Europeans arrived in Africa as explorers, missionaries, traders or settlers, they typified Beach's environmental pioneers encountering a 'new, misunderstood environment' (Beach, T. 1998, p. 400). They knew nothing about the soils, vegetation or climates they encountered, and had little idea of ecosystem interactions. Their traditional agricultural, livestock and forest management systems had been developed in humid, temperate, glacially rejuvenated Europe, yet the Europeans were certain of these systems' suitability to tropical and sub-tropical Africa. For some, the African landscapes appeared to be deficient and, to most, the indigenous land use systems seemed inefficient or unproductive. Whether they arrived with settlers, missionaries, merchants or officials, European agricultural and land management technologies and practices changed vegetative cover and altered soil properties. Although the Europeans carried out many of the same activities Africans had before them, they did so on a larger scale, with more invasive techniques and powerful technologies, and their economic system demanded that mostly closed cycling systems be opened.

a. Agricultural Production and Soil Conservation

The 'differentness' of African soils became apparent soon after European land use practices were applied. The inherently low soil fertility and weak soil structures of most West African soils confounded efforts to increase yields through the application of 'superior' European management techniques (Richards 1986 provides West African examples). Adams (1992, pp. 103, 105), among others, describes some of the larger agricultural projects that failed, including the Tanganyika (modern Tanzania) Groundnut Scheme in which there was an attempt to use surplus World War Two Sherman tanks to clear stumps and plough the land. Soil compaction clearly resulted from this endeavour. Over time, the plantation managers of West Africa, settled farmers and ranchers of East and Southern Africa, and various government agencies introduced agricultural chemicals – fertilisers, pesticides, insecticides, nematicides – as well as agricultural equipment and machinery. Some of these changed soil chemical and physical properties temporarily, while others had long-term effects. A review of the African agricultural modernisation and development literature is, however, beyond the scope of this chapter.

In Southern Africa, it was the erodible nature of the soils that initially attracted attention. Soil erosion was first described around European establishments: on mission stations and agricultural experiment stations, as well as on farms, plantations and ranches (Roland 1938, pp. 7, 8; Tempany 1949, pp. 63–4; Ross 1963, p. 11; Beinart 1984, p. 54; Nyamapfene 1987, p. 2; Showers 1989, p. 267; Showers 2005, p. 143). European farming practices included ploughing large acreages of land and planting annual crops in rows. Ploughing increases soils' susceptibility to erosion because it creates channels in fields in which water can collect or flow, reduces soil organic matter levels, weakens aggregated soil particles, compacts subsoil to form a dense layer (plough pan), can physically shift soil particles downslope, and results in a more exposed soil surface than does hoeing.[13] European livestock management resulted in overstocked pastures and the concentration of animals around corrals and watering holes, which caused soil compaction (see discussion of compaction above under African grazing systems). These land use practices were associated with soil erosion (Hailey 1938, p. 1080; Tempany 1949, p. 1; Jacks and Whyte 1938, p. 71; Ross 1963, p. 12; McIlwane 1941 in Nyampafene 1987, p. 2). As Africans adopted European practices, erosion began to occur on their land as well.

Soil erosion in Southern Africa cannot be discussed in isolation from the great disturbance in the landscape caused by the arrival of new groups of land users from Europe who perceived much of the region as unused or empty, and claimed it for their own purposes. As the number of settler farmers increased, the land available to Africans decreased. Treaties signed in the late nineteenth century effectively transferred most of the land used by the Swazi people in the Transvaal Republic (north of the Vaal River in modern South Africa) to settlers (Swaziland Col.An. Rept. 1956, p. 17), and the rich farmland of the Basotho (modern Lesotho) to the Afrikaner farmers of the Orange Free State (modern Free State Province, South Africa) (Germond 1967, p. 308). Large blocks of land with the highest rainfall in the Bechuanaland Protectorate (modern Botswana) were declared Crown Lands available for European settlement (Tlou and Campbell 1984, p. 181). Laws passed in South Africa in 1913 and 1936 crowded the African population onto 13 per cent of the territory, giving the tiny European population control of the remaining 87 per cent of the land (Platzky and Walker 1985, p. 92), including the most favourable agricultural regions of the country. In Southern Rhodesia (modern Zimbabwe), Native Reserves were formally created for Africans in 1913 on marginal land with infertile soils and low rainfall (Mpofu 1987, p. 2), again leaving the more favoured land in the hands of European settlers.

The result was the thorough disruption of indigenous land use systems. Although details and dates vary from nation to nation, the overall form is of systems that had supported societies for generations and centuries, if not millenia, had their land bases seriously reduced. Indigenous land use systems were extensive, requiring large amounts of land. When confined to a smaller area, the African systems no longer functioned well (Basutoland Dept. Ag. An. Rept.1930, p. 9; Nyampafene 1987, pp. 1, 2; Swaziland Col. An. Rept. 1946, p. 14; Swaziland Col. An. Rpt. 1950, p. 7; Swaziland Col. An. Rept. 1952, p. 52; Weinmann 1975, p. 200). First

---

[13] For references and discussion of the consequences of ploughing on rangeland and of erosion associated with ploughing see Showers 2005, pp. 19–22.

there was no surplus, and then not enough for self-sufficiency. This process resulted in increasing – and inevitably unbearable – pressures on the landscape; pressures that resulted in the cultivation of marginal land, overuse of pastures, deforestation and soil erosion. In many areas accelerated soil erosion had been an unknown and unnamed concept. By the end of the nineteenth century, 'overcrowding' or 'over-population' of people and their livestock was identified as the major cause of erosion in most countries of southern Africa. A land problem created by the arrival of European settlers led to the erosion problem.

The earliest discussion of accelerated soil erosion occurred in South Africa's Cape Colony (Beinart 1984, pp. 54, 55). Apprehension about soil conditions was not unique to this colony. In the adjacent Protectorate of Basutoland (modern Lesotho), missionaries and government officials had expressed particular concern about gullies along roads and paths, and on government and missionary land (Showers 1989; Showers 2005). Further north in the colony of Southern Rhodesia (modern Zimbabwe) and the Nyasaland Protectorate (modern Malawi), officials worried about erosion on European farms. Tobacco and cotton fields and tea estates were reportedly particularly under threat (Nyasaland Dept. Ag. An. Rept. 1924, p. 5). In Nyasaland by the mid-1910s, both Europeans and Africans believed that each other's land use practices had caused an ecological crisis (Mlia 1987, p. 4), including an increase in soil erosion.

During the 1920s regional awareness of soil erosion as a problem that could be controlled and prevented resulted in a number of government and professional publications, conferences and exchange among government officers in South Africa, Southern Rhodesia (modern Zimbabwe) and Nyasaland (modern Malawi) (Report of Select Committee 1914; Torrance 1919; Showers forthcoming). The release of the South African Drought Investigation Commission's report in 1923 confirmed that European land management practices were causing the destruction of the soil and hydrological regimes, and that erosion was a national problem.

The 1920s also marked the beginning of systematic and sustained investigations of the southern African environment. Experiment stations were established initially to identify the best crops, varieties, and management practices for European settlers. Regional departments of agriculture became increasingly interested in identifying practices to limit 'surface wash' (sheet erosion) because it was understood to be linked to losses of soil fertility. Sheet – not gully – erosion was the major concern in most places (Showers forthcoming). A government research programme to characterise Nyasaland's soils was implemented in the mid-1920s. Beginning in 1924, soils were surveyed and samples were taken for chemical and physical analysis. The results were used to construct an erosivity index (susceptibility of a soil to erode). Both the physical and chemical properties of major Nyasaland soils and the erosivity index were published in the 1930 Department of Agriculture Annual Report (Nyasaland Dept. Ag. An. Rept. 1925, p. 26; 1930, p. 21). It was also noted that soils with a distinct textural change were most susceptible to erosion. In the 1930s and 1940s increasing official concern about soil erosion was reflected in an increase in government and professional publications; research to understand erosion processes; and programmes for farmers and ranchers, in an attempt to prevent erosion, or mitigate its consequences by restoring eroded land (Showers 1989; Showers forthcoming).

Europeans settled as farmers and ranchers in modern South Africa, Swaziland, Botswana (Bechuanaland), Zimbabwe (Southern Rhodesia) and Malawi (Nyasaland). Most were literate, and had access to both regional publications and professional and social contacts with government officials. However, these land users did not appear to share official concern about soil erosion's threat to the future of the landscape. Although European farmers in Southern Rhodesia were visited by the Agricultural Engineer to advise on soil conservation in the late 1920s, in 1938 only two districts had over 40 per cent of the cultivated land protected by contour ridges (S. Rhodesia An. Rept. 1929, p. 12; S. Rhodesia An. Rept. 1930, p. 39; S. Rhodesia An. Rept. 1931, p. 13; S. Rhodesia Dept. Ag. and Lands An. Rept. 1939, p. 28; S. Rhodesia Dept. Ag. and Lands An. Rept. 1939, p. 3). The 1939 report of the Commission to Enquire into the Preservation of the Natural Resources stated that soil erosion was still a serious problem on European farmland (McIlwaine 1939). By the end of 1940, less than three per cent of white farmers in South Africa had elected to participate in government programmes to build conservation works on their land because 'only a minority of farmers and a very small percentage of townspeople' were concerned about soil erosion (Chief, Div. Soil and Veld Cons. 1941; Ross 1963, p. 19). Despite the creation of a Natural Resources Board in Swaziland, meetings, and advice to employ conservation techniques, the government had to resort to fines to enforce reclamation and erosion control measures on European farms in the mid-1950s (Swaziland An. Rept. 1953, p. 15; Swaziland An. Rept. 1955, p. 15; Swaziland An. Rept. 1958, p. 17).

African farmers were equally uninterested in the Departments of Agriculture's soil conservation techniques. Their lack of enthusiasm for the various methods was noted in South Africa, Basutoland, Swaziland and Southern Rhodesia (Engineer to Chief Native Commissioner 1941; Basutoland Dept. Ag. An. Rept. 1937/38, p. 76; Basutoland Dept. Ag. An. Rept. 1938/39, p. 77; Basutoland Dept. Ag. An. Rept. 1948, p. 31; Weinmann 1975, p. 206; Swaziland Col. An. Rept. 1950, p. 7; Swaziland Col. An. Rept. 1951, p. 5; Swaziland Col. An. Rept. 1954, p. 4). However, African land users did not have a choice about the adoption of these conservation practices. Colonial officials decided what would be best for 'African land' and for Africans. Each colonial government devised a programme that it felt suited its colony's unique conditions. Most included some kind of engineering structures, like diversion ditches and terraces (which were referred to as contour banks, ridges or bunds), grass strips and mechanisms for the regulation of grazing.

In the 1940s and 1950s major state-sponsored soil conservation campaigns were implemented throughout Africa, particularly in British Africa, and most especially in those regions with European settlers. While most programmes were optional for Europeans, they were mandatory for Africans. Three basic approaches were taken: structure building (contour banks, bunds, terraces), vegetation management, and livestock/grazing control. The purpose of structure building was to change the surface hydrology of the soils, since soil erosion was thought to result from surface wash (sheet erosion). Basutoland, where the first state-sponsored national soil conservation programme in Africa was implemented, went from having an estimated 10 per cent of its landscape threatened by gullies when the programme began in 1936, to being one of the most gullied nations in the world by the 1970s. Conservation structures, untested for local conditions and modified to minimise costs, were implemented by non-professionals throughout the lowlands. Previously dispersed overland flow was collected and concentrated, creating rills, gullies and tunnelling (internal erosion), ultimately changing the hydrology of the soils (Showers 1989; Showers 2005). In Southern Rhodesia, beginning in the 1950s, the soil was disturbed as conservation works were dug and, as in Basutoland, erosion was increased by their inappropriate design. The mandatory conservation programme was so offensive to the rural population that it was one of the reasons given for supporting the guerrilla war to create independent Zimbabwe (Showers 1994). Kenyan farmers also resisted terracing in the late 1930s and 1940s (Carswell 2003, pp. 3, 4; Mackenzie 1998). In Tanganyika (modern Tanzania), a programme of bench terrace construction in the 1950s produced riots, and hillside ridges were so fiercely resisted that they had to be abandoned (Carswell 2003, pp. 3, 4). Vegetative approaches to soil conservation were less disruptive – to soil and to societies. Grass strips, which mimic the function of savanna vegetation, had been mandatory in Swaziland's 1949 conservation programme, but were embraced by farmers as essential for crop production and continued voluntarily, with modifications, in the late twentieth century (Showers 1994; Osunade and Reij 1996, p. 4). Where vegetative and structural measures were consistent with existing local practice, they were accepted, as in Kigezi, south-western Uganda (Carswell 2003).

Soil conservation structures clearly changed soil hydrology and, in some cases, permanently transformed or destroyed soil bodies. Construction disturbed soil profiles, and increased erosion removed soil. Soil water relations were changed when water ponded behind structures causing soils to become waterlogged. The new source of soil water increased internal drainage of the soils. In some instances erosion by tunnelling (piping) was initiated when newly developed rills and gullies provided outlets for water moving within the soil. In Basutoland, soils became irreversibly drier as they were drained by the developing network of tunnels (or pipes) and gullies.

Africans have not been passive in the face of the social, economic and environmental changes that occurred with the arrival of Europeans and their technologies. They made changes in their land use practices. In many places, as fallowing periods were constrained or eliminated, they were replaced by manuring. During the 1990s this process was studied in the semi-arid region of West Africa (including significant parts of Burkina Faso, Chad, Gambia, Mali, Mauritania, Niger and Senegal), where soils are inherently low in plant nutrients, soil organic matter and water retention capacity (Williams 1999, pp. 15, 17). Manure and fallow, not fertiliser, were the major means of maintaining soil fertility. Manure was applied either by hand-spreading on fields, or by corralling stock on a field at night. Except on small fields, manure was placed on specific spots within a field deemed to need nutrients rather than the whole field. Over time, each area of the field received a manure application. The farmers' belief that manure had long-term effects and was not needed each year (Williams 1999) has been confirmed in studies such as by Powell et al. (1998). Research has shown that manure augments soil organic matter contents, raises soil pH, improves cation exchange capacity (CEC) and water-holding capacity of soils. When used in combination with inorganic fertiliser, especially nitrogen, it reduces the fertiliser's negative effects, particularly acidification, and increases removal of nutrients other than those supplied by the fertiliser (Williams 1999, p. 15). However, not every farmer was found to have access to as much manure as s/he would like (Powell et al. 1998, p. 260; Williams 1999, p. 19).

When livestock are kept in stalls, as is advocated for modern production systems, instead of corralled on a field as in traditional systems, urine is lost through runoff from the barn and in seepage. The remaining manure must be handled, stored and transported before spreading on fields. Researchers warn that the result of such a switch in management practices could reduce the amounts and types of nutrients available for recycling and increase nutrient losses, thus jeopardising the long term productivity of the soil (Powell et al. 1998, p. 260)

b. Livestock Management

European methods of livestock management involved concentrated use of grazing areas. Fences contained the animals, and there were limited locations for the stock to drink. This approach produced overstocked pastures and heavy animal traffic around corrals and watering holes – all of which reportedly caused soil erosion (Hailey 1938, p. 1080; Tempany 1949, p. 1; Jacks and Whyte 1938, p. 71; Ross 1963, p. 12). As Europeans claimed land for farms and ranches, first in southern Africa and then in East Africa (Kenya in particular), African grazing areas were reduced and pastures became overgrazed. Ward et al. (2000) addressed concerns about 'land degradation' on a communal ranching area in arid Namibia. Otjimbingwe's mean annual rainfall of 165.4 mm precludes crop production; livelihoods could only come from livestock management. Using archival materials (including photographs) and oral histories as well as soil and vegetational sampling, the researchers concluded that it was rainfall, not stock numbers or management, that determined the condition of the landscape. Even around watering points used for 150 years, changes in soil quality (as measured by soil nitrogen, phosphorus, organic carbon, and water-holding capacity by bioassay) could not be detected (Ward et al. 2000, p. 351).

c. Architecture/Urbanisation

European urban areas in Africa were established for reasons of commerce, transportation or administration, and also grew up around mining and industrial enterprises. Some were expansions of existing African settlements, while others were completely new. As with African settlements before them, European towns affected soil by covering it, quarrying it for use in construction, and stimulating the creation of roads and paths. Many European buildings were constructed with brick or cement, which required local quarrying operations. Waste management also affected soils.

Transportation to and from urban areas posed problems of compaction and drainage. As unimproved roads became too rutted, wagons – and later automobiles and trucks – created new tracks in the grass alongside the existing one. In this way, small paths became large, rutted, eroding spaces (Kruger 1882 in Germond 1967, p. 407; Assistant Engineer, 1934). Faulty road and railway drainage systems also caused gully erosion in South Africa, Basutoland and Southern Rhodesia (Watt 1913; Stewart 1917; S. A. Dept. Ag. and For. 1941, p. 18; Hailey 1938, p. 1058; Jacks and Whyte 1938, p. 70; Pentz 1940, p. 4; Proc. Conf. Soil Erosion 1945; Showers 1989, pp. 270, 275; Phiri 2003, p. 12; Showers 2005, pp. 143, 150–1) and railway construction's demand for wooden sleepers encouraged deforestation and subsequent erosion, especially in savanna regions that did not have many trees.

Municipal waste disposal – domestic and industrial – was generally understood to be a process of removal. Treatment was rarely a consideration. Sewage and industrial effluents were dumped in streams and the ocean; abandoned quarries were used as disposal sites, and, in some semi-arid regions, liquid waste was simply dumped on the ground. Most urban areas were built with septic, rather than centralised water borne, sewage systems. The septic systems overwhelmed the soils, resulting in subsoil pollution (for discussion and tables of African urban and industrial waste disposal and pollution see Showers 2002).

Phiri (2003) assessed the environmental impact of the establishment and growth of the Southern Rhodesian town of Umtali (modern Mutare, Zimbabwe). Established as a base from which gold mining operations could be expanded, Umtali displaced an indigenous African population to other, less productive land, which ultimately became overused. As mines were dug, soil was disturbed directly and indirectly. To prospect for gold, miners burned the grass for easier access to the earth, and no attention was paid to maintaining soil cover against erosion. The combined mining and urbanisation processes led to deforestation: trees were cut for timbers in the mines, sleepers for the expanding railway track, and for construction. Erosion resulted (Phiri 2003, p. 6).

As mining operations encroached on the original site of Old Umtali, the town was moved and expanded to New Umtali. Soils were again disturbed as buildings were built, and roads laid. A new town with new and better buildings required more wood and of higher grade. Brick was a desired building material, so large scale soil excavation began for brick making. The new town's roads were not well constructed, and drained badly. Domestic, milling and mining wastes posed disposal problems that were never properly resolved. Soil disruption and erosion, waste accumulation, and pollution accompanied Umtali's development. As modern Mutare, it faced the same problems in 1995 – although on a different scale – that Old and New Umtali had in the late nineteenth and early twentieth centuries.

d. Mining

It was the 1867 discovery of diamonds in northern Cape Colony, South Africa that triggered the 'mineral revolution', despite coal having been found in Natal Colony in the 1840s (Wilson and Thompson 1971, p. 11). Large-scale European mining in Africa began with diamonds, but it was institutionalised as a component of economic development when the Witwatersrand was proclaimed a gold mining area in 1886 (Wilson and Thompson 1971, p. 13). Although Southern Africa has Africa's greatest concentration of non-petroleum minerals, by the end of the twentieth century every African country had large- and small-scale mines (Griffiths 1984, p. 126). In the 1920s and 1930s mining operations began around the continent, including small-scale gold mines in Cameroon, mines for iron and diamonds in Sierra Leone, diamonds on the west coast of southern Africa, and copper in Zambia (Desmet and Cowling 1999, p. 35; Grove, A.T. 1970, p. 109; Mobbs 1998; Lewis and Berry 1988, p. 369). A further wave of mine development occurred in the 1960s and 1970s, as African nations gained independence. Mining activities disturb surface soils, and create sources of pollution, dispersed by water and wind to surrounding soils undisturbed by actual mining operations (Lewis and Berry 1988, p. 367).

When minerals are quarried, the unwanted material surrounding the minerals (tailings) is discarded in piles. This can involve a considerable amount of soil; northern Botswana's rich Orapa Mine produces 0.89 carats of diamonds for every ton of soil excavated (Lewis and Berry 1988, p. 367). In South Africa, until the 1920s, tailings from the increasingly deep and urban Witwatersrand gold mines were deposited in dry form on so-called 'sand dumps'; after 1921 the deposits were slurries on 'slimes dams' (Groves 1974, p. 296). By the mid 1970s the Witwatersrand alone had 1,200 ha of sand dumps and 6,800 ha of slimes dams (Groves 1974, p. 296). Until stabilisation measures began to be taken in the mid-1960s, dust blew at such a rate from these dams and dumps that visibility was sometimes reduced to 10 metres (Groves 1974, p. 296). The dust settled on soils in surrounding areas. Tailings piles also produced sulphuric acid, which travelled in rainwater runoff to surrounding soil and water courses. On South Africa's arid southwest coast, opencast mining for diamonds since 1929 has resulted in large overburden dumps and other disturbed areas (Desmet and Cowling 1999, p. 35). In contrast to the acidic (low pH) tailings from gold mines, these overburden dumps are composed of dune sands and marine deposits with high pHs and high salt contents (Desmet and Cowling 1999, p. 35). Left on their own, the slimes dams, sand dumps and overburden dumps were not completely revegetated. In the mid-1960s attention turned to stabilising the urban tailings piles, but only recently has attention turned to the rural arid overburden dumps, some of which have remained bare since the 1970s (Desmet and Cowling 1999, p. 36). By 1990 South Africa had 450 mine dumps (Koch et al. 1990, p. 49). For photographs of mines in the Orange River catchment, including coastal diamond mines, see the South African Department of Water Affairs Orange River website at http: //www.dwaf.gov.za/orange/Low_Orange.htm.

In Northern Rhodesia (modern Zambia), copper mining began in 1921. From 1921 to 1957, when hydroelectricity became available, the surrounding forest was cleared to fuel the thermal power stations supplying the mines. Some 118,857 hectares were cleared between 1947 and 1964. Not only did this disrupt the local shifting cultivation system by removing land for fallow, but it also changed the water balance of the soil (Lewis and Berry 1988, pp. 370, 371). The mines themselves produced huge open pits and tailings piles, and the refining process released copper and lead into the air, which settled on the surrounding landscape (Lewis and Berry 1988, p. 371).

Southern Africa's mining industry is, perhaps, the biggest and best documented, but the processes are the same around the continent. Mining operations make holes and generate waste, they require large amounts of electricity and water for operation, and they require a population of workers. Mines are, as Phiri (2003) pointed out, an agent of urbanisation, and create all of the environmental effects of any urban centre.

e. Trade

African local, regional and intercontinental importing and exporting has an ancient history. Ceramics, metal ore and finished metal products were widely circulated. Livestock played a significant role in transport and as objects of trade. Shaw (1995) discussed camels in Roman North Africa and the Sahara, and Kreike (2004) outlined the extensive local and long-distance cattle trade in the late-eighteenth and nineteenth centuries carried out by the Ovakwanyama, residents of the Ovambo flood plain (modern northern Namibia and southern Angola). In contrast to pre-colonial economies in Africa, those constructed by Europeans placed increasingly heavy pressures on soils. With the sale of agricultural and livestock products, nutrient cycling systems were broken, and replacements were not always made. As industrial processes increased the demand for both the amount and kind of minerals, prospecting and mining expanded, magnifying the effects on soil discussed above. Urban and industrial areas required power, which resulted in rich alluvial soils being submerged by hydroelectric dams (Showers 2001). Finally, increased trade required increased infrastructure – from transportation to urbanisation – with its attendant soil effects.

AFRICAN SOILS AS HISTORICAL BODIES

African soils are historical bodies. They enter the twenty-first century with a long history of interaction with climate, topography, parent material, time and organisms – the factors of soil formation (Figure 6.3) (Buol et al. 1975; Buol et al. 2003, pp. 122–3). It is interactions with organisms, especially human organisms, that is the least understood, and the source of many of Roe's (1991, 1995) 'crisis narratives'. Can people be considered as participants in essential nutrient cycles, like mound-building termites or deep-rooted grasses cycling nutrients from down in the soil profile to surface deposition as litter? Do they participate in soil formation – creating new soil bodies or contributing new properties to existing soil profiles? Do they participate in soil destruction – the permanent change of soils. And if so, do new soils result?

Answers to all of these questions seem to be both yes and no, depending on location and time. Since human beings have lived all over the African continent for thousands of years, it is difficult to imagine soil conditions without any human interaction. Based on the information presented, in pre-European Africa, arguments can be made for both savanna and forest dwellers as being participants in fairly closed nutrient cycling systems that did not disrupt soils. Human interventions in these nutrient cycling systems certainly modified temporarily, if they did not transform permanently, soil properties. Roads and paths caused soil compaction that transformed soil water relations locally. Soil creation could arguably have occurred in those scattered locations where terraces or other surface water diversion devices resulted in the collection of soil particles, or where organic matter (as household refuse or deliberate fertilisation) was consistently added to a specific place. Soil destruction would most obviously have occurred where quarrying for minerals or special soils resulted in large holes. Although perhaps significant locally, quarrying is unlikely to have had a regional impact.

It is the use of fire in shifting cultivation or livestock management that provokes great debate. From a soil perspective, the significance of fire is its heat, and the result of burning is the return of nutrients from vegetation to soil in the form of ash. As mentioned above, fires that burned quickly did not affect soil below the top 5 cm. under either forest or savanna. Soil properties did change, however, when wood was deliberately piled to create an intense heat, as in the ***chitemene*** system. The cation exchange capacity was increased, and could persist for as long as 16 years. Burning in this instance could be considered to be a temporary transforming process – for the increased CEC would not persist long after cessation of the burning sequence. From a soil chemical perspective, burning produced ash that released cations and anions into the soil solution. This affected the acidity of soils (making them less acid), and the nutrient availability to plants. The pH change, which affects soil microbial communities as well as plant roots, is temporary. However, the consequence of burning that receives the most attention is the loss from the soil-plant system of carbon, nitrogen and sulphur. If these losses can be compensated for over time by plants, micro-organisms, or rainfall and lightning (nitrogen), then burning can be seen as another aspect of nutrient cycling, without net loss.

There is no question that with the arrival of Europeans and their land use practices and technologies, soils changed dramatically in some places. European settlers and plantation owners displaced Africans from land, thus limiting previously unbounded extensive land use systems and increasing human and livestock populations on the remaining land. This concentration had implications for the stability of fallow systems and the stimulation of soil erosion, as discussed earlier. The newly arrived technologies were untested on African soils (although arguments have been made for the pre-European rejection of ploughs in favour of less soil-disturbing 'digging sticks' by Africans living on structurally fragile soils of West Africa). Whether employed by Europeans or Africans, the new agricultural technologies brought change. Certainly the soil conservation ideas imported from the United States in the early twentieth century proved to be highly destructive to many of the soils they were meant to protect. (See Showers 2005 and forthcoming for discussion of soil conservation technology transfer from USA to southern Africa.)

Each landscape was affected differently, but in places, soil textures, structures, chemistry and hydrology were changed – some irreversibly. As population densities increased and urban areas were created or expanded, soil cover was altered; ploughs, monocrops and continuous cultivation lowered soil organic matter levels and nutrient supplies; and agricultural chemicals (synthetic fertilisers, pesticides, herbicides) changed soil chemistries and microbial populations temporarily or permanently. Soil hydrologies were changed by mining operations, compaction and cover associated with urban areas and roads, and by gullies. Soils lost surfaces through erosion. Mines and industries released toxins to the soil environment, some of which persist for years, as do many agricultural chemicals. Soil was the object of concern only when it presented a problem, such as when it eroded, produced lower than expected yields, or failed to accept all of the waste products deposited on or into it. Soil conservation was seen as a technology that could remedy land shortage and allow the persistence of certain European practices, rather than an expression of concern about soil bodies themselves. It is obvious that soil properties, and in some places entire soil bodies, were changed substantially in the twentieth century.

A true history of African soils will have to be many histories, with each region's soils considered. Documentation will be needed to show the mechanism(s) by which a particular soil body has been changed. Analysis must consider whether the changes identified were permanent or temporary in nature, as well as whether they are reversible or irreversible. A tool for distancing researchers from the logics of dominant narratives (crisis or otherwise) is historical environmental impact assessment (HEIA), which focuses attention on landscape dynamics and facilitates the use of data from a wide variety of sources (Showers 1996). Given the tremendous paucity of information about most of the African continent, local soil knowledge systems ought to be major sources. Even when there are simply no data, and no sources can be imagined, the temptation to extrapolate or generalise must be resisted. Clear statements of gaps are more useful than projections and assumptions. 'No information' is a kind of data, sometimes the only legitimate kind.

The twenty-first century dawns on an Africa with different soil conditions than at the start of the twentieth century, and knowledge of Africa's soils remains incomplete – including its history in detail. Proper histories of African soils could make a useful contribution to addressing the very real soil concerns of the twenty-first century.

REFERENCES

Abrahams, P.W. 1997. 'Geophagy (Soil Consumption) and Iron Supplementation in Uganda'. *Tropical Medicine and International Health,* **2**(7): 617–23.

Adams, W.M. 1992. *Wasting the Rain: Rivers, People and Planning in Africa*. Minneapolis: University of Minnesota Press.

Alemayehu, Million. 1996. 'Traditional Ditches in Northern Showa, the Ethiopian Highlands'. In Chris Reij, Ian Scoones and Camilla Toulmin, *Sustaining the Soil: Indigenous Soil and Water Conservation in Africa* (London: Earthscan), pp. 163–9.

Amborn, Hermann. 1989. 'Agricultural Intensification in the Burji-Konso Cluster of South-Western Ethiopia'. *Azania,* **24**: 71–83.

Asrat, Kebede, Kederalah Idris and Mesfin Semegen. 1996. 'The "Flexibility" of Indigenous Soil and Water Conservation Techniques: A Case Study of the Harerge Highlands, Ethiopia'. In Chris Reij, Ian Scoones and Camilla Toulmin, *Sustaining the Soil: Indigenous Soil and Water Conservation in Africa* (London: Earthscan), pp. 156–62.

Assistant Engineer. 1934. Answers to Question 16, A.W. Pim Commission of Inquiry, 1944. Assistant Engineer, Department of Public Works. Public Records Office DO 119/1051.

Aufreiter, S., R.G.V.Hancock, W.C. Mahaney, A. Stambolic-Robb and K. Sanmugadas. 1997. 'Geochemistry and Mineralogy of Soils Eaten by Humans'. *International Journal of Food Sciences and Nutrition*, **48**: 293–305.

Basutoland. *Department of Agriculture Annual Report for the Year*: 1930; 1937; 1938; 1948.

Batterbury, Simon, Timothy Forsyth and Koy Thompson. 1997. 'Environmental Transformations in Developing Countries: Hybrid Research and Democratic Policy'. *The Geographical Journal*, **163**(2): 126–32.

Beach, D.N. 1984. *Zimbabwe before 1900*. Gweru: Mambo Press.

Beach, Timothy. 1998. 'Soil Catenas, Tropical Deforestation and Ancient and Contemporary Soil Erosion in the Petén, Guatemala'. *Physical Geography*, **19**(5): 378–405.

Beinart, William. 1984. 'Soil Erosion, Conservationism and Ideas About Development: a Southern African Exploration'. *Journal of Southern African Studies*, **11**(1): 52–83.

Birmingham, D.M. 2003. 'Local Knowledge of Soils: The Case of Contrast in Côte d'Ivoire'. *Geoderma*, **111**(3–4): 481–502.

Boardman, John. 1998. 'An Average Soil Erosion Rate for Europe: Myth or Reality?' *Journal of Soil and Water Conservation*, **53**(1): 46–50.

Bolinder, M.A., D.A. Angers, M.Giroux, and M.R. Laverdière. 1999. 'Estimating C Inputs Retained as Soil Organic Matter from Corn (*Zea mays* L.)'. *Plant and Soil*, **215**(1): 85–91.

Buckle, Colin. 1978. *Landforms in Africa: An Introduction to Geomorphology*. London: Longmans.

Buol, S.W., F.D. Hole and R.J. McCracken. 1973. *Soil Genesis and Classification* [1st edn]. Ames: Iowa State University Press.

Buol, S.W., R.J. Southard, R.C. Graham and P.A. McDaniel. 2003. *Soil Genesis and Classification*, 5th edition. Ames: Iowa State University Press.

Carswell, Grace. 2003. 'Soil Conservation Policies in Colonial Kigezi, Uganda: Successful Implementation and an Absence of Resistance'. In William Beinart and Joann McGregor (eds) *Social History and African Environments* (Oxford: James Currey), pp. 131–54.

Casey, Joanna. 2005. 'Holocene Occupations of the Forest and Savanna'. In Ann Brower Stahl (ed.), *African Archaeology: A Critical Introduction* (Malden, MA: Blackwell Publishing), pp. 225–48.

Cassell, D.K. and R. Lal. 1992. 'Soil Physical Properties of the Tropics: Common Beliefs and Management Restraints'. In R. Lal and P.A. Sanchez, *Myths and Science of Soils of the Tropics*, Soil Science Society of America Special Publication No.29 (Madison: Soil Science Society of America), pp. 61–90.

Chambers, Robert. 1983. *Rural Development: Putting the Last First*. London: Longman.

Chami, Felix A. 1996. 'The Archaeology of the Rujigi Region since 1987 to 2000: Coastal and Interior Dynamics from AD 00 – 500'. In Felix Chami, Gilbert Pwiti and Chantal Radimilahy (eds), *People, Contact and the Environment in the African Past* (Dar es Salaam: DUP [1996], c. 2001), pp. 7–20.

Chief, Division of Soil and Veld Conservation. 1941. To Secretary for Agriculture and Forestry, No.R.1435/91, September 1941. South African Archives LDB R4250 Vol. 1.

Childs, S. Terry and Eugenia W. Herbert. 2005. 'Metallurgy and its Consequences'. In Ann Brower Stahl (ed.), *African Archaeology: A Critical Introduction* (Malden, MA: Blackwell Publishing), pp. 276–300.

Coupland, R.T. 1979. 'Natural Temperate Grasslands: Use and Management'. In R.T. Coupland (ed.), *Grassland Ecosystems of the World* (Cambridge: Cambridge University Press), pp. 107–11.

Davis, Diana K. 2005. 'Potential Forests: Degradation Narratives, Science and Environmental Policy in Protectorate Morocco, 1912–1956'. *Environmental History*, **10**: 211–38.

Denyer, Susan. 1978. *African Traditional Architecture*. New York: Africana Publishing Company.

Desmet, P.G. and R.M. Cowling. 1999. 'Patch Creation by Fossorial Rodents: A Key Process in the Revegetation of Phytotoxic Arid Soils'. *Journal of Arid Environments*, **43**(1): 35–45.

Dokuchaev Central Soil Museum, 2000–2003. *V.V. Dokuchaev (1846–1903)*. http: // www.soilmuseum.by.ru/english/dokuchaev_eng.htm.

Dreschel, P., D. Kunze and F. Penning de Vries. 2001. 'Soil Nutrient Depletion and Population Growth in Sub-Saharan Africa: A Malthusian Nexus?' *Population and Environment*, 22(4): 411–423.

Eggert, Manfred K.K. 2005. 'The Bantu Problem and African Archaeology'. In Ann Brower Stahl (ed.), *African Archaeology: A Critical Introduction* (Malden, MA: Blackwell Publishing), pp. 301–26.

Encyclopaedia Britannica. 1929. 'Laterite'. Vol. 13: 740.

Endfield, Georgina H. and David I. Nash. 2002. 'Drought, Desiccation and Discourse: Missionary Correspondence and Nineteenth-Century Climate Change in Central Southern Africa'. *The Geographical Journal*, **168**(1): 33–47.

Engineer to Chief Native Commissioner. 5 November 1941. No. 115/8. South African Archives, NTS 15/276 Vol. II.

Eswaran, Hari. 1988. 'Taxonomy and Management Related Properties of the 'Red Soils' of Africa'. In *The Red Soils of East and Southern Africa*, Proceedings of an International Symposium, Harare, Zimbabwe, 24–27 February 1986, Rapport Manuscript. IDRC-MR170e (Ottawa: International Development Research Centre), pp. 1–23.

Eswaran, H., F.H. Beinroth, J. Kimble. and T. Cook. 1992. 'Soil Diversity in the Tropics: Implications for Agricultural Development'. In R. Lal and P.A. Sanchez, *Myths and Science of Soils of the Tropics*, Soil Science Society of America Special Publication No.29 (Madison: Soil Science Society of America), pp.1–16.

Faniran, A. and O. Areola. 1978. *Essentials of Soil Study: With Special Reference to Tropical Areas*. London: Heinemann.

FAO/Unesco.1977. *Soil Map of the World, 1: 5 000 000*, Vol. 6, Africa. Paris: Unesco.

Germond, R.C. 1967. *Chronicles of Basutoland*. Morija: Morija Sesuto Book Depot.

Gifford-Gonzalez, Diane. 2005. 'Pastoralism and its Consequences'. In Ann Brower Stahl (ed.), *African Archaeology: A Critical Introduction* (Malden, MA: Blackwell Publishing), pp. 187–224.

Giles-Vernick, Tamara. 2002. *Cutting the Vines of the Past: Environmental Histories of the Central African Rain Forest.* Charlottesville: University Press of Virginia.

Goldammer, J.G. and C. de Ronde (eds). 2004. *Wildland Fire Management Handbook for Sub-Sahara Africa.* Frieburg: Global Fire Management Centre/Oneworld Books.

Gradwell, M.W. 1968. 'Compaction of Pasture Topsoils under Winter Grazing'. In *Transactions 9th International Congress of Soil Science (Adelaide Australia, 1968)* (New York City: American Elsevier Publishing Co.), vol.3, pp. 429–35.

Griffiths, Ieuan Ll. 1984. *An Atlas of African Affairs.* New York: Methuen.

Grove, A.T. 2000. 'The African Environment, Understood and Misunderstood'. In Douglas Rimmer and Anthony Kirk-Greene (eds), *The British Intellectual Engagement with Africa in the Twentieth Century* (New York: St. Martins Press), pp. 179–206.

Grove, A.T. 1970. *Africa South of the Sahara,* 2nd edition. London: Oxford University Press.

Grove, A.T. and J.E.G. Sutton. 1989. 'Agricultural Terracing South of the Sahara'. *Azania,* **24**: 113–22.

Grove, Richard. 1989. 'Scottish Missionaries, Evangelical Discourses and the Origins of Conservation Thinking in Southern Africa 1820–1900'. *Journal of Southern African Studies,* **15**(2): 163–87.

Groves, J.E. 1974. 'Reclamation of Mining Degraded Land'. *South African Journal of Science,* **70** (10): 296–9.

Guyer, Jane and Samuel M.E. Belinga. 1995. 'Wealth in People as Wealth in Knowledge: Accumulation and Composition in Equatorial Africa'. *Journal of African History,* **36**(1): 91–120.

Hailey, Lord. 1938. *An African Survey: A Study of Problems Arising in Africa South of the Sahara.* London: Oxford University Press.

Hamza, M. Ait. 1996. 'The Mastery of Water: SWC Practices in the Atlas Mountains of Morocco'. In Chris Reij, Ian Scoones and Camilla Toulmin (eds) *Sustaining the Soil: Indigenous Soil and Water Conservation in Africa* (London: Earthscan), pp. 44–7.

Haynes, R.J. and G.S. Francis. 1993. 'Changes in Microbial Biomass C, Soil Carbohydrate Composition and Aggregate Stability Induced by Growth of Selected Crop and Forage Species Under Field Conditions'. *Journal of Soil Science,* **44** (4): 665–75.

Heusch, B. 1981. 'Sociological Constraints in Soil Conservation: A Case Study, the Rif Mountains, Morocco'. In R.P.C. Morgan (ed.), *Soil Conservation: Problems and Prospects* (London: John Wiley and Sons), pp. 419–24.

Hilson, Gavin. 2002. 'Small-Scale Mining in Africa: Tackling Pressing Environmental Problems with Improved Strategy'. *Journal of Environment and Development,* **11**(2): 149–74.

Hiol Hiol, François; Dieudonné Ndoum Mbeyo and François Tchala Abina. 1996. 'Traditional Soil and Water Conservation Techniques in the Mandara Mountains, Northern Cameroon'. In Chris Reij, Ian Scoones and Camilla Toulmin, *Sustaining the Soil: Indigenous Soil and Water Conservation in Africa* (London: Earthscan), pp. 191–201.

Hoffman, Irene, Dirk Gerling, Usman B.Kyiogwom, and Anke Mané-Bielfeldt. 2001. 'Farmers' Management Strategies to Maintain Soil Fertility in a Remote Area in Northwest Nigeria'. *Agriculture, Ecosystems and Environment*, **86**: 263–75.

Hopkins, A.G. 1973. *An Economic History of West Africa.* London: Longman.

Igbokwe, E.M. 1996. 'A Soil and Water Conservation System Under Threat: A Visit to Maku, Nigeria'. In Chris Reij, Ian Scoones and Camilla Toulmin, *Sustaining the Soil: Indigenous Soil and Water Conservation in Africa* (London: Earthscan), pp. 219–27.

Ikpe, F.N., J.M. Powell, N.O. Isirimah, T.A.T. Wahua and E.M. Ngodigha. 1999. 'Effects of Primary Tillage and Soil Amendment Practices on Pearl Millet Yield and Nutrient Uptake in the Sahel of West Africa'. *Experimental Agriculture*, **35**(4): 437–48.

Jacks, G.V. and R.O. Whyte. 1938. *Erosion and Soil Conservation*, Herbage Publication Series Bulletin No. 25. Aberystwyth: Imperial Bureau of Pastures and Forage Crops.

Jacobs, Nancy. 2003. *Environment, Power and Injustice: A South African History.* Cambridge: Cambridge University Press.

Johns, T. and M. Duquette. 1991. 'Detoxification and Mineral Supplementation as Functions of Geophagy'. *American Journal of Clinical Nutrition*, **53**(2): 448–56.

Johnson, Douglas H. and David M. Anderson.1988. *The Ecology of Survival: Case Studies from Northeast African History.* Boulder, CO: Westview Press.

Kass, Donald C., H. David Thurston, and Kenneth Schlather. 1999. 'Sustainable Mulch-Based Cropping Systems with Trees'. In Louise E. Buck, James P. Lassoie and Erick C.M. Fernandes (eds), *Agroforestry in Sustainable Agricultural Systems* (Boca Raton, FL: CRC Press), pp. 361–79.

Killingray, David. 2000. Colonial Studies. In Douglas Rimmer and Anthony Kirk-Greene (eds), *The British Intellectual Engagement with Africa in the 20th Century* (New York: St. Martins Press).

Kjekshus, Helge. 1996. *Ecology Control and Economic Development in East African History: The Case of Tanganyika 1850–1950*, 2nd impression with new introduction. London: James Currey.

Klieman, Kairn A. 2003. *'The Pygmies were our Compass'. Bantu and Batwa in the History of West Central Africa, Early Times to c. 1900 C.E.* Portsmouth, N.H.: Heinemann.

Koch, Eddie, Dave Cooper, and Henk Coetzee. 1990. *Water, Waste and Wildlife: The Politics of Ecology in South Africa,* Penguin Forum Series. London: Penguin Books.

Krasilnikov, Pavel V. and Joe A. Tabor. 2003. 'Perspectives on Utilitarian Ethnopedology'. *Geoderma*, **111**: 197–215.

Kreike, Emmanuel. 2004. *Re-creating Eden: Land Use, Environment and Society in Southern Angola and Northern Namibia.* Portsmouth, NH: Heinemann.

Krenzer, E.G., Jr., Chaw Foh Chee, and J.F. Stone. 1989. 'Effects of Animal Traffic on Soil Compaction in Wheat Pastures'. *Journal of Production Agriculture*, **2**(3): 246–9.

Lal, R. and P.A. Sanchez. 1992. *Myths and Science of Soils of the Tropics*, Soil Science Society of America Special publication No. 29. Madison: Soil Science Society of America.

Laris, Paul. 2004. 'Grounding Environmental Narratives: The Impact of a Century of Fighting Against Fire in Mali'. In William G Moseley and B. Ikubolajeh Logan (eds). *African Environment and Development: Rhetoric, Programs and Realities* (Burlington VT: Ashgate Publishing), pp. 63–86.

LaViolette, Adria and Jeffrey Fleisher. 2005. 'The Archaeology of Sub-Saharan Urbanism: Cities and their Countrysides'. In Ann Brower Stahl (ed.), *African Archaeology: A Critical Introduction* (Malden, MA: Blackwell Publishing), pp. 327–52.

Lewis, L.A. and L. Berry. 1988. *African Environments and Resources*. Boston: Unwin Hyman.

Lipman, J.G. 1933. 'A Quarter Century of Progress in Soil Science'. *Journal of the American Society of Agronomy*, 25(1): 9–25.

Lloyd, P.C. 1967. 'Osifekunde of Ijebu'. In Philip D. Curtin (ed.), *Africa Remembered: Narratives by West Africans* (Madison: University of Wisconsin Press), pp. 217–88.

Mackenzie, A. Fiona D. 1998. *Land, Ecology and Resistance in Kenya, 1880–1952*. Edinburgh: Edinburgh University Press.

MacKenzie, John M. 1988. *The Empire of Nature*. Manchester: Manchester University Press.

McCarty, M.K. and A.P. Mazurak. 1976. 'Soil Compaction in Eastern Nebraska After 25 Years of Cattle Grazing Management and Weed Control'. *Journal of Range Management* 29(5): 384–6.

McIlwaine, R. 1939. Report of the Commission to Enquire into the Preservation of the Natural Resources of the Colony. Salisbury: Government Printer.

Mapunda, Bertram B. 2003. 'Iron Technology and Deforestation: Myths and Realities'. Paper presented at the International Conference on Forest and Environmental History of the British Empire and Commonwealth, Centre for World Environmental History, University of Sussex, March 19–21, 2003. Forthcoming in Richard Grove and Vinita Damodaran (eds), *Empire and Environment: The British Empire and the Natural World, 1600 – 2000* (New Delhi: Oxford University Press).

Marean, Curtis W. and Zelalem Assefa. 2005. 'The Middle and Upper Pleistocene African Record for the Biological and Behavioural Origins of Modern Humans'. In Ann Brower Stahl (ed.), *African Archaeology: A Critical Introduction* (Malden, MA: Blackwell Publishing), pp 93–129.

Marshall, Alfred. 1938. *Principles of Economics*, 8th edition. New York: The Macmillan Company.

Mazzucato, Valentina and David Niemeijer. 2000. *Rethinking Soil and Water Conservation in a Changing Society: A Case Study in Eastern Burkina Faso*, Tropical Resource Management Papers No. 32, Department of Environmental Sciences, Erosion and Water Conservation Group. Wageningen: Wageningen University and Research Centre.

Mlia, J.R. Ngoleka. 1987. 'History of Conservation in Malawi'. In *History of Soil Conservation in the SADCC Region*, Report No. 8. Maseru: SADCC Soil and Water Conservation and Land Utilization Programme.

Mobbs, Philip M. 1998. The Mineral Industry of Cameroon – 1998. Http: //minerals.usgs.gov/minerals/pubs/country/1998/9204098.pdf.

Moore, Henrietta L. and Megan Vaughn. 1994. *Cutting down Trees: Gender, Nutrition and Agricultural Change in the Northern Province of Zambia, 1890–1900*. Portsmouth, NH: Heinemann.

Mpofu, T.P.Z. 1987. 'History of Soil Conservation in Zimbabwe'. In *History of Soil Conservation in the SADCC Region*, Report No. 8. Co-ordination Unit. Maseru: SADCC Soil and Water Conservation and Land Utilization Programme.

Nachtergaele, F.O. undated. From the Soil Map of the World to the Digital Global Soil and Terrain Database, 1960–2002. http://www.itc.nl/rossiter/Docs/WRB/SoilMapWold.pdf

National Geographic Society. 1990. *National Geographic Atlas of the World*. Washington, D.C.: National Geographic Society.

Neumann, Katharina. 2005. The Romance of Farming: Plant Cultivation and Domestication in Africa. In Ann Brower Stahl (ed.), *African Archaeology: A Critical Introduction* (Malden, MA: Blackwell Publishing), pp. 249–75.

Niemeijer, David and Valentina Mazzucato. 2003. 'Moving beyond Indigenous Soil Taxonomies: Local Theories of Soils for Sustainable Development'. *Geoderma*, **111**: 403–24.

Nyamapfene, K.W. 1987. *A Short History and Annotated Bibliography on Soil and Water Conservation*, Report No. 12. Maseru: Coordination Unit, SADCC Soil and Water Conservation and Land Utilization Programme.

Nyasaland. *Department of Agriculture Annual Report for the Year*. 1924, 1925, 1930. Zomba: Government Printer.

Nye, P.H. and D.J. Greenland. 1960. *The Soil under Shifting Cultivation*, Technical Communication No. 51. Harpenden: Commonwealth Bureau of Soils.

Oliver, Roland. 1991. *The African Experience: Major Themes in African History from Earliest Times to the Present*. New York: IconEditions.

Osunade, M. and C. Reij. 1996. 'Back to the Grass Strips': A History of Soil Conservation Policies in Swaziland. In C. Reij, I. Scoones and C. Toulmin (eds), *Sustaining the Soil: Indigenous Soil and Water Conservation in Africa* (London: Earthscan).

Payton, R.W., J.F. Barr, A. Martin, P. Sillitoe, J.F. Deckers, J.W. Gowing, N. Hatibu, S.B. Naseem, M. Tenywa and M.I. Zuberi. 2003. 'Contrasting Approaches to Integrating Indigenous Knowledge about Soils and Scientific Soil Survey in East Africa and Bangladesh'. *Geoderma*, **111** (3–4): 355.

Pendleton, Don. 1982. 'Range Use and Management of Great Plains Soils'. In Andrew R. Aandahl (ed.), *Soils of the Great Plains* (Lincoln: University of Nebraska Press), pp. 122–7.

Pentz, J.A. 1940. *Soil Erosion Survey of the Reclamation Area in Natal*, Veld Management and Pasture Research Series – Memoir No. 2, Science Bulletin No. 212, Union of South Africa Department of Agriculture and Forestry. Pretoria: Government Printer.

Phiri, Christopher. 2003. 'Environmental Aspects of Umtali's Urbanisation: 1890–1923: An Analysis'. Paper presented at the International Conference on Forest and Environmental History of the British Empire and Commonwealth, Centre for World Environmental History, University of Sussex, March 19–21, 2003. Forthcoming in Richard Grove and Vinita Damodaran (eds), *Empire and Environment: The British Empire and the Natural World, 1600 – 2000* (India: Oxford University Press).

Platzky, Laurine and Cheryl Walker. 1985. *The Surplus People: Forced Removals in South Africa.* Johannesburg: Ravan Press.

Powell, J.M., S. Fernández-Rivera, P. Hiernaux, and M.D. Turner. 1996. 'Nutrient Cycling in Integrated Rangeland/Cropland Systems of the Sahel'. *Agricultural Systems*, **52**(2/3): 143–70.

Powell, J.M., F.N. Ikpe, Z.C. Somida and S. Fernández-Rivera. 1998. 'Urine Effects on Soil Chemical Properties and the Impact of Urine and Dung on Pearl Millet Yield'. *Experimental Agriculture*, **34**(3): 259–76.

Proceedings of a Conference on Soil Erosion caused by Roads and Railways, October 1945. Pretoria. South African Archives LDB R 4250/1.

Report of the Select Committee on Droughts, Rainfall and Soil Erosion together with proceedings of the committee, minutes of evidence and appendix. 1914. South African Parliament, Senate, 4th Session, 19 June 1914. Cape Town: Government Printer.

Richards, Paul. 1985. *Indigenous Agricultural Revolution: Ecology and Food Production in West Africa.* London: Hutchinson.

Roe, Emery. 1991. 'Development Narratives, or Making the Best of Blueprint Development'. *World Development*, **19**(4): 287–300.

Roe, Emery. 1995. 'Except-Africa: Postscript to a Special Section on Development Narratives'. *World Development*, **23**(6): 1065–9.

Roland, J.W. 1938. *Pasture Research in South Africa*, Department of Agriculture and Forestry, Progress Report No. 1. Pretoria: Government Printer.

Ross, J.C. 1963. *Soil conservation in South Africa: A Review of the Problem and Developments to Date*, Department of Agriculture Technical Services. Pretoria: Government Printer.

Russell, E.W. 1973. *Soil Condition and Plant Growth*, 10th edition. London: Longman.

Sayer, Jeffrey A., Caroline S. Harcourt and N. Mark Collins (eds), 1992. *The Conservation of Tropical Forests: Africa.* New York: Simon and Schuster.

Schantz, H.L. and C.F. Marbut. 1923. *Vegetation and Soils of Africa*, American Geographical Society Research Series No. 13, National Research Council and American Geographical Society. New York: AMS Press (Reprinted 1971).

Scoones, Ian. 1997. 'The Dynamics of Soil Fertility Change: Historical Perspectives on Environmental Transformation from Zimbabwe'. *The Geographical Journal*, **163**(2): 161–9.

Shaw, Brent D. 1995. 'The Camel in Roman North Africa and the Sahara: History, Biology and Human Economy'. In Brent D. Shaw, *Environment and Society in Roman North Africa* (Aldershot: Variorum), pp. 663–721.

Shahack-Gross, Ruth, Fiona Marshall and Steve Weiner. 2003. 'Geo-Ethnoarchaeology of Pastoral Sites: The Identification of Livestock Enclosures in Abandoned Maasai Settlements'. *Journal of Archaeological Science*, **30**(4): 439–59.

Showers, Kate B. 1989. 'Soil Erosion in the Kingdom of Lesotho: Origins and Colonial Response', 1830s–1950s, *Journal of Southern African Studies* **15**(2): 263–86.

Showers, Kate B. 1994. Early Experiences of Soil Conservation in Southern Africa: Segregated Programs and Rural Resistance. Working Paper No. 184, African Studies Center. Boston: Boston University.

Showers, Kate B. 1996. 'Soil Erosion in the Kingdom of Lesotho and Development of Historical Environmental Impact Assessment'. *Ecological Applications*, **6**(2): 653–64.

Showers, Kate B. 2001. 'Dams and Power in Southern Africa'. In C.M. Miller, M. Cioc and K.B. Showers (eds), *Water and the Environment Since 1945: A Cross-Cultural Perspective*, History in Dispute Series, vol. 7 (Detroit: St. James Press), pp. 236–41.

Showers, Kate B. 2002. 'Water Scarcity and Urban Africa: An Overview of Urban-Rural Water Linkages'. *World Development*, **30**(4): 621–48.

Showers, Kate B. 2003. 'Origins of Soil Conservation in British Africa, Pt. I: Southern Africa through the 1920s'. Paper presented at the International Conference on Forest and Environmental History of the British Empire and Commonwealth, Centre for World Environmental History, University of Sussex, March 19–21, 2003. Forthcoming in Richard Grove and Vinita Damodaran (eds), *Empire and Environment: The British Empire and the Natural World, 1600 – 2000* (India: Oxford University Press).

Showers, Kate B. 2005. *Imperial Gullies: A History of Soil Erosion and Conservation in the Kingdom of Lesotho*. Athens: Ohio University Press.

Sillitoe, P. 1998. 'Knowing the Land: Soil and Land Resource Evaluation and Indigenous Knowledge'. *Soil Use and Management* **14**: 188–93.

Soil Survey Staff. 1951. *Soil Survey Manual*, USDA Handbook No. 18, Bureau of Plant Industry, Soils and Agricultural Engineering, USDA. Washington DC: Government Printing Office.

South African Department of Agriculture and Forestry. 1941. Report of the Departmental Soil Erosion Committee. South African Archives LDB R4250 Vol. 1.

Southern Rhodesia. Annual Report for the year: 1929, 1930, 1931. Salisbury: Government Printer.

Southern Rhodesia Department of Agriculture and Lands Report. 1939. Salisbury: Government Printer.

Stahl, Ann Brower. 2005a. Introduction: Changing Perspectives on Africa's past. In Ann Brower Stahl (ed.) *African Archaeology: A Critical Introduction* (Malden, MA: Blackwell Publishing), pp. 1–23.

Stahl, Ann Brower (ed). 2005b. *African Archaeology: A Critical Introduction*. Malden, MA: Blackwell Publishing

Stromgaard, Peter. 1986. 'Early Secondary Succession on Abandoned Shifting Cultivators' Plots in the *Miombo* of South Central Africa'. *Biotropica*, **18**(2): 97–106.

Stewart, G.A. 1917. Roads of Basutoland. Report, 4 May 1917. In Colonial Office: High Commission for South Africa, Original Correspondence. Despatches: Basutoland 1917. Public Records Office CO 417/593.

Sutton, J.E.G. 1984. 'Irrigation and Soil-Conservation in African Agricultural History with a Reconsideration of the Inyanga Terracing (Zimbabwe) and Engaruka Irrigation Works (Tanzania)'. *The Journal of African History*, **25**(1): 25–41.

Sutton, J.E.G. 1989. 'Towards a History of Cultivating Fields'. *Azania*, **24**: 98–112.

Swaziland. *Colonial Annual Report for the Year:* 1946; 1950; 1951; 1952; 1953; 1954; 1955; 1956; 1958.

Tanner, C.B. and C.P. Mamaril. 1959. 'Pasture Soil Compaction by Animal Traffic'. *Agronomy Journal*, **51**(6): 329–31.

Temato, P. and K.R. Olson. 1997. 'Impacts of Industrialized Agriculture on Land in Bafou, Cameroon'. *Journal of Soil and Water Conservation*, **52**(6): 404–5.

Tempany, Harold A. 1949. *The Practice of Soil Conservation in the British Colonial Empire*, Technical Communication No. 45. Harpendon: Commonwealth Bureau of Soil Science.

Temu, A.E.M. and S. Bisanda. 1996. 'Pit Cultivation in the Matengo Highlands of Tanzania'. in Reij, Chris, Ian Scoones and Camilla Toulmin (eds) *Sustaining the soil: Indigenous soil and water conservation in Africa* (London: Earthscan), pp. 145–50.

Thompson, Robert Farris and Joseph Cornet. 1981. *The Four Moments of the Sun: Kongo Art in Two Worlds*. Washington, DC: National Gallery of Art.

Tlou, Thomas and Alec Campbell. 1984. *History of Botswana*. Gaberone: Macmillan Botswana Publishing Co.

Torrance, W. 1919. *Observation on Soil Erosion*, Union of South Africa Department of Agriculture Bulletin 4. Pretoria: Government Printer.

Tráore, P.A. 1994. 'Constraints on Small-Scale Mining in Africa'. *Natural Resources Forum*, **18**(3): 207–12.

Trustees of Boston University. n.d. *How Big is Africa?* Outreach program. Boston: Boston University African Studies Center.

Twerdoff, D.A., D.S. Chanasyk, E. Mapfumo, M.A. Naeth, and V.S. Baron. 1999. 'Impacts of Forage Grazing and Cultivation on Near-Surface Relative Compaction'. *Canadian Journal of Soil Science*, **79**(3): 465–71.

Van der Vijver, C.A.D.M., P. Poot, and H.H.T. Prins. 1999. 'Causes of Increased Nutrient Concentrations in Post-Fire Regrowth in an East African Savanna'. *Plant and Soil* **214**(1–2): 173–85.

Vansina. Jan. 1990. *Paths in the Rainforests: Toward a History of Political Tradition in Equatorial Africa*. London: James Currey.

Ward, David, Ben T. Ngairorue, André Apollus and Hermanus Tjiveze. 2000. Perceptions and Realities of Land Degradation in Arid Otjimbingwe, Namibia. *Journal of Arid Environments*, **45**: 337–56.

Wardell, D. Andrew, Thomas Theis Nielsen, Kjeld Rasmussen and Cheikh Mbow. 2004. 'Fire History, Fire Regimes and Fire Management in West Africa: an Overview'. In J.G. Goldammer, and C. de Ronde (eds), *Wildland Fire Management Handbook for Sub-Sahara Africa* (Frieburg: Oneworld Books/Global Fire Management Centre), pp. 350–81.

Warren, S.D., T.L. Thurow, W.H. Blackburn and N.E. Garza. 1986. 'The Influence of Livestock Trampling under Intensive Rotation Grazing on Soil Hydrologic Characteristics'. *Journal of Range Management*, **39**(6): 491–5.

Watt, W. Martin. 1913. The Dangers and Prevention of Soil Erosion. *Rhodesia Agriculture Journal* X: 5.

Weinmann, H. 1975. 'Agricultural Research and Development in Southern Rhodesia, 1924–1950'. Salisbury.

Williams, Martin. 2003. 'Changing Land Use and Environmental Fluctuations in the African Savanna'. In Thomas J. Bassett and Donald Crummey (eds), *African Savannas, Global Narratives and Local Knowledge of Environmental Change* (London: James Currey), pp. 31–52.

Williams, T.O. 1999. 'Factors Influencing Manure Application by Farmers in Semi-Arid West Africa'. *Nutrient Cycling in Agroecosystems*, **55**(1): 15–22.

Wilson, Monica and Thompson, Leonard (eds). 1971. *The Oxford History of South Africa, Vol. II: South Africa 1870–1966*. New York: Oxford University Press.

WinklerPrins, Antionette M.G.A. 1999. 'Local Soil Knowledge: A Tool for Sustainable Land Management'. *Society and Natural Resources*, **11**(7): 151–66.

WinklerPrins, Antoinette M.G.A. and Jonathan A. Sandor. 2003. 'Preface'. *Geoderma*, **111**: 165–70.

7

Prolegomena to a History of Soil Knowledge in Europe

Verena Winiwarter

A GUIDE TO THIS CHAPTER

This chapter aims at introducing you to the history of soil knowledge as part of agricultural knowledge from Antiquity to the fourteenth century with an outlook to later times. I have imagined readers to have no detailed knowledge of European Classical or Medieval history, and I provide translations or paraphrases of any Greek and Latin texts cited. Further, while I assume that the terms 'knowledge', 'experience' and 'experiment' are taken to be completely clear notions in a reader's disciplinary surroundings, I assume that readers have not previously been concerned with a possible difference between 'history of ideas' and 'archaeology of knowledge'. I further assume that readers have no detailed concept of how knowledge transmission and dissemination in the age prior to the printing press worked, including, for example, how long it took a scribe to copy a treatise of say, 10,000 words, how much parchment he needed for that, and how many sheep or goats were needed to produce this parchment. While I shall not explain any of this in detail, it should be borne in mind that a book was an extremely costly item, and that access to books, while limited, was possible and was a matter of display of wealth and erudition. Over the course of time, the largest part of the material that was ever produced has been destroyed, so we work from very sketchy evidence. There is not a single case in which an 'original' manuscript of an agricultural treatise from Antiquity still exists. Only traces remain of such papyri, the usual writing material of this period. Usually, we deal with several (different, and often corrupt or reworked) later copies of these works. Any copy we know of represents a lot of effort and time, and thus by itself speaks to the relevance of the text for those who made it or ordered it to be made.

I start with sketching a 'universe' in which agriculture is a dominant concept in society, so dominant that agricultural metaphors are as commonplace as, for example, today's allusions to information technology if one wants to explain how the human brain works. Then I show how problematic the notion of 'knowledge' is for me, and try to define what I mean by it. I am concerned with the authority of knowledge (i.e. what is taken to be true at a given time and place, and why), which brings me to use a concept for the history of soil knowledge which was laid out by Foucault for (scientific) knowledge in general. Next I introduce many authors who wrote agricultural works, giving a very brief outline of their life and times. I find it important to make visible the manifold conjunctions between the works, and I do this using two examples, which I discuss in some detail. A timetable I have compiled may be used in conjunction with this chapter for an overview. Finally, I have chosen two examples to discuss how soil knowledge has evolved over time. One example is the body of soil terminology that was developed in Antiquity, the other is the question of 'suitable' soil. The article does not talk in any detail about agricultural literature in languages other than Latin or Greek; therefore, while I give some names and dates, I do not give any detail on works later than the fourteenth century. As I start around 700 BC, the time span is still some 2000 years.

Should readers be familiar with all I have assumed them not to be, I hope that they will still find the composition of the material and the unknown detail I unearthed interesting.

GLIMPSES OF AN AGRICULTURAL UNIVERSE

> Villicandi enim scientia adeo super omnes pollebat, quae videlicet in aedificiis construendis, iumentis et pecoribus educandis, agris serendis aliarumve rerum rusticarum constare videtur quacunque cultura, quam tamen non usu constat eum didicisse, sed arte, ut pene haec eo nemo curaret solertius nec his felicius abundaret.
>
> [With reference to knowledge in the field of agriculture, this means e.g. in construction of buildings, in breeding cattle and smaller animals, in treating fields or in any work of other rural matters, he was far ahead of all others. Almost no-one had more skills and better success than him. Yet his knowledge did not stem from tradition, but from studies, and thus almost nobody worked more circumspectedly and had better success than him.] (Chapter 8 from the *Life of Bishop Benno II of Osnabrueck*, written 1090–1100, my translation of the German translation in Kallfelz 1973).[1]

Bishop Benno II was a learned and well-travelled man, although – or even because – he was no member of the aristocracy. In the eleventh century, *ministeriales* like him could climb up the social ladder considerably. Benno was born in 1020 in Swabia, and studied in Strasbourg and at the Reichenau, the famous monastery

[1] Thanks to Christoph Sonnlechner for alerting me to this source.

situated in Lake Constance. He made a pilgrimage to the Holy Land, was a member of the cathedral school of Speyer, then became governor of the cathedral school of Hildesheim, after which he was appointed chief administrator (*vicedominus*) of the Goslar royal estate. After a short interlude in Cologne he became Bishop of Osnabrueck by appointment of King Henry IV. The bishop worked in a diplomatic mission during the uprising of the Saxons 1073–75. He founded the Benedictine monastery of Iburg, where he spent the last years of his life, and died in 1088. His agricultural knowledge was found worth mentioning in the *Vita*, the life of Benno, written by the abbot Norbert of Iburg shortly after Benno's death.

There is no way of knowing what Benno's sources were, whether he was a clever and practical person who outdid his contemporaries by means of his own thinking, or whether he had read any of the agricultural manuals of Antiquity, which were available, if scarce, at his time. Actually, one could also understand the juxtaposition in the text 'non usu … sed arte' in this way: that he was above the usual in his doing.

Benno's life is part of the unwritten history of agricultural knowledge which begins in Europe with one of the earliest preserved works of Greek poetry, Hesiod's *Works and Days*, and continues to the present day. The medieval part of this history is the least well known, but no detailed knowledge exists for any time and any part of Europe.

Knowledge about agriculture, of which soil knowledge is part, can be found from Ancient Greek times onwards in agricultural textbooks, in manuals and agricultural calendars, in texts by natural philosophers and theologians and, later on, by biologists and natural scientists. Some bits and pieces can also be found in other texts such as charters, legal texts and administrative (e.g. land taxation) records and in topographical descriptions which contain information about soils. The *Vita* of Bishop Benno is a case in point illustrating the wealth of sources – but it also serves as an example of the difficulties one can expect to encounter in locating them.

Humans living in agricultural societies possess knowledge about how to produce harvests. They have competence in dealing with domesticated animals and plants and their competitors and parasites, they are acquainted with various aspects of their doing, and through trial and error they possess means to recognise information as being correct; therefore one can summarise that they are in possession of the three basic constituents of knowledge, competence, acquaintance and recognition of correctness (Mattey, 2002).

While I shall deal with the concept of knowledge in the next section, and with the agricultural textbooks in the main part of this chapter, I shall first demonstrate the context of the more specialised soil knowledge by alluding to the mentality of agricultural societies. It is commonplace that Greek and Roman mythology show peoples' concern for agriculture, with Demeter (Ceres) as the goddess of agriculture having 12 minor gods connected with agricultural operations such as ploughing or storing grain worshipped with her; with Dionysos (Bacchus) as the god of wine; and with Artemis (Diana) as the goddess of hunting, to name but the most prominent examples – such is the world-view one expects of a society based on solar energy harvested via terrestrial biomass. Caesar, after his calendar reform, had so-called ***menologiae rusticae*** made, stones which detailed the new calendrical order and listed important agricultural tasks to be performed (Winiwarter, 1992). To assume that agricultural textbooks were written and used seems not too far-fetched under such circumstances, when culture and agriculture are obviously closely connected.

One is less well aware of the close connection of agriculture and culture in the Christian Middle Ages. Anyone who could read during the Middle Ages would probably first and foremost read the Bible. The Old and New Testament make use of agricultural metaphors, and many of the stories are embedded in the world-view of agricultural people. One will immediately think of the tense relationship between pastoralists and those who plant field crops as referred to in the story of Cain and Abel, as well as in the story of Joseph and his brothers. But there is more detailed reference than that. I will give some examples, without trying to be exhaustive. Psalm 65 is a good example.

> Visitasti terram et inebriasti eam;
> multiplicasti locupletare eam.
> Flumen Dei repletum est aquis;
> parasti frumenta illorum,
> quoniam ita parasti eam.
> Sulcos eius irrigans, glebas eius complanans;
> imbribus emollis eam, benedicis germini eius.
> (Vulgate)

> You care for the land and water it;
> you enrich it abundantly.
> The streams of God are filled with water
> to provide the people with grain,
> for so you have ordained it.
> You drench its furrows
> and level its ridges;
> you soften it with showers
> and bless its crops.
> (Translation of the New International Version)

Note that the translation has 'ridge' for ***glaeba***, which actually means clod. All translations are approximations, which is why I also give the Vulgate text, because this is closer to the text that medieval monks would have read.

At the end of his lament, Job evokes the utmost possible pain of the peasant: thorny bushes instead of wheat, and weeds instead of barley shall spring, if he had ever committed any evil with this land. It is significant that this is the end of his lament. A peasant can go no further than putting his land at stake.

> si adversum me terra mea clamat et cum ipsa sulci eius deflent
> si fructus eius comedi absque pecunia et animam agricolarum eius adflixi
> pro frumento oriatur mihi tribulus et pro hordeo spina finita sunt verba Iob

> [if my land cries out against me and all its furrows are wet with tears,
> if I have devoured its yield without payment or broken the spirit of its tenants,
> then let briers come up instead of wheat and weeds instead of barley.
> The words of Job are ended.] Job 31: 38–40.

Turning to the New Testament, Matthew 13, 3–9, Luke 8, 4–9 and Mark 4, 3–9 all tell the same story of the difference between sowing on fertile ground and among thorns as an allegory of the word of God and its reception by the human soul. An often cited biblical story is the one in Matthew 13, 24–31, where not so much the soil itself, but the problem of weeds serves as an allegory of evil and its subsequent fate (the weeds are to be collected and burnt). Weeds are used in allegory also in Hosea 10, 4: 'loquimini verba visionis inutilis et ferietis foedus et germinabit quasi amaritudo iudicium super sulcos agri' [They make many promises, take false oaths and make agreements; therefore lawsuits spring up like poisonous weeds in a ploughed field].

To end up unburied, and be used as manure seems to be one of the worst threats one can issue, as can be found, for example, in Jeremiah 16, 4:

> 'mortibus aegrotationum morientur non plangentur et non sepelientur in sterquilinium super faciem terrae erunt et gladio et fame consumentur et erit cadaver eorum in escam volatilibus caeli et bestiis terrae'

> [They will die of deadly diseases. They will not be mourned or buried but will be like refuse lying on the ground. They will perish by sword and famine, and their dead bodies will become food for the birds of the air and the beasts of the earth]. Note that what is translated as 'refuse' here is *stercor*, usually translated as 'dung'.

Latin Christian writers in Late Antiquity and in the Middle Ages were familiar with the Bible and used it in sermons or wrote about its interpretation. Interestingly, they used several words to describe features of the soil which are not found in the Bible itself. Latin was the spoken language in large parts of the Christian Ecumene until the thirteenth century, and one cannot tell from where the authors took their vocabulary, but whether it was from their reading of ancient works on agriculture or from everyday spoken language, either way proves the ubiquitous nature of agriculture in the spiritual life of these writers. Again, I shall give only very few, very specific examples.

The church father Hieronymus (Jerome) lived c. 350–420. He was born in Dalmatia, had travelled in the Middle East, and then lived in Trier and Rome.

He spent his last years in a monastery he had founded. The most famous thing about him is his translation of the Bible into Latin, for which he also undertook philological studies. In addition, he wrote several commentaries on books of the Bible. In the *Commentary on Isaiah,* he uses vocabulary that cannot be found in the Bible, talking about infertile swamps, which are muddy and silty, in contrast to fertile irrigated fields. The commentary relates to Isaiah 14, 22–23, the biblical text of which is given below.

> et consurgam super eos dicit Dominus exercituum et perdam Babylonis nomen et reliquias et germen et progeniem ait Dominus
> et ponam eam in possessionem ericii et in paludes aquarum et scopabo eam in scopa terens dicit Dominus exercituum (Vulgate)
>
> ['I will rise up against them,' declares the LORD Almighty. 'I will cut off from Babylon her name and survivors, her offspring and descendants,' declares the LORD . 'I will turn her into a place for owls and into swampland; I will sweep her with the broom of destruction,' declares the LORD Almighty.] (Translation of the New International Version)

> ubi non est ager irriguus, qui afferat fructus diuersorum seminum, sed paludes infertiles, et limosae ac lutosae, in quibus caeno gaudentia reptant animalia.
>
> [where there is not an irrigated field, which produces fruit from diverse seeds, but infertile swamps, muddy and silty, where animals are creeping rejoicing in mud.] (Hieronymus, *Commentarii in Isaiam,* 6, 14, 23, trans. VW).

None of the terms *limosus, lutosus* or *irriguus* are from the biblical text, as one can see (nor are they found anywhere else in the Vulgate). Hieronymus tries to explain what one is to understand by the biblical 'swampland' and its owls using terminology he must have known from elsewhere.

But Hieronymus was not the only one to use such metaphor. His contemporary, Rufinus of Aquileia, likens the care of the soul in all aspects to agriculture, mentioning ploughing, thorough watering, breaking of the soil (*scindere*) and furrowing, so that the seed of charity, hope and justice can grow:

> Agricola terrae suae putandus est ille, qui campos animae suae cordis que sui noualia indesinenter uerbi dei aratro et scripturarum uomere scindit et sulcat, qui plantaria fidei et caritatis ac spei et iustitiae de israhel fontibus rigat et omnem agriculturae disciplinam in animae suae rure deprendit
>
> [Take him as a farmer of his fields who ceaselessly tears and carves his unbroken land with the plough of the Lord's word and of the Scripture, who irrigates the plantations of Faith, Love and Hope and Justice from the springs of Israel, and who gathers all knowledge of agriculture in the lands of his soul.] (Rufinus, *De benedictionibus patriarcharum,* II, 14, trans. VW).

Caesarius of Arles, 470–542, also an important church father, wrote – among many other things – sermons to promote Christian values, practices and beliefs among the pagans, Jews and Christians of southern 'France' (cf. Klingshirn 1994). He used agricultural analogies abundantly, comparing the care of the soul with the cultivation of the fields. He compared eradicating evil from the soul to weeding a field, stating that if the evil is not uprooted, the good seed cannot sprout (Caesarius of Arles, *Sermones*, p. 32). Likewise, he described the care of the soul in an agricultural sequence: 'Thorns and stones have to be removed, then the soul can be ploughed, and after that, the legitimate seed can be laid in the furrow' (Caesarius of Arles, *Sermones*, p. 34).

While one could give more examples from his sermons and other works, I will move on to the years around the first millennium, when Heriger, the Abbot of Lobbes (Belgium) wrote a history of the diocese of Liège, the *Gesta pontificum Tungrensium et Leodiensium*. In it, he compares a field and the soul, pointing out that even a fertile field cannot bring fruit without culture, likewise without learning, meditation, self-restraint and prayer monastic life will decline: 'ut enim ager quamuis fertilis sine cultura non potest esse fructuosus, sic sine doctrina animus' [as a field, even if fertile, cannot bear fruit without cultivation, the spirit cannot do without teachings] (Heriger of Lobbes, *Gesta pontificum Tungrensium et Leodiensium*, 55, p. 188, trans. VW).

It would be possible to give numerous further examples, and it would surely be interesting to investigate the vocabulary and the interdependence between the Bible and the various authors more thoroughly, but this cannot be pursued here.

Ulrich Meyer has pointed out the importance of the concept of the *oikos* (in German 'das ganze Haus') and the origin of home economics for Christian Antiquity and the Early Middle Ages (Meyer, 1998). His study emphasises the importance of good housekeeping in moral and material matters alike as a duty of the *pater familias*. The analogy often drawn between the care of the fields and the care of the soul is consistent with this world-view. Meyer's second point is even more important: he explains how the highly literate and educated church fathers after long struggles solved the problem of pagan knowledge in their Christian context. If the Bible was God's product, a Holy Scripture, how could one dare to use pagan texts for its interpretation? Jerome (as cited above) is an excellent example of the solution that was found: the Bible came to be understood as consisting of layers of meaning, of which the uppermost level, the spiritual one, was indeed only to be interpreted within Christian doctrine. But to understand the spiritual content, the lower levels of meaning had to be perfectly understood, for which, as these were profane matters, pagan knowledge could indeed be expected to be of help. Jerome chose to explain Isaiah with words he could not have taken from the Bible, because obviously he expected them to have explanatory power on the level of factual content.

The *oikos*, the household, as the core unit of society dominated Western thought for centuries. One should also bear in mind that the only economic occupation Roman Patricians were allowed to pursue was land ownership. Making money from money was considered improper or even forbidden in many agricultural societies. Even if such restrictions were often more theoretical than real, the education of heirs to an estate needed to comprise agricultural matters to ensure their legal income. A copy of John Loudon's famous *Encyclopaedia of Agriculture* of 1883 (Loudon 1883[1839]) has a dedication that allows us to glimpse the thinking about proper education of landowners which still existed in the late nineteenth century:

> To Evan Gordon Macpherson [...] on the occasion of his attaining his maturity, With every good wish from A Macpherson, Kingussie, 15th June, 1889 (the signatory is probably Alexander MacPherson, who was then provost of Kingussie).

While I have only briefly sketched the importance of agriculture in the world-view of agricultural societies, what I hope has become clear now is that with Antiquity, the Middle Ages and Early Modern Times in Europe alike, we are dealing with societies whose culture is inextricably tied to agriculture. This should be borne in mind when judging the reception of specialised agricultural texts by a readership whose literacy is usually underestimated, especially for the Early Middle Ages (McKitterick 1988), and whose ways of knowing what they were doing are quite unknown.

CONCEPTUAL CONSIDERATIONS FOR THE STUDY OF AGRICULTURAL KNOWLEDGE

Traditional Knowledge, Sustainability and Authority

Soil knowledge has developed in the context and as part of agricultural knowledge. It is under this presupposition that the quote about Bishop Benno becomes relevant to the search for soil knowledge, and it is in this context one needs to develop the tools for studying soil knowledge.

No history of agricultural knowledge exists, on which a more specific soil study could rest. Güntz (1897) made a compilation of Latin and German texts, which, while extremely useful, is merely a list of treatises rather than a history of them. Much the same is true for McDonald (1908), who lists the English works on agriculture.

Historians of science have neglected agriculture, because it is not a proper scientific field. Historians of agriculture have paid more attention to the economy of production than to the theoretical knowledge on which agriculture rested. They have often taken the agricultural manuals as sources for reconstruction of agriculture at a given time and place, but have not studied the evolution of knowledge as such. Famous historians of agriculture are divided over the question of whether the agricultural manuals influenced the practice of agriculture. Duby was rather positive about an influence, whereas Adriaan Verhulst denies it completely (Verhulst 1989, p. 1). Even Verhulst admitted that a thorough study of the works would be necessary to decide about their practical relevance, and in the years since the publication of his short piece, no such work has appeared. John Ziman has pointed out that the interrelation of theory and practice cannot be caught by a simple theoretical model (Ziman 1976, p. 35), and the present study, which is meant as study of ideas about soil, starts by analysing texts rather than aiming directly at the question of practices.

Within environmental history, the history of ideas has a long tradition. Clarence Glacken's seminal volume ***Traces on the Rhodian Shore: Nature and Culture in Western thought from Ancient Times to the End of the Eighteenth Century*** is a classic, frequently reprinted and widely used (Glacken 1967). A distinctive feature of environmental history is its connection to environmental sciences or, more globally, its interest in being relevant to relations between society and the environment. Concepts from biology or from environmental sciences have played an important part in the development of the field. 'Equilibrium', 'resilience' and 'adaptation' are prominent examples of biological notions used for environmental history narratives. The notion of 'sustainability' or 'sustainable development' is the narrative basis for many such studies, which ask if a particular society in a given place was sustainable or not, and if so, why. The critique of sustainability has also been a topic for environmental historians (Worster 1988; Winiwarter 2001). This chapter is an attempt to address an ongoing debate in the sustainability context. In preparation for the Johannesburg World Summit, The International Council of Scientific Unions (ICSU) issued a report on ***Science, Traditional Knowledge and Sustainable Development*** (ICSU 2002). Traditional knowledge therein is distinguished from 'pseudo-science', and a partnership between science and traditional knowledge is envisaged to foster sustainable development. The role of written knowledge from Europe before the so-called 'scientific revolution', or works that appeared after it but had no connection to it, is quite unclear in this paper, which does not mention historical knowledge at all.

In the ICSU paper the boundaries between science, pseudo-science and traditional knowledge are drawn in a quest to determine the authority of knowledge (what statements are taken to be true and relevant at a given point). The ICSU position is ***strictu sensu*** ahistorical, and has to be differentiated in historical perspective. As the paper shows, questions of epistemology and questions of authority arise in conjunction with soil knowledge.

This elaboration on soil knowledge history is therefore written as a Foucaultian 'archaeology' rather than as a history of ideas. A Foucaultian 'archaeological' inquiry, in essence, asks what conditions foster particular statements that come to be taken as true, alluding to the way memory is conceptualised in cognitive theory. The appearance of one statement rather than another is, of course, not a simple question for Foucault. A statement is not an isolated 'utterance', but 'always belongs to a series or a whole'. '[I]t is always part of a network of statements' (Foucault 1972, p. 99) which compose a discourse. Reflexively, the discursive formation – 'a relatively autonomous system of serious speech acts in which [a given statement] was produced' (Dreyfus and Rabinow 1982, p. 49) – sets the context in which constitutive statements are held to make 'serious sense', to be 'true'. And truth, for Foucault, is no more than an 'ensemble of rules ... [and] a system of ordered procedures for the production, regulation, distribution and operation of statements' (Foucault 1980, pp. 132–3).

Later on, Foucault has worked with the notion of 'genealogy', which does not intend to displace the archaeological 'dig', but to broaden the scope of inquiry. It seeks to trace the 'descent' and 'emergence' (Foucault 1984) of new discursive formations; to trace a discourse's lineage across the path of contradictions and logical discontinuities – the 'accidents, chance, passion, petty malice, surprises ... and power' (Davidson 1986, p. 224) that foster new discursive formations (Rice 1992).

Foucault's emphasis on the non-linear character of knowledge as produced in discursive formations is important as a concept because it works against the false impression of a body of accumulating knowledge, its quality improved by succession towards truth (which would then mean: current scientific concepts). Secondly, he emphasises that writing history is an act of rewriting it, (re-écriture) rather than an act of reconstruction: the history I offer here is my rewriting, and should be understood as part of the discourse rather than as outside of it. Especially if one deals with pre-modern times, the vagaries of the archival situation, the availability, visibility and invisibility of parts of the body of knowledge for authors who wanted to contribute to these discourses, should be taken into account when judging the relationship of a text to the 'existing' contemporary literature. If, for example, Petrus de Crescentiis, whom I shall present later, did not cite Columella directly, but only Palladius, this could mean that he did so by choice, or it could mean that only Palladius was available to him. Many such questions will unfortunately remain without answer.

François Jacob has written a fascinating account of the development of biology, making excellent use of Foucault's concept of an archaeology of knowledge (Jacob 1972). According to Jacques LeGoff, Jacob's book can be considered the only successful 'archaeology of knowledge' he had ever encountered (LeGoff et al. 1990, p. 42). How the aims of such a project differ from a type of historiography where the history of knowledge is taken as factual, has been explained by Athar Hussain: 'Archaeology does not seek to trace the intentions, motives, etc., of the speaking subject. It is not the discovery of the primordial germ of a discourse. It is the *re-writing* (*reécriture*) of a discourse, i.e., a rule-governed transformation of something which has already been written' (Hussain 1971, p. 105).

Discontinuities and epistemic cleavages are major features of soil knowledge systems, just as much as continuities and traditions are. Facts about nature – the nature of soils in this particular case – that are provided in these knowledge systems are not to be mistaken as mirror images of a presumed reality. Ethnological studies in particular have shown how important such a reflexive approach is when interpreting concepts of nature and world-views (recently Gingrich and Mader 2002), albeit for different source types.

To tell a long story briefly, the authors, their aims and ramifications and the networks they are part of are all important for understanding their works. Ideological presuppositions have to be taken seriously, instead of discrediting knowledge because of its ideological context. Cognition is considered as historically bounded, and as being variable in its structure.

KNOWING ABOUT (SOIL) KNOWLEDGE

Dealing with 'knowledge', I must clarify what 'knowledge' means, what I assume to be particulars of soil knowledge. One needs to be precise in differentiating between 'experience' and 'practice', and again in discerning them from 'experiment'. For the latter, Hans-Jörg Rheinberger's work on the history of natural science experiment, which discusses the relations of 'experience' and 'experiment', can be a useful guideline (Rheinberger 2001).

At the dictionary level, knowledge seems to be a well-defined term, it being 'a clear and certain perception of something, the act, fact, or state of knowing, understanding' (Webster 1983, p. 1007). Knowledge does include practical experience and skill and comprises information, and thus seems to denote a body of facts and principles accumulated by mankind over the course of time. In such a conceptual frame, knowledge can be stored in media other than the human mind, and re-processed from storage, and it is apparent that knowledge, being an entity which can be accumulated, grows with time. But there are also epistemological considerations, questions about how we can know, basically, if knowledge is empirical or innate, and thus if knowledge is a reference to an outside world which can be known or if knowledge is confined to the individual mind. Having posed such questions, the idea of knowledge as something that can be accumulated as an entity independent of the medium becomes somewhat less convincing. To define knowledge accommodating such considerations, one can conceptualise it as the matrix of impressions within which an individual situates newly acquired information (Clarke 2001). Whenever the term knowledge is used in this chapter, I refer to this latter concept. It is clear that one would then need to define 'information'. One way is to define it as 'the quantity of data in a message', which then yields the problem of defining what data are. The other dictionary definition, information being 'an accretion to knowledge' is circular. Information will therefore be defined as data (observation-based impressions on the human senses) which under given circumstances are considered as different from noise (i.e. all the impressions on the senses which are not used in interpretation of the world from which they emanate). After this excursion into information theory, it should have become clear that 'knowledge' rests on a set of decisions made by humans on how to proceed with sensory perceptions. It is such decision-making processes that interest me as historian of knowledge, though without any hope of being able to study them directly. A 'second order' type of knowledge-generating process can be studied, however: the web of communication within which decisions about noise versus data, about interpretation of data and about strategies of communication itself are made and can be traced – or at least re-written – by the historian.

The next question is about the relationship between knowledge and experience. One should note that experience and experiment both come from the same Latin word, ***experior***, a verb meaning 'try' or, as it still means in English today, experience something. ***Experimentum***, also a Latin word, accordingly means a 'trial' or 'test'.

Experience, according to dictionary wisdom, is an activity that includes training, observation or practice and personal participation. In the dictionary the notion bears a strong connection to the life-world, as experiences are lived through or happen to one (Webster 1983, p. 645). If one accepts this definition, the process(es) by which experience is transformed into knowledge remain(s) unclear. The transformation of experience into language, be it oral or written, is critical and needs attention. In such a context, the notion of 'tacit knowledge', which is used to describe the unspoken or implicit in an action which the actors are unable to formulate in language, might be helpful. Another distinction along these lines it the one between 'embodied knowledge' and 'theoretical knowledge'. Knowing-how or embodied knowledge is characteristic of the expert, who acts, makes judgments and so forth without explicitly reflecting on the principles or rules involved. The expert works without having a theory of his or her work; he or she just performs skilfully without deliberation or focused attention. Knowing-that, by contrast, involves consciously accessible knowledge that can be articulated and is characteristic of a person learning a skill through explicit instruction, recitation of rules, attention to his or her movements, etc. While such declarative knowledge may be needed for the acquisition of skills, the argument goes, it is no longer necessary for the practice of those skills once the novice becomes an expert in exercising them, and indeed it does seem to be the case that when we acquire a skill we acquire a corresponding understanding that defies articulation. That an exhaustive equation of tacit knowledge with pre-theoretical, skilled expertise cannot be maintained becomes particularly clear when we consider that one widely accepted paradigm of tacit knowledge is to be found in language competence. In contrast to other varieties of tacit knowledge, knowledge of language is not understood to constitute a skill, and thus to consist in a capacity to do something but rather is a properly cognitive capacity, and therefore defined in terms of mental states and structures that are not always or reliably manifested in behaviours or performances. There is thus reason to suppose that at least some – though by no means all – forms of tacit knowledge can behave like ordinary dispositions to believe, and accordingly can be brought to awareness given the proper circumstances. We might say then that these kinds of tacit knowledge are tacit to the extent that they are initially inaccessible to the person to whom they are attributed, but that given the proper conditions, this inaccessibility can be converted to the kind of accessibility enjoyed by our ordinary knowledge (Barbiero n.d.).

In the transformation of (tacit) knowledge from practical experience and from the ubiquitous metaphorical context found in agricultural societies into a specialised agricultural text (whether a manual or a poem, if its theme is agriculture it will be considered an agricultural text), it is changed and reformed in the conversion, and becomes part of a body of texts dealing with agriculture, texts with diverse aims and audiences, written from different experiences and by authors with different skills in the transformation process described above.

The theme of their works, agriculture, nevertheless poses similar writing problems for all of the authors. These are reflected in similarities in their texts. But the notion of 'similarity' is not a precision tool. To formulate the dissection plan in Foucault's terminology, all those texts should be considered as belonging to one discursive field. Within it, one needs to differentiate between several forms of difference/similarity: linguistic analogy (which means the ability to be translated), logical identity (equivalence) and homogeneity of the statement. Changes in each of these three dimensions can be indicative of the formation of a new discursive practice, and it is such discursive practices that I will try to distinguish from each other. But synchronic and diachronic comparisons will be needed for that, as several discursive practices can co-exist at the same time. I shall not analyse the entirety of the works I refer to, but confine myself to those parts that are closely linked to soils.

While I consider it important to distinguish between knowledge, practice, experience and the like, in analysing agricultural treatises one has to bear in mind that the differentiation we make today between practice and theory and between 'experience' and 'experiment' are relatively new (post-Baconian or even post-eighteenth century) developments, at least in the English language. Therefore, one must be careful not to misinterpret these two synonyms as being different from each other in any particular way when one finds them in the sources.

AN INTRODUCTION TO MAJOR AGRICULTURAL WORKS AND THEIR VIEW OF SOILS

Table 7.1. A selection of extant agricultural treatises of classical antiquity and beyond

Author	Title	Lifetime	Comments
Hesiod	*Erga kai hemerai*	Around 700 BC	First Greek author who is known by name
Xenophon	*Oikonomikós*	430–354 BC	
M. Porcius **Cato**	*De agri cultura*	234–149 BC	First work of prose in Latin
M. Terentius **Varro**	*Res rusticae*	116–27 BC	
P. **Vergilius** Maro	*Georgica*	70–19 BC	Poem
L. Iunius Moderatus **Columella**	*De re rustica*	1st cent. AD	Most detailed work
C. **Plinius** Secundus the Elder	*Naturalis historiae*	AD 23–79	Vol. 17 and 18 of the Encyclopaedia
Anonymous	*Liber de arboribus*	2nd cent. AD.	Transmitted with Columella
Gargilius **Martialis**	*De arboribus pomiferis*	3rd cent. AD.	Only parts extant
Rutilius Taurus Aemilianus **Palladius**	*Opus agriculturae*	4th or 5th cent. AD	Agricultural calendar
Isidor of Seville	*Etymologiarum sive originum libri XX, Liber XVII: De rebus rusticis*	6th/7th century AD	One book of his Encyclopaedia
(H)rabanus Maurus	*De rerum naturis*	Work written between 842 and 846	Theological, abundant references to agriculture
Walafrid Strabo	*Liber de cultura hortorum*	809–849	Horticultural poem
Wandalbert of Prüm	*De mensium duodecim nominibus signis culturis aerisque qualitatibus*	813–870	Agricultural calendar poem
Several authors	*Geoponika*	10th century AD	Collection; in Greek
Albertus Magnus	*De vegetabilibus et plantis*, 7 books	1193–1280	Part of commentaries on Aristotle
Vincent of Beauvais	*Speculum maius. Speculum doctrinale, liber sextus: De arte Oeconomica*	1190–1264	Part of his Encyclopaedia
Walter of Henley	*Treatise on husbandry*	written c. 1270	First English work

Author	Title	Lifetime	Comments
Petrus de Crescentiis	*Ruralia commoda*	c. 1300	
Konrad of Megenberg	*Yconomica (lib I, tractatus tertius, cap 38. De agricola)*	1309–1374	
Konrad of Heresbach	*Rei rusticae libri quattuor* (1570)	1508–1576	Latin
Johannes Coler	*Oeconomia* (1591ff.)		German, 14 print runs
Philipp Jacob von Grünthal (and family members)	*Haushaltungsbüchl*	1580s–1630s	German
Martin Grosser	*Anleitung zu der Landwirtschaft*	1640	German
Abraham von Tumbshirn	*Oeconomia*	1616–	German
John Claudius Loudon	*Encyclopaedia of Agriculture* (1829 ff)	1783–1843	English

All extant works mentioned in this text can be found in Table 7.1, which is ordered according to the age of the texts. Many more must have existed, if we trust the authors who mention many names of experts of whom this reference is the sole trace we have. One of the most mysterious works is a treatise written by a Phoenician named Mago. As he is sometimes called 'the father of agricultural textbooks' I find it important to show that the reference to his work is reference to an unknown. We have only very blurred traces of it – but to follow them is extremely interesting. Mago's work is said to have comprised 28 volumes. It was rescued during the fall of Carthage (146 BC) so it must be older than that, and the Roman Senate even commissioned its translation, as Pliny the Elder reports in the book on agriculture in his encyclopaedia, where Mago is mentioned alongside with other non-Latin writers:

> Igitur de cultura agri praecipere principale fuit etiam apud exteros, siquidem et reges fecere, Hiero, Philometor, Attalus, Archelaus, et duces, Xenophon et Poenus etiam Mago, cui quidem tantum honorem senatus noster habuit Carthagine capta, ut, cum regulis Africae bibliothecas donaret, unius eius duodetriginta volumina censeret in Latinam linguam transferenda, cum iam M. Cato praecepta condidisset, peritisque Punicae dandum negotium, in quo praecessit omnes vir clarissimae familiae D. Silanus.

> [To give instruction on agriculture was a matter of importance also for foreigners, because even kings dealt with it, such as Hieron,[2] Philometer Attalos,[3] Archelaos[4], and military leaders like Xenophon, and even Mago of Carthage, whom our senate paid such respect after the capture of Carthage, that while it distributed the library among the kings of Africa, it had only the 28 books of this man translated into Latin, although M. Cato had already written such a work. The senate confided this work to men experienced in the Punic language; among whom D. Silanus, a man from a distinguished family, stands out] (Plin. n.h. XVIII, V, 22f., trans. VW).

Two of the prominent Roman writers on agriculture, Varro and Columella, inform us about the subsequent history of Mago's work. Varro gives an account that shows great interest in Mago's work and hints at the fact that the translator was not in fact a mere translator, but added Greek knowledge and left out part of the original work.

> All these are surpassed in reputation by Mago of Carthage, who gathered into 28 books, written in the Punic tongue, the subjects they had dealt with separately. These Cassius Dionysius of Utica translated into Greek and published in twenty books, dedicated to the Praetor Sexti(li)us. In these volumes he added not a little from Greek writers whom I have named, taking from Mago's writing an amount equivalent to eight books. Diophanes, in Bithynia, further abridged these in convenient form into six books, dedicated to king Deiotarus. (Varro, rust. 1,1,10, transl. Loeb Classical Library, pp. 165, 167)

Varro compresses a history of about 100 years into these few sentences. Carthage was destroyed in 146 BC, and the translation was made soon thereafter, around 140 BC. The abridged six-volume edition can be dated to around 60 BC because one can date King Deiotarus to this time. The Greek translation was made in between, and it has been suggested that it was commissioned in connection with the colonisation projects of C. Gracchus, because in several of them Greek would have been the only common language of the people assembled from different parts of the Empire (Mahaffy 1889/90, pp. 33f).

Columella, who refers to the same incident, is far less precise: 'As a matter of fact, Diophanes of Bithynia epitomised in six abridged volumes the treatise of the entire work of Dionysius of Utica, who translated in many prolix volumes the treatise of the Carthaginian Mago' (Col. r.r. 1, V, 10, transl: Loeb Classical Library, Col. r.r. Book 1, p 33).

Neither the translation Columella refers to nor its abbreviated version have been passed on to us. So, unfortunately, we have no knowledge of the scope of this work, but Mago's book was obviously important to the authors we do know

---

[2] Tyrant of Syracuse

[3] King of Pergamon

[4] King of Macedonia

of. Mago is of particular interest because through the references to his work the geographical extension of the Mediterranean world in which agricultural knowledge was exchanged can be anticipated. The active interest by the Senate is a very important proof of the high value agriculture had, and the complicated history of translation, expansion and compression makes visible the discursive formation to which these authors belonged.

The oldest works (also older than Mago's treatise) mentioned in Table 7.1 are by Greek authors. Hesiod predates all others by centuries. In his poem (there was no prose writing in the sixth century BC) he placed agriculture in the context of myth. He recounts the generations of beings created by the gods, humans being the fifth generation. He then goes on to describe the works that have to be done in the yearly cycle, for which the stars serve as guides. The rise and setting of the Pleiades and of Arcturus are used as indicators of time for sowing, reaping or other work, the themes being organised according to an astronomical calendar. His list of things to do does not include any reference to specific types of soil. But Hesiod is very specific about the type of wood best suited for constructing a plough; he recommends having a spare plough in case one breaks; he gives details about the age of oxen best suited to do the work and how to go about it; and he has a lot to say about the right timing of ploughing, planting vines, cutting wood and other agricultural tasks. He generally recommends getting new land under the plough – enlarging the agricultural area was a good thing to him.

The other important Ancient Greek treatise, Xenophon's, is the first extant example of another narrative structure, the form of a dialogue between Socrates, Kritobulos and others, among them the landowner Isomachos as an expert. The *Oikonomikos*, as this book has been called, has some information on soil qualities, albeit not very differentiated. The basic idea is that soil and plant are an interactive system: 'For they tell us that to be a successful farmer one must first know the nature of the soil.' 'Yes, and they are right', I remarked; 'for if you don't know what the soil is capable of growing, you can't know, I suppose, what to plant or what to sow' (Xen., *Oikon.*, 16, 2). Soil can be 'fat' or 'lean' (17, 8), dry or moist (e.g. 16, 11), and by means of the plants that grow on a piece of ground one can judge the quality of this ground (16, 3) (such plants are called indicator plants today). Xenophon, like Hesiod, is concerned about the right timing for ploughing and sowing (17, 5). Manuring is known as a means of improving the soil. Some editors hold that particular reference is made to green manure (*Oikon.*, 16, 12), as is the application of salt-free, dry or wet substances to clean soil from salt (20, 12), a form of land improvement. The multifaceted knowledge of agriculture does not easily yield to narration, and Xenophon uses the form of the dialogue in a very elegant and convincing way to convey his information. However, Xenophon is mainly concerned with the economic output of the farm and with its management, and less with details of agricultural operations. We might consider his rendering of soil quality as the most basic knowledge one can expect in agriculture.

Hesiod's and Xenophon's are the two most widely known Greek agricultural works; the third one that is widely known is a Byzantine collection of older works (which we only know via this collection) called *Geoponika*. Much discussion has been going on about this collection, which was apparently compiled in the tenth century AD, but comes from works dating back at least to the sixth century. Most probably it was compiled by an unknown writer for the Emperor Constantine VII Porphyrogenitus, to whom the work was formerly ascribed. It is based mainly on a collection made in the sixth or seventh century by Cassianus Bassus, who used two earlier writers, Anatolius Vindanius (or Vindonius) of Berytus, and Didymus of Alexandria (both of the fourth or fifth century). The most recent – and excellent – summary and discussion can be found in a study on Byzantine gardening (Rodgers 2002). In his study Rodgers also shows the close connection between Latin and Greek and Arabic texts (see Figure 7.1), again pointing to the cross-fertilisation of several traditions, such as we have already seen when discussing Mago's work.

Due to their expansion from the seventh century onwards, Arabs became a major political and cultural factor in the Mediterranean world. Their flourishing realms on the Iberian Peninsula became cultural contact zones of great importance. There is at least one famous Arab treatise on Agriculture (Ibn Al' Awwâm's *Kitab al-filaha*). We know next to nothing about the author, who was an Arab from Al Andalus and wrote his treatise in the middle of the twelfth century. He compiled his work using agricultural and veterinary medicine works of Greek and Latin origin, which he presumably had available in Arab translation, putting them together without much comment. But selection is also a creative process of knowledge generation, and as Karl Butzer (1994 and 1993) has shown, he is part of the multi-lingual network of Mediterranean agricultural authors who all knew many others' works and used them according to their own selection criteria and preferences with their own experience as a guideline for selection. Mago's work was translated on behalf and for the Roman Senate. The *Geoponika* was another work of public interest, as far as the interest of the aristocracy of mid-tenth century Byzantium can be called public. The overall value placed on agriculture can be seen in the preface to one of the manuscripts of the *Geoponika*:

> Knowing that the state consists of three elements – the army, the clergy, and agriculture – you have devoted no less care to the latter, which is best able to preserve human life. That which several of the ancients discovered through their study and experience of cultivating the land and their care of plants, the seasons, the methods and terrain suitable to each, and furthermore the discovery of water and the construction of buildings, their implantation and orientation, all these and many other important things, the greatness of your genius and the depth of your spirit have gathered together, and you have offered to all a work that is generally useful.

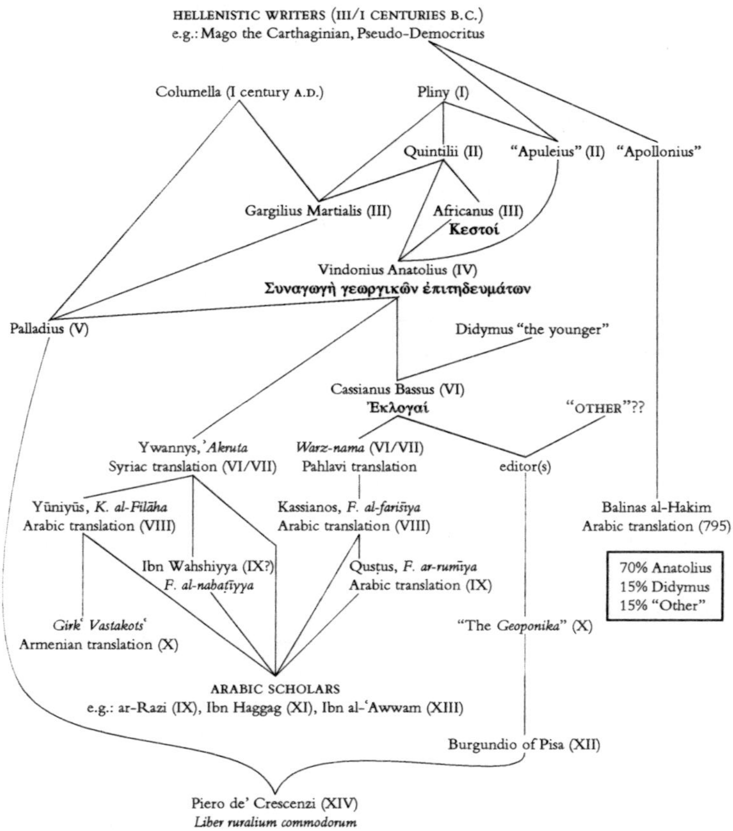

Figure 7.1. The connections between texts on agriculture in Arab, Latin and Greek, according to Rodgers (2002, p.163)

> Immediately, he who applies himself to the fruit of your labours is able to recognise exactly what his existence consists of, and he may observe in perfect order that which is both useful and necessary, what the basis of human life is and that on which he lavishes all his care. He can see not only what is necessary, but also superfluous things, conducive solely to the enjoyment of his eyes and his sense of smell. (Prooimion 6–9, trans. Lefort 2002).

Like the above-mentioned works, the *Geoponika* also contains information on soils, e.g. in Book 2, Chapters 9–11 and in Book 12, Chapter 3, which talks about soils for vegetables. In Chapter 12, 2 we encounter soils in the garden:

> (12, 2: 3) The person who wants to excel in growing garden plants must take forethought for good seeds, suitable soil, water, and manure. (4) Good seeds will produce offspring like themselves. Suitable and fertile soil will guard what is entrusted to it. Water will make the vegetables grow larger through nurture. Manure makes the soil more friable, so that it receives water more readily, to make space for the roots and to allow the foliage to sprout (trans. Rodgers 2002, p. 173).

While today we have another theory as to why manure works, the passage contains some interesting conceptual remarks on soils. That there is fertile and infertile soil was probably the most common knowledge about agriculture, and that soil and plant have to match to yield good results was also 'standard wisdom'. Friability as a distinct soil quality is interesting in terms of soil knowledge.

The *Geoponika* has been widely translated into, among other languages, Arabic, Latin, German and French (in 1545), but no modern translation of the whole work is available. Rodgers at least has a table of contents, which is helpful (Rodgers 2002, p. 166).

In Figure 7.2, I have tried to convey that when dealing with 'agricultural writers' we are dealing with a synthetic category which encompasses works whose theme might be (in part or in general) agriculture, but which are part of several different 'genres'. I have to emphasise that my categories are haphazard (as all such schemes are). Connections exist from within the agricultural literature to other realms of knowledge. Whereas I have treated works on (home) economics such as Xenophon's as part of the agricultural realm, I have excluded important sources for the agricultural tradition, especially for the learned treatises, viz. botanical books. The most important and influential botanist of antiquity was Theophrastus (371–286 BC), a pupil of Aristotle. He was also in charge of the first existing botanical garden. Two of his works still exist: ***Historia plantarum*** (A History of Plants) and ***De causis plantarum*** (On the Reasons of Vegetable Growth). Quite often, the way an ancient text is classified and studied is determined more by modern scholarship than by the distinctions drawn by the texts' authors themselves. A few examples of the agricultural tradition will elucidate this point. Albertus Magnus, although he deals with agriculture as much as other compilers of encyclopaedias did, has mainly been encompassed into the history of science, and not into the history of agriculture. The famous quotation which makes him the predecessor of scientific rigour, 'Experimentum solum certificat in talibus' (Experiment is the only safe guide in such investigations, ***De vegetabilibus***, VI, tr. ii, i), actually comes from his botanical treatise, which might as well be considered part of his agricultural studies. As mentioned above, at the time he was writing, no sharp distinction existed between experiment and experience, so the historians of science might be overstating their case. In contrast to him, Petrus de Crescentiis, who draws heavily on the work of Albertus, is counted among the agricultural texts. Historiography has its mantraps.

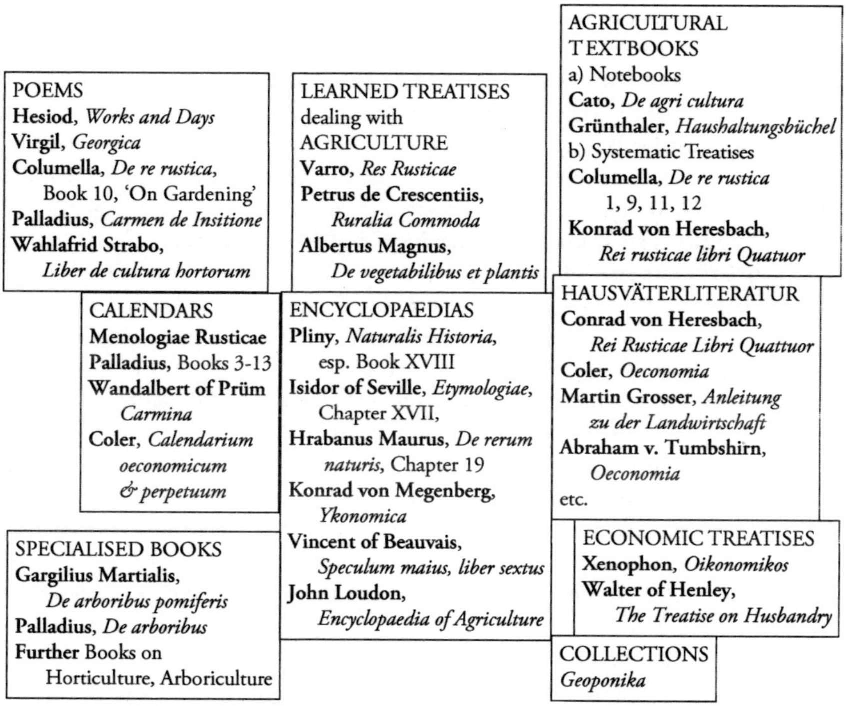

Figure 7.2. Works Containing Information on Soils

Works containing information on soils in agricultural context can tentatively be grouped into several categories, albeit none of the given categories is exclusive. In some cases I have sorted parts of works into different categories, in other cases this was not possible

Along the same lines, the theological interests of Hrabanus Maurus are certainly the most important aspect of his work, but it would certainly be wrong not to treat him also as someone who used agricultural knowledge. So Figure 7.2 is also an attempt at bringing together works that have been studied under quite different presuppositions, treating them as the precursors to economy (Xenophon), natural science (Albertus Magnus), bucolic poetry (Virgil), etc., and not trying to compose a larger picture of the many ways in which agricultural knowledge was present in these works and thus in learned circles.

These historiographic considerations have brought us to the beginning of the fourteenth century AD. Now we must go back 1600 years, to the second century BC. The first prose work ever written in Latin was an agricultural notebook. This fact is another sign of the importance of agricultural knowledge in solar-based societies.

Cato the Censor, a distinguished Roman patrician, left a work about agriculture which has more the character of a journal than of a systematic treatment of the subject. But the author is aware of the importance of appropriate timing, saying, 'Opera omnia mature conficias face. Nam res rustica sic est, si unam rem sero feceris, omnia opera sero facies' [See that you carry out all farm operations betimes, for this is the way with farming: if you are late in doing one thing you will be late in doing everything.] (Cato, agr. V, 7). Cato, however, does not give any kind of calendar of tasks. A century after Cato another distinguished Roman patrician, Varro, wrote the treatise ***Res Rusticae*** (On Agriculture) in the style and form in which a learned man of that time would write, as a dispute among three estate owners discussing a variety of subjects. He was very concerned with etymological aspects of agricultural vocabulary, a part of his interest we have come to belittle. The book was written late in his life, and was meant as a manual for his young wife, Fundania, so that she would be able to oversee the estates after his death. This might be purely narrative construction yet it is interesting, because Varro understood his book as practical advice. The work consists of three books: one on animal husbandry, one on agriculture proper, and one on all types of horticultural activities and the breeding of fowl. In the first book he talks about the two measures of time – one being the annual solar cycle, the other being the monthly lunar cycle – and he elaborates at some length the various periods into which the year can be divided. (Varro, rust. I, XXVIII f.). First, Varro divides the year into eight periods and explains what should be done in each, then he talks about the lunar cycle. His general remark that there is a planting season appropriate for each kind of seed shows that he, too, is aware of the importance of timing (Varro, rust. I, XXXIX). Varro makes an interesting remark about the grafting of plants, saying that some that were formerly grafted in spring are now grafted in summer. Obviously there was some development of agricultural techniques of which Varro takes account (Varro, rust. I, XLI).

Agriculture played a very important role in ancient Roman life; high-ranking individuals were concerned with and about it. So, it is no wonder that even a work of verse was written by a very prominent Roman poet, Virgil. His *Georgics* are again divided into books dealing with the various subjects of agriculture, as is Varro's work. In the first book, Virgil explains the main tasks of fieldwork. The second part of the first book (204–310) is an agricultural calendar, and the third part is a meteorological treatise. His second book is about trees, the different kinds of soils and climates, and the work to be done in an orchard. The last part of this book is in praise of rural life. The third book is about both large and small animals, and the fourth is about bee-keeping, beginning with the legendary creation of the bees and continuing with a dramatic mythological story, in which no practical agricultural knowledge is presented. Virgil is unsurpassed in the amount of manuscripts that have survived. He was certainly very widely known in later times. He also must be credited with being the founder of the agricultural poem, in which he had two important followers. The books by Columella and Palladius each contain a part in verse. Columella's tenth book on gardening and Palladius's *Carmen de insitione* (On Grafting) were written (as the authors let us know) in homage to the Georgics. The *Hortulus* of Walafrid Strabo, a poem on the plants of the garden written by the learned Abbot of the Reichenau monastery (formerly one of the teachers of Charlemagne), is an early medieval example of this literary tradition, which has had importance in maintaining the status of agriculture as part of culture ever since.

By modern standards, the most scientific ancient work ever written on agriculture in pre-industrial times is surely Columella's treatise on agriculture (*De re rustica*). We do not know much about him. He lived in the first century AD, was born in Spain and passed through a typical Roman career. He had several estates, and as we can infer from his invention of a new implement for grafting of trees, he must have taken part in the practical running of them. His agricultural work was written late in his life. It consists of twelve books, amongst which the one on horticulture is written in verse, as mentioned above. The other books are organised by topics; one entire book (Book 2) is on soils, but the work also contains an agricultural calendar using astronomic markers (e.g. the rise of the Pleiades) to give dates.

Within the socio-political entity of the Imperium Romanum, several more specialised treatments of agricultural matters were undertaken, and agriculture also became part of the encyclopaedic attempts. We know of specialised works on viticulture only by way of their mention by Columella, but several treatises on veterinary medicine and at least fragments of a work on arboriculture written in the third century AD survive. While veterinary medicine will not be dealt with here, Gargilius Martialis' work on tree culture, which contains detailed information on soils, is an important part of the discursive formation. Unfortunately, we know nothing about the author of this book.

Pliny the Elder, who was not primarily an agricultural writer, did write the very famous encyclopaedia *Naturalis Historia*, summing up the knowledge of his age, including in Books 17 and 18 a great deal of agricultural knowledge and a calendar of agricultural tasks. Pliny, his Latin full name Gaius Plinius Secundus, was born in AD 23 in Novum Comum, Transpadane Gaul, and died on 24 August AD 79 in Stabiae, near Mount Vesuvius, while studying the eruption of the volcano. Pliny came from a prosperous family, and could pursue studies in Rome. He embarked on a military career, serving in Germany, and on his return to Rome he possibly studied law. He later became procurator in Spain. His devotion to his studies and his research technique were described by his nephew, Pliny the Younger. Upon the accession of Vespasian in AD 69, Pliny returned to Rome and assumed various official positions.

We know of another encyclopaedia, by Aulus Cornelius Celsus, of which only the medical part survived. His work also contained chapters on agriculture.

In late antiquity a man named Rutilius Taurus Aemilianus Palladius, writing somewhere in southern France, completed a treatise which is most important not for the new knowledge it presents but for the new structure by means of which knowledge is presented. Palladius was a ***vir inlustris***, a person of highest social status, had estates in Italy and Sardinia and presumably came from southern Gaul. We do not know if he was Christian or pagan; his work does not allow any conclusions about that. He wrote at a time when rural life had again become desirable among the literate. Palladius wrote the first agricultural work which is almost entirely structured by time, in calendar style. Only two introductory books precede the next twelve books, all of which deal with agricultural work on a monthly basis. Palladius not only built on existing treatises but he also was able to utilise the calendar tradition, which was a very important element of ancient Roman culture.

It is notable that Palladius was able to organise all knowledge he possessed into a calendar with the sole exception of soils, to which he devoted the second book of this work.

Not all agricultural knowledge is on paper. Following the calendar reform under Julius Caesar, stone steles were made, informing the rural population about the new calendar and about the various agricultural tasks to be performed in the respective months. Nineteen of these ***menologiae rusticae*** have survived and can be found in collections of ancient Roman inscriptions.

Whereas the calendrial way of presenting knowledge is of great practical value and can thus be found today in many popular works, the tradition of order by topic is still rigidly applied in 'scientific' books on agriculture, with ever more specialised themes, this being a characteristic of scientific studies in general. Mixed organisation, as implemented by Columella or Pliny, is – at least not as a usual form – no longer found today. But mixed organisation, taking into account both the networked character of nature and the need of the head of the household to have a guide at hand in which to look up the tasks for each month of the year, is the principle in the agricultural treatises from the thirteenth to the seventeenth centuries, and thus forms an important part of the tradition.

Table 7.1, long as it may be, is by no means complete. The major omissions, however, are not works that I have left out, such as the Middle English ***Craft of Grafting*** by Nicholas Bollard, or the entire Francophone literature, but translations of the Latin writers, among them most prominent translations of Palladius (e.g. into English in the 1350s and again in 1442, cf. Keiser 1998). If one added the translations, a more complete picture of the sheer amount of available information would emerge. The interest in agricultural matters which is proved by translations can only be hinted at – a detailed inventory is beyond the scope of this study. To see even more, one would have to add to this picture the transmission and number of manuscripts. The oldest manuscript of Columella, for instance, is from 830 and was written in the monastery of Corbie in France. From such evidence, a full picture of the amount of available information on agriculture, hence on soils, could be constructed.

Isidore of Seville, Bishop of Seville in the sixth century, wrote a relatively short encyclopaedia, which was enormously influential in the Middle Ages. Many of the Latin authors he mentions became more widely known through his work, and his treatment of agriculture, brief as it may be, discusses the ploughing sequence, speaks about manuring, and introduces fertile and 'fat' soils. I have already introduced Walafrid Strabo, who wrote a poem about 24 plants in his garden, which rests on the botanical work of Dioscórides, but was written from practical experience. He was a pupil of Hrabanus Maurus, also mentioned above. The poem of the third Carolinginan monk who dealt with agriculture, Wandalbert of Prüm, monk and Prior of the abbey of Prüm in Germany in the ninth century, deals with the names of the months and gives information about agricultural operations in these months. Of the three, Hrabanus Maurus has the most explicit information on soil treatment, albeit merely to explain the care of the soul.

Between the 850s and the 1250s, no new work on agriculture was written – or at least, none has survived, nor do we know about any indirectly. But this does not mean that no agricultural knowledge was available, or that nobody was interested in it. As the language of the literate was still Latin, the old works could be used, and several of them have been copied during this period. For example, there are tenth-century manuscripts of Palladius known, which means that the dissemination process did not stop.

With Albertus Magnus we proceed into the thirteenth century. Albertus Magnus wrote what has been called one of the most eminent contributions to botany, seven books, of which the sixth is an alphabetical list of about 400 plants, for which he also gives the therapeutic uses. The seventh book is on agriculture and horticulture. Albertus was born around 1200 in Lauingen in Suabia, and died in 1280 in Cologne. His work and study led him to several German towns, such as Hildesheim and Regensburg, and also to Padua and Paris. He was a teacher of the famous Thomas Aquinas, led a very rich and active life as a Dominican friar and bishop, and one can hardly believe the sheer amount of his works. He was beatified in 1622, and canonised as a saint in 1931. His work has been used by, among others, Vincent of Beauvais and Petrus de Crescentiis. Vincent (1184–1264) was a Dominican friar, librarian and teacher at the court of King Louis IX of France, and the author of a multi-volume encyclopaedia of which agriculture is a part. It is remarkable that Petrus de Crescentiis (Pier de' Crescenzi) was a lay person. He came from a wealthy and high-ranking family, and participated in the public life of Bologna. For the greater part of his life he worked in several northern Italian cities as a legal advisor. His work is dedicated to King Charles II of Sicily and Naples. For Vincent as an encyclopaedist, and for Pier de' Crescenzi as a compiler and writer of a learned agricultural treatise, Albertus Magnus was the state-of-the-art work on which to base their own thinking. One often forgets that the conscious use of expert knowledge is a creative act, and in this case in particular, one can see that the choice was made according to a quite 'modern' principle, taking the latest and most advanced work.

Konrad of Megenberg, the last person to introduce, is again a cleric. He, like so many others I have mentioned, travelled through Europe and was actively engaged in the politics of his day. His work encompassed the *Yconomia*, an encyclopaedic treatment of all possible branches of home and public economics. It includes a lengthy passage on soil treatment and gives enormous detail on soil qualities, and it is also mainly based on Albertus Magnus. Although Vincent of Beauvais made use of the work of Albertus, his treatment of agriculture is mainly based on Palladius, abridged and blended into the larger concept of his ***Speculum***. Palladius's second book, which is entirely on soils, is cut to almost half the original length. This must have been a conscious decision, and is worth studying in detail. Some of the authors mentioned have recently been treated thoroughly in the context of literary studies in a dissertation on Latin agricultural writers (Rex 2001). None of the medieval works exists in a modern translation, although all are edited and thus available for study.

Of all the detail on soils these works contain, I can only mention very few examples here. Albertus Magnus explains why one should plough thoroughly, because the nutrient content of the field increases in the lower layers. This is one of the first systematic mentions of the multi-layered nature of the soil (but see Col. r.r. II, 2,21, who mentions the layer below the humic upper soil as important), and also of the understanding that nutrients are distributed unevenly. For Konrad of Megenberg, the care of the fertility of fields is like the care of the medical doctor for a patient. He uses this metaphor when he states in the chapter on land improvement that some kinds of soil cannot be improved: 'Sed terra sicca salsa et amara numquam accipit medicinam' [But a dry, salty and bitter earth does not ingest medicine] (trans. VW).

SOIL TERMINOLOGY AND SOIL CONCEPTS

In Table 7.2a I have compiled important Latin vocabulary used to describe soils by the authors of Roman Antiquity. The table is a compilation of the treatises of Cato, Varro, Columella, Pliny and Palladius. Table 7.2b gives a list of soil qualities which are found in Petrus de Crescentiis, but not in the classical authors. Table 7.2 is not based on the details given in Appendix II of Kenneth Douglas White's seminal volume on Roman farming, but has been compiled independently. The reader is referred to White (1970) for separate lists of soil vocabularies for Cato, Varro, Columella and Pliny. Table 7.2c gives a list of Latin soil nouns.

Table 7.2a. Soil classification in Roman Antiquity

Grain size

(h)arenosus .... sandy (fine sand)
sabulosus .... sandy (coarse sand)
glareosus .... gravelly
saxus .... rocky

Density and Structure

crassus .... dense
densus .... compacted
gluttus .... mellow
putris .... crumbly
gravis .... heavy
levis .... light
solutus .... loose
spissus .... dense
subactus .... worked through
tenuis .... delicate

Humidity

aquosus .... watery
aridus .... arid
(h)umidus .... wet
limosus .... muddy
siccus .... dry
sucosus .... juicy

Colour

niger .... black
pullus .... dark
rubicundus .... reddish
rubidus .... dark-red (or redbrown)

Taste

amarus .... bitter
salsus .... salty

Fertility

laetus .... lush
macer .... meagre
pinguis .... fat
sterilis .... infertile
(note: *infecundus* [infertile] is used only by Pliny.)

[Temperature]

aestivus .... summerly (hot)
calidus .... hot
frigidus .... cold
sitiens .... thirsty*
(*probably not with reference to water)

Special Properties

cariosus .... carious
creteus .... clayey
cretosus .... chalky
lutosus .... loamy
rimosus .... riven

Table 7.2b. New soil-related terminology used by Petrus de Crescentiis (numbers refer to the books)

Term	Meaning	Book
adustius	bronzy	II, 17, 2
caenulentus	boggy	II, 19, 5
infecundus	infertile	II, 17, 3
lapidosus	stony	II, 20, 6
melencolicus	melancholic	II, 16, 7
nemorosus	shady	II, 20, 6
pessimus	bad	II, 16, 3
porosus	porous	II, 16, 2
pulverulentus	dusty	II, 16, 3
sativus	fertile,fat	II, 16, 2
subcaeneus	consumed from below	II, 19, 5
uliginosus	wet	II, 19, 6

Table 7.2c Latin soil terminology compiled from agricultural works in antiquity

Nouns of general content	
ager, agri, m.	field, plot
arvum, i, n.	arable land
humus, i, f.	soil (humus)
gla(e)ba, ae, f.	clod
solum, i, n.	soil
terra, ae, f.	earth
tractus, us, m.	land
Nouns to specify types of soil (or related matter, such as fertiliser)	
argilla, ae, f.	clay
carbunculus, i, m.	carbuncle
cinis, eris, m.	ash
creta, ae, f.	chalk
limus, i, m.	mud
limus levis	cow dung
marra, marga, ae, f.	marl
sabulum, i, n.	sand, gravel
stercus, oris, n.	manure
uligo, inis, f.	swamp

White has sorted the vocabulary into the categories 'mineral content', 'moisture or dryness', 'structure and texture', 'warm and cold', 'richness or leanness' and 'heavy or light'. My suggestion in Table 7.2a is to sort them in a slightly different way. I have tried to build categories compatible with qualities of soil which modern soil science discerns. Therefore, I have compiled all words that are about 'grain size', 'density and structure', 'humidity' and 'colour', which correspond to modern categories, and 'fertility', 'taste', 'temperature' and 'special properties', which do not, or not directly translate into modern categories

Columella set up a systematic soil classification system, which rests on dichotomies: soil can be dry or wet (*siccus vel umidus*), dense or loose (*spissus vel solutus*), and fat or lean (*pinguis* or *macer*). His interest is the overall soil quality for agriculture, and for such an assessment these three qualities are surely encompassing and accurate.

He did not use the systematic treatment of different minerals, as can be found in Varro. This is remarkable, and clearly shows that decisions about classification systems are fundamental and should not be taken for granted. Varro uses a theory about soils which holds that the different types are generated by mixing of eleven kinds of (mineral) substances. The passage in Varro (rust., I, 9. 2–3) reads:

> In illa enim cum sint dissimili vi ac potestate partes permultae, in quis lapis, marmor, rudus, arena, sabulo, argilla, rubrica, pulvis, creta, glarea, carbunculus, id est

> quae sole perferve ita fit, ut radices satorum comburat, ab iis quae proprio nomine dicitur terra, cum est admixta ex iis generibus aliqua re, cum dicitur aut cretosa sic ab illis generum discriminibus mixta.
>
> [For there are many substances in the soil, varying in consistency and strength, such as rock, marble, rubble, sand, loam, clay, red ochre, dust, chalk, ash, carbuncle (that is, when the ground becomes so hot from the sun that it chars the roots of plants); and soil if it is mixed with any part of the said substances, is e.g. called chalky, as well as according to other differences as mixed] (translation compiled from Loeb Classical Library and the German translation by Flach).

Note that ***sabulum*** is translated as loam in Loeb, whereas Flach has potter's clay. Translation is interpretation, as I have said before. Besides, the Roman authors were not trying to be consistent with all their predecessors, and each work has its distinct use of words. As another example, we cannot judge for sure, if ***cretosus*** and ***creteus*** are synonyms, and overall, the difference between clayey and chalky seems to be blurred.

Temperature is not to be mistaken for degrees Fahrenheit or Centigrade. A hot soil for Columella should not be given too much manure, as it would be burned, whereas a cold soil needs a lot of it. The Roman scholars used words derived from Greek philosophy, where hot and cold were properties not simply denoting temperature, but more the temperament, as of a human being. A cold soil then has to be understood more like one of cold temper, one with too much 'phlegm', a phlegmatic soil. Of course manuring means adding chemical energy to a soil, so we can understand Columella today using the energy concept. Whether he himself had a concept of manure as added energy, one cannot say.

Of the special properties, ***cariosus*** will be discussed in more detail. The idea of the notion is clear for Columella, a practitioner, and he is able to explain it to us, unlike his predecessors, above all Cato, who only warns you to touch it with no explanation at all. ***Terra cariosa*** is not a soil, but a state of soil which can arise from inappropriate treatment. Soil that is worked when the upper layer is wet from rain and the lower layers have not been humidified becomes 'cariosa', it loses its fertility for years through ploughing under adverse conditions.

Word and concept come from the ***rustici*** – the rural people – as Columella informs his readers, and are probably understood by those scholars that have practical experience. The various authors are more or less successful in explaining it – a typical case of tacit knowledge. It is not understood by Pliny, the encyclopaedist, as can clearly be seen in his text. He misunderstands ***terra cariosa*** as a type of soil, whereas for Columella it is a result of poor treatment and a (reversible) condition (Winiwarter 2000).

One must not forget that at least 900 years (from Palladius's time onwards) lie between Petrus de Crescentiis and his ancient predecessors. While some of the words given in Table 7.2b have to be considered as mirroring developments of the Latin language (such as ***uliginosus***), and some are mere synonyms, some, such as ***melencolicus*** or ***subcaeneus*** give proof of new theoretical concepts. It would be interesting to trace the first appearance of, for example, ***ager melencolicus***, a description of soil quality which can only be understood in the context of humour theory, meaning a cold and dry complexion. Petrus de Crescentiis wrote for a literate audience and his book, in my opinion, is an attempt at developing agriculture into a truly 'scientific' undertaking. He uses the (impractical) alphabetical classification system, which he takes from the best and most recent scientific work he can get hold of, and works hard to press the different qualities of soils into the fourfold scheme of the humour-theory. So wet and dry, and cold and hot have to be seen as meta-categories, under which the more detailed, practical denominations come to be subsumed. A complete comparative vocabulary of soil terminology by author as well as a thorough study of the influence of humour theory on soil concepts could yield a more detailed picture of the evolution of conceptual knowledge. Such work is planned but will have to be done in years to come.

WHAT IS A SUITABLE SOIL?

Xenophon was the first to make clear that there are suitable soils, suitable for a specific plant or, more general, under specific circumstances of cultivation (Xen. 16, 2). In the centuries after him, authors tried to define what makes a soil good and/or suitable. Two kinds of differentiation are developed: one is a more accurate determination of soil qualities, in order to be able to give descriptions which are inter-subjective; the second takes into account that plants will grow optimally at a particular site, but might still be grown elsewhere, developing a plant-relative quality measure. I have dealt elsewhere in detail with the soil test methods described to enable readers to assess this quality (Winiwarter 2000 and 1999). A list with short descriptions of them can be found in Table 7.3. This chapter will give a few examples to trace the development of the quality description.

Columella's classification system has already been introduced above. His assessment of general quality (quot sint genera terreni – 'land types') rests on these categories, and thus he is able to define the kind of land least suited for agriculture, as well as the best one:

> Nullum deterius habetur genus, quam quod est siccum pariter et densum et macrum; quia cum difficulter tractetur, tum ne tractatum quidem gratiam refert; nec relictum pratis vel pascuis abunde sufficit.
>
> [No kind is considered worse than that which is at the same time dry, stiff (compacted, VW), and lean; for not only is it worked with difficulty, but even when worked it makes no recompense, and when left idle it is not altogether adequate for meadows or for grazing land] (Col. r.r. 2, 2, 7, trans. Loeb Classical Library, Vol. I, pp. 112–13).

Table 7.3. Soil test methods in Roman Antiquity

Test for Salinity and Bitter Taste

Important for wine-growing. One takes clear, fresh water and mixes it thoroughly with a soil sample. Once the mixture is filtered, through an unglazed earthenware or a sieve as used in wine-making [Virgil] – details vary depending on the author – one can cautiously taste the water, which will have taken on the taste of the soil. (Pall. Op.Agr. I, V, 3; Verg. Georg. II 238–47)

Test for Soil Fertility (Structural Stability – Density)

[Not applicable to black soils, according to Columella.] After digging a hole into the ground, one tries to refill the hole. If the earth has increased in volume, leaving a small hill, it is fertile. If the volume has not changed and the hole can be filled evenly, the earth is of middle quality. The soil quality is meager if the soil volume has decreased due to digging and the earth leaves a trough (Col. r.r. II, 2, 18–19; Verg. Georg. II 226–37).

Test for Cohesion

A similar test is still practised today, although in a somewhat more elaborate form. One takes a small clod of soil, adds a few drops of water to moisturise it and then moulds it in the palm of the hand. If it becomes sticky and clings to the skin, the soil is of good quality (explained in terms of 'natural humidity' and 'richness'). (Verg. Georg. II 250; Col II, 2, 2, 18; also Pall. Op.Agr. I, V, 3). Pliny remarks that as potters clay, which is infertile, also shows this stickiness, the test is inconclusive (Plin. n.h. XVII, 27)

Test for Fertility

Nowadays it is known that the typical smell of earth is due to the activity of fungi, in particular the Actinomyces. Fungi play a crucial role in soil biology, and a soil that smells is obviously a healthy soil even by modern standards. Pliny (Plin. n.h. XVII, 39) already mentions soil smell as an indicator of quality, which according to him is strongest when rain wets a surface that has dried out, and also becomes stronger when the soil is worked.

Indicator Plant Species

Pristine land-cover also serves as a good indicator, and several indicator species are named that allow one to distinguish between sweet and saline grounds, and especially between grounds fit for grain or not. Columella gives a long list of plants (Col II, 2, 20); Pliny remarks generally on the possibility of plant cover as a soil quality indicator (Plin n.h. XVIII, 34) and refers to a list of plants that Cato had already given. Columella, who is generally the most cautious author, warns against restricting oneself to the use of plant cover as the sole indicator.

Ideoque maximos quaestus ager praebet idem pinguis ac putris, quia cum plurimum reddat, minimum poscit; et quod postulat, exiguo labore atque impensa conficitur. Praestantissimum igitur tale solum iure dicatur.

[And on this account, a field which is both rich and mellow (crumbly, VW) yields the greatest returns, because in producing most it demands least, and what it does require is supplied with trifling labour and expense. Such a soil may therefore with justice be called the very best] (trans. Loeb Calssical Library, Vol. I, p. 111).

Similar assessments can be found in all Roman authors. In Walafrid Strabo's ninth-century AD work on gardening, general descriptions of good and bad land open the poem:

Ruris enim quaecumquae datur possessio, seu sit
Putris, harenoso qua torpet glarea tractu,
Seu pingui molita graves uligine foetus
Collibus erectis alte sita, sive jacenti
Planitie facilis clivo, seu vallibus horrens,
Non negat ingenuos holerum progignere fructus,
si modo non tua cura gravi compressa veterno,

[Whatever kind of land you might own, be it
that on crumbly sand infertile gravel it might rest
be it that from fat humidity it might yield heavy fruit
high up on towering hills, on the gently sloping plain or furrowed with valleys
nowhere does it refuse to produce its peculiar plants
if only your care does not exhaust itself in laming laziness.]
(translation with changes after the German adaptation by Stoffler, VW)

While Columella uses ***putris*** as a notion for a good piece of land, Strabo has it to describe the looseness of sand which will fall apart in crumbles and not stick together. This example shows that a general Latin soil vocabulary is not possible. One will always have to account for the specific way each author uses the words.

The plant-relative description system will be demonstrated using the Chestnut tree, for which we also have Gargilius Martialis as a reference.

Ea pullam terram et resolutam desiderat; sabulonem humidum vel refractum tofum non respuit; opaco et septentrionali clivo laetatur; spissum solum et rubricosum reformidat.

[It likes a black and loose soil: does not refuse a damp, gravelly soil or crumbling tufa; delights in a shady slope with a northern exposure; and fears a heavy soil that is full of red ochre.] (Col. r.r. 4, 33, 2, trans. Loeb Classical Library, Vol. I, p. 457). Note: The Loeb edition translates 'pullus' as black here, I would prefer 'dark'.

quaerit solum facile nec tamen harenosum, maximeque sabulum umidum aut carbunculum vel tofi etiam farinam, quamlibet opaco septentrionalique et praefrigido situ, vel etiam declivi. recusat eadem glaream, rubricam, cretam omnemque terrae fecunditatem. (Plin. n.h. XVII, xxxiv, 147)

Solum castanea desiderat molle et subactum; non tamen arenosum patitur et sabulum. Si umor adfuerit, pulla terra vel maxime prodest: sed et carbunculus, itemque tofus optime servit diligenter infractus. Spissum rubricosum locum non vult: vix provenit in argilla: in glarea cretaque nec nascitur. (Gargilius Martialis, *De Hortis*, IV. 5, 314–318)

Amant solum molle et solutum, non tamen harenosum. In sabulone proueniunt, sed humecto. Nigra terra illis apta est et carbunculus et tofus diligenter infractus. In spisso agro et rubrica uix prouenit: in argilla et glarea non potest nasci. (Pall. Op.agr., XII, 19)

Amant solum molle et solutum, non tamen arenosum; et in sabulone proveniunt, sed humecto. Terra nigra illis est apta et carbunculus et tofus diligenter infractus. In agro spisso et rubrica vix proveniunt: in argilla et glarea non possunt nasci. (Petrus de Crescentiis, *Ruralia Commoda*, V, 6, 1)

If one compares the five statements, which are in general quite similar, one can still see differences, as well as different degrees of detail. What becomes clear is that the soil a chestnut tree likes is definitely ***not*** sandy or gravelly, and that a good, fertile, soft soil is best suited for chestnuts. The second thing the authors agree on is that a dense soil (***spissus***) is not suitable for chestnuts. All authors but Columella agree that ***argilla*** (clay) is hardly (***vix***) suitable, if we consider that Pliny uses the word '*creta*' which also means clay in this context. The only two statements which are completely identical (only changes in word order) are those of Palladius and Petrus de Crescentiis. It is sure that the latter used Palladius as his source and found no reason to change any of the substance. Between them and Gargilius Martialis, only a few words are different: while the later texts have ***solutus***, Gargilius gives ***subactus*** and instead of ***amare*** uses ***desiderare*** for 'to like'. Palladius used Gargilius, but left out the dependency on humidity which his source gives: the part 'Si umor adfuerit' is left out in Palladius's chapter, and hence, in Petrus de Crescentiis.

The stability with which knowledge about the suitable soils for chestnuts was preserved over about 1200 years, is astounding. The manuscripts (as far as the editions I use give information on them) are not corrupt for the cited parts of texts, so one can conclude that the similarity was not produced by the editors but is original.

So one could infer that the suitable soil for a chestnut tree was well established in Antiquity and should not be matter of much dispute thereafter. But the later treatises do not support this easy conjecture. I have at my disposal only the 1632 printing of Johannes Colers ***Oeconomia*** from 1591. He has a chapter on the chestnut, in which he also gives information about his source:

> Sie bekommen gerne auff steinigten festem Lande / und an kalter lufft / doch kann ihnen warme lufft auch nicht viel schaden. Auf festem und lehmigen Lande bekommen sie selten/ daß sie Früchte trügen. *Barthol. anglic.lib.* 17 cap.88.
>
> [They readily grow on stony, dense land / and in cold air/ but warm air cannot do them much harm. On dense and loamy ground it seldom happens / that they bear fruit] (transl. VW).

Bartholomaeus Anglicus was a Franciscan friar in the first half of the thirteenth century, who wrote an encyclopaedia (*De proprietatibus rerum*), which was used until the sixteenth century. It was also translated into several vernacular languages. Exchange between agricultural textbooks and encyclopaedias worked both ways, from the specialised to the general, and vice versa. Pliny, Albertus Magnus, Konrad von Megenberg and others have to be viewed as part of the transmission network of agricultural knowledge.

Coler obviously speaks about two kinds of soil, which he describes with the same word *fest* (dense). Coler gives no further detail, so we cannot assess what his distinction between dense and stony on one hand and dense and loamy on the other hand really means, but altogether his opinion is contrary to all the previously cited works, which were in agreement about loose and soft soil as being well suited. But there is agreement as to loamy conditions not being suitable.

Loudon's *Encyclopaedia* of 1883 has more detail. A table shows soils suitable for trees. Chestnut trees are considered suitable for topsoils consisting of 'flinty and gravelly loam', with a subsoil composed of 'chalk at four feet with deep gravelly loam and a few flints', they are also recommended for 'gravelly and chalky loams' with the identical subsoil, and for 'sandy gravel' and 'sandy loam', with 'gravelly and sandy loam' as subsoil.

Consulting the internet to find out what current wisdom has to offer, one finds: 'The most suitable soil for Chestnut trees is a sandy loam, with a dry bottom, but they will grow in any soil, provided the subsoil be dry' (http://www.botanical.com/botanical/mgmh/c/cheswe59.html). Also: 'Geeignete Böden verfügen über lockeres, tiefes, feuchtes, karbonat- und salzarmes Erdreich' [suitable lands have loose, deep, humid soil devoid of carbonate and salt] (http://www.frujorge.com/sorayafruits/aleman/castanas.htm trans. VW). And finally, 'Sie verträgt Schatten, meidet Kalk, braucht nährstoffreiche, tiefgründige Böden, mildes Klima' [(The chestnut) tolerates shade, avoids lime, and needs nutrient-rich and deep soils] (http://www.biozac.de/biozac/capvil/Cvcastan.htm trans. VW). The very last quote bears more similarity to the classical authors than Coler or Loudon or the other two internet resources. What can one conclude from this example?

The history of soil knowledge is indeed full of contradictions, (mis-)interpretations and at times seems circular. The difficulty in describing a phenomenon like the soil in inter-subjective categories is still discernible today. But this is all the more reason to respect the cognitive and linguistic achievements of the first writers on soils.

CONCLUDING REMARKS

Reflected in the title 'Prolegomena' is the notion that this chapter will be a prologue to future studies on the history of soil knowledge. There is not much point in writing a conclusion to the beginning of something, but I can sum up my initial arguments. I have attempted to show that 'knowledge', and in particular the authority granted to knowledge, are concepts which are best understood as interactive. I have attempted to make clear the amount in numbers and the wide diffusion of the body of texts, which are part of the interaction of producing knowledge, and have tried to contextualise the texts within the overall mentality of agricultural societies. Historiography has not conceptualised texts about soils as one body to be studied in its entirety so far, so by re-writing I construct a new knowledge system.

I have tried to make clear that the transformation from the perception of a handful of soil into a text describing it is complicated and by no means easy. Terminology cannot be taken for granted but must be understood as the result of a conscious attempt of transformation of sensory perception. Other than handfuls of soil, texts are a primary agent of interaction in the textual network. I have tried to make the discontinuities and vagaries of textual transmission visible using the examples of Mago and of the *Geoponika*. I have also tried to show the influence of the requirements of the narrative on the conceptualisation of soil knowledge, and I have tried, finally, to show processes of differentiation and synthesis of texts at work.

I hope it has become clear that the sheer amount of available information prevents any general picture within the limited space of this article. An attempt to describe the entirety of textual tradition would have to proceed in several (interrelated) stages. One would need to show the interrelation between the classical texts and other texts now lost that the authors refer to. Then one would have to make visible the dissemination of classical knowledge by listing all manuscripts and their life-histories, their probable place(s) of origin and of use, their travels and the reworking of their texts, paying attention to what was excluded or included in the process. It would be necessary to trace all extant translations of classical texts, and relate them to manuscript traditions, and to investigate the use of classical texts in several genres of literature such as encyclopaedias and specialised works on subjects such as agriculture, natural sciences or economics. The matter is complicated by the fact that use of existing information can be direct or indirect, derived from an 'original' or only from a later user of it. The later the authors write, the more material they (theoretically) have at their disposal, so the question arises why authors grant authority to a writer they use, or whether they would have had the choice of using others but did not. Sometimes some, if sketchy, evidence on that can be unearthed. One would be looking for breaks and interruptions, for misunderstanding and for adaptation, as well as taking into account the question of availability and abilities of the authors. The examples I gave are certainly a part of this larger picture – whether and, if so, how their interpretation would change within the sketched larger framework, will have to be left for future study.

ACKNOWLEDGEMENTS

Thanks (as so often) go to Herwig Weigl, who not only read this text, but also helped with translations, and thus made it a lot better; to Karl Brunner, who helped locating the Church Father references; and to Martin Schmid, who made valuable suggestions on how to structure the piece. Funding by the Austrian Academy of Sciences, APART (Austrian Program on Advanced Research and Technology) Grant 10916, 'Precious Dirt Underfoot. A History of Soil Knowledge' enabled the preparation of this chapter.

SOURCES

Albertus Magnus. *De vegetabilibus libri VII, historiae naturalis pars, XVIII*, ed. Ernst Meyer and Karl Jessen. Berlin: G. Reimer, 1867.

Caesarius of Arles. *Caesarii Arelatensis opera, I: Sancti Caesarii Arelatensis sermones*, ed. G. Morin, 2nd edn (Corpus Christianorum, Series Latina, Vol. 103). Brepols: Turnhout, 1953.

Cato, agr. = M. Porcius Cato, *De agri cultura*, Latin/English, ed. William Davis Hooper and Harrison Boyd Ash, Loeb Classical Library. Cambridge MA: Harvard University Press 1934.

Col. r.r. = L. Iunius Moderatus Columella, *De re rustica*, Latin/German, ed. Will Richter et al. Munich: Tusculum, Vol. 1 1981, Vol. 2 1982, Vol. 3 1983.

Gargilius Martialis (Q. Gargilii Martialis). *De hortis*, ed. Innocenzo Mazzini, Opuscula philologica 1. Bologna: 1988.

Heriger of Lobbes. *Gesta pontificum Tungrensium et Leodiensium*, Monumenta Germaniae Historia, Scriptores 7, ed. G. Pertz (Hannover: 1846), pp. 162–238.

Hieronymus (Jerome). *Commentariorum in Esaiam libri XII-XVIII.* In *Esaiam parvula adbreviatio*, ed. M. Adriaen, G. Morin, (Corpus Christianorum Series Latina Vol. 73A). Brepols: Turnhoult, 1963.

Konrad von Megenberg. *Werke, Ökonomik*, ed. Von Sabine Krüger. In MGH, *Staatsschriften des späteren Mittelalters*, Vol. 3. Stuttgart: 1973.

Pall. Op.agr. = Rutilius Taurus Aemilianus Palladius, *Opus Agriculturae*, ed. Robert H. Rodgers. Leipzig: 1975.

Pall. Op.agr. = Rutilius Taurus Aemilianus Palladius, *Opus Agriculturae* I and II Latin/French, ed. Martin René. Paris: 1976.

Petrus de Crescentiis. *Ruralia Commoda: das Wissen des vollkommenen Landwirts um 1300*, ed. Will Richter, foreword by Reinhilt Richter-Bergmeier, Editiones Heidelbergenses 25–27, 30. Heidelberg: C. Winter 1995–2002.

Plin. n.h. XVII = C. Plinius Secundus [the elder], *Naturalis Historiae Liber XVII*, Latin/German, ed. and trans. Roderich König and Joachim Hopp. Munich: Artemis & Winkler/Tusculum, 1994.

Plin. n.h. XVIII = C. Plinius Secundus [the elder], *Naturalis Historiae Liber XVIII*, Latin/German, ed. and trans. Roderich König, Joachim Hopp and Wolfgang Glöckner. Munich: Artemis & Winkler/Tusculum, 1995.

Rufinus. *Rufin d'Aquilée. Les Bénédictions des Patriarches*, Sources chrétiennes, 140, ed. Manlio Simonetti. Paris, 1968.

Varro, rust. = M. Terentius Varro, *Res Rusticae*, Latin/German, ed. and trans. Dieter Flach, Gespräche über die Landwirtschaft. Darmstadt: Vol. 1 1996, Vol. 2 1997.

Verg. Georg. = P. Vergilius Maro (Virgil), *Georgica*, Latin/German, ed. J. and M. Götte et al. Munich: Tusculum, 5th edn. 1987.

Vulgata. http://www.vatican.va/archive/bible/nova_vulgata/documents/nova-vulgata_vetus-testamentum_lt.html (Jerome's translation of the Greek and Hebrew Scriptures into the common language, Latin, was completed in 405. It was recognised as authoritative during the Council of Trent (1546) and became the official Bible of the Roman Catholic Church. The widespread use of the Vulgate is also recognisable in its influence in early modern Bible translations, such as the Authorised, or King James, Version. The Vulgate continues to be of scholarly use today in the study of the textual transmission of the Bible and in the historical study of Christian theology. The Vulgate used is the fourth edition of the *Biblia Sacra iuxta vulgatam versionem*. It was originally published in 1969, and this fourth edition was released in 1994.)

Xenophon. Oikonomikós. Xenophon, Ökonomische Schriften. ed. and transl. Gert Audring, Schriften und Quellen der Alten Welt. Berlin: Akademie-Verlag, 1992.

REFERENCES

Barbiero, Daniel. n.d. 'Tacit Knowledge'. In Chris Eliasmith (ed.), *Dictionary of the Philosophy of Mind*, http://www.artsci.wustl.edu/~philos/MindDict/tacitknowledge.html

Butzer, Karl W. 1993. 'The Classical Tradition of Agronomic Science: Perspectives on Carolingian Agriculture and Agronomy'. In P.L. Butzer and D. Lohrmann, *Science in Western and Eastern Civilization in Carolingian Times* (Basel: Birkhäuser), pp. 540–605.

Butzer, Karl W. 1994. 'The Islamic Traditions of Agroecology: Crosscultural Experience, Ideas and Innovations'. *Ecumene*, 1: 7–49.

Clarke, Roger. 2001. 'Knowledge'. http://www.anu.edu.au/people/Roger.Clarke/SOS/Know.html

Davidson, Arnold I. 1986. 'Archaeology, Genealogy, Ethics'. In David Couzens Hoy (ed.), *Foucault: A Critical Reader* (New York: Basil Blackwell), pp. 221–33.

Dreyfus, Hubert L. and Paul Rabinow. 1982. *Michel Foucault: Beyond Structuralism and Hermeneutics*. Chicago: University of Chicago Press.

Foucault, Michel. 1972. *The Archaeology of Knowledge*, trans. A.M. Sheridan Smith. London: Tavistock.

Foucault, Michel. 1980. *The History of Sexuality. Vol. 1.* New York: Vintage.

Foucault, Michel. 1984. 'Nietzsche, Genealogy, History.' In Paul Rabinow (ed.), *The Foucault Reader* (New York: Pantheon), pp. 76–100.

Gingrich, Andre and Elke Mader (eds). 2002. *Metamorphosen der Natur*, Sozialanthropologische Untersuchungen zum Verhältnis von Weltbild und natürlicher Umwelt, series editors Marianne Nürnberger, Ulrike Davis-Sulikowski and Wolfgang Kraus. Vienna/Cologne/Weimar: Böhlau.

Glacken, Clarence J. 1967. *Traces on the Rhodian Shore. Nature and Culture in Western Thought from Ancient Times to the End of the Eighteenth Century.* Berkeley/Los Angeles/London: University of California Press.

Güntz, Max. 1897. *Handbuch der landwirtschaftlichen Literatur. Mit biographischen Notizen und 38 Bildnissen von Autoren.* 2 vols. Leipzig 1897–1902, reprinted Vaduz: Topos 1977.

Hussain, Athar. 1971. 'A Brief Resumé of the Archaeology of Knowledge'. *Theoretical Practice*, 3/4: 104–7.

International Council of Scientific Unions. 2002. *Science, Traditional Knowledge and Sustainable Development*, ICSU Series on Science for Sustainable Development No. 4. Paris: ICSU.

Jacob, François. 1972. *Die Logik des Lebenden. Von der Urzeugung zum genetischen Code.* Frankfurt: S. Fischer. (English, *The Logic of Life: A History of Heredity*, trans. Betty E. Spillmann. Princeton University Press, 1993.)

Kallfelz, Hatto (trans.). 1973. 'Vita Bennonis II. Episcopi Osnaburgensis auctore Norberto abbate'. In *Lebensbeschreibungen einiger Bischöfe des 10.–12. Jahrhunderts* (= Ausgewählte Quellen zur deutschen Geschichte des Mittelalters, Freiherr vom Stein-Gedächtnisausgabe Vol. XXII [Darmstadt: Wissenschaftliche Buchgesellschaft], pp. 363–441).

Keiser, George R. 1998. *Works of Science and Information*, A Manual of the Writings in Middle English 1050–1500, series editor Albert E, Hartung, XXV. New Haven: The Connecticut Academy of Arts and Sciences.

Klingshirn, William E. 1994. *Caesarius of Arles. The Making of a Christian Community in Late Antique Gaul.* Cambridge: Cambridge University Press.

Le Goff Jacques, Roger Chartier and Jacques Revel (eds). 1990. *Die Rückeroberung des historischen Denkens. Grundlagen der neuen Geschichtswissenschaft.* Frankfurt: Fischer.

Lefort, Jacques. 2002. 'The Rural Economy, Seventh-Twelfth Centuries'. In Angeliki Laiou (ed.), *The Economic History of Byzantium: From the Seventh through the Fifteenth Century*, Dumbarton Oaks Studies 39. Washington: Dumbarton Oaks Research Library.

Loudon, J[ohn]. C[laudius]. 1883[1839]. *An Encyclopaedia of Agriculture: Comprising the Theory and Practice of the Valuation, Transfer, Laying Out, Improvement, and Management of Landed Property, and of the Cultivation and Economy of the Animal and Vegetable Productions of Agriculture.* 8th edn., 1883. London: Longmans, Green, and Co.

Mahaffy, J.P. 1889/90. 'The Work of Mago on Agriculture'. *Hermathena*, 7: 29–35.

Mattey, George J. 2002. 'Lecture Notes, Lehrer's *Theory of Knowledge*, 2nd edn., Chapter 1, The Analysis of Knowledge' http://www-philosophy.ucdavis.edu/phi102/tkch1.htm

McDonald, Donald. 1908. *Agricultural Writers from Sir Walter of Henley to Arthur Young, 1200-1800.* New York: Burt Franklin, reprinted 1968.

McKitterick, Rosamond. 1988. *The Carolingians and the Written Word.* Cambridge/New York: Cambridge University Press.

Meyer, Ulrich. 1998. *Soziales Handeln im Zeichen des 'Hauses'. Zur Ökonomik in der Spätantike und im früheren Mittelalter*, Veröffentlichungen des Max-Planck-Instituts für Geschichte, 140. Göttingen: Vandenhoeck und Ruprecht.

Rex, Hannelore. 2001. Die lateinische Agrarliteratur von den Anfängen bis zur frühen Neuzeit. Phil. Diss. Bergische Universität - Gesamthochschule Wuppertal.

Rheinberger, Hans-Jörg. 2001. *Experimentalsysteme und epistemische Dinge Eine Geschichte der Proteinsynthese im Reagenzglas.* Wallstein. (English, *Toward a History of Epistemic Things: Synthesizing Proteins in the Test Tube.* Stanford University Press, 1997).

Rice, John S. 1992. 'Discursive Formation, Life Stories, and the Emergence of Co-Dependency: 'Power/Knowledge' and the Search for Identity'. *Sociological Quarterly*, **33**(3): 337–64.

Rodgers, Robert. 2002. 'Kepopoia: Garden Making and Garden Culture in the *Geoponica*'. In Anthony Littlewood, Henry Maguire and Joachim Wolschke-Bulmahn (eds), *Byzantine Garden Culture.* Washington: Dumbarton Oaks Research Library.

Verhulst, Adriaan E. 1989. 'Agrarian Revolutions: Myth or Reality?' *Sartoniana* 2: 71–5.

Webster's New Twentieth Century Dictionary. 1983. Simon and Schuster: New York.

White, Kenneth Douglas. 1970. *Roman Farming*, Aspects of Greek and Roman Life. London: Thames and Hudson.

Winiwarter, Verena. 1992. 'Landwirtschaftliche Kalender im frühen Mittelalter. Überlegungen zum Fortleben antiken Bildungsgutes und zu seinen Funktionszusammenhängen'. *Medium Aevum Quotidianum*, 27: 33–55.

Winiwarter, Verena. 1999. 'Böden in Agrargesellschaften: Wahrnehmung, Behandlung und Theorie von Cato bis Palladius'. In Rolf Peter Sieferle and Helga Breuninger (eds), *Natur-Bilder. Wahrnehmungen von Natur und Umwelt in der Geschichte* (Frankfurt: Campus), pp. 181–221.

Winiwarter, Verena. 2000. 'Soils in Ancient Roman Agriculture: Analytical Approaches to Invisible Properties'. In H. Novotny and M. Weiss (eds.), *Shifting Boundaries of the Real: Making the Invisible Visible* (Zurich: vdf Hochschulverlag), pp. 137–56.

Winiwarter, Verena. 2001. 'Nachhaltigkeit. Zur Paradoxie von Theorie und Praxis'. In Görg, Bernhard, *Innovation und Verantwortung. Werte, Wege und Aufgaben für das 21. Jahrhundert* (Vienna: Holzhausen), pp. 205–19.

Worster, Donald. 1988. '"Dust Bowl". Dürre und Winderosion im amerikanischen Südwesten'. In Rolf Peter Sieferle, *Fortschritte der Naturzerstörung* (Frankfurt: Suhrkamp), pp. 118–57.

Ziman, John. 1976. *The Force of Knowledge. The Scientific Dimension of Society.* Cambridge: Cambridge University Press.

8

Nutrient Flows in Pre-Modern Agriculture in Europe

Robert S. Shiel

INTRODUCTION

Europe is fortunate in that, with the exception of the far north and highland regions, there is sufficient warmth and moisture to grow wheat, barley, oats or rye with good reliability provided the soil has an adequate supply of nutrients and is not excessively wet. Along the Atlantic fringe drying and ripening of the crop may be made difficult by the wet climate. In the southern tip of Spain, Greece and Italy drought is common but there is usually sufficient winter moisture for barley or millet to be grown. The range of fruit, vegetable and cereal crops and animals that can be produced in any location varies but, with the above exceptions, there is always a range of potential food products available. Nearly half the soil is cambisols, which are inherently fertile, but one quarter is podzols and luvisols, which contain fewer nutrients and may be excessively acidic. Under the widespread natural deciduous woodland, these soils changed fertility slowly and accumulated a large reserve of nitrogen; soils in those areas with coniferous woodland contained fewer nutrients. After clearance of the woodland, nutrients were lost rapidly by increased leaching, soil erosion and offtake in crops; crop yield under a monoculture decreased to a low level within one human generation and the soils would not then yield adequate crops to repay the effort of cultivation. The soil nutrient status recovers more slowly, if the natural vegetation is allowed to re-establish, than it decreases during exploitation and other land will have to be used during this recovery phase. Such shifting cultivation depends on having a large excess of cultivable land on which recovery proceeds, and if this is not available then techniques to maintain fertility are required. These include the application of wood ash, the use of animal manures and the recycling of human waste. It may also be possible to 'import' material such as marl to raise the pH rapidly. Several of these methods maintain fertility on the manured area but result in slow fertility loss on the large areas of grassland and woodland exploited for manure or ash; fallowing increases crop yield but leads to a more rapid decrease in soil organic matter and nitrogen. The intentional use of legumes and deep rooting plants, respectively, introduce external nitrogen and make other nutrients more available within the root zone of the plant. The sophisticated rotations of the eighteenth century maintained fertility much better than had the earlier rotations but still depended on large amounts of animal manure or ash from outside the rotated area if the yields were to be commercially attractive and sustainable. The modern concept of 'organic farming' eschews inputs of water soluble and most exotic materials and is able to maintain yields at a substantial proportion of those obtained using conventional farming methods. It does however make use of external natural fertilising materials such as rock phosphate and limestone, which were not widely available or used in the distant past, but remain endemically short of plant nutrients. The tradition of recycling urban waste 'night soil', which could replace much of the nutrient loss, is less acceptable today on aesthetic grounds and much of that collected is now contaminated with toxic industrial wastes.

In this paper I will examine the original pattern and properties of soils in Europe before the initiation of widespread land management as we recognise it today and will then look at the natural and human-induced alterations to those soils. This will involve looking at the sustainability of earlier land use systems and will be based, where possible, on sound scientific evidence from field research and experiments. The soils of Europe have for long been modified by the human population. In some other parts of the world soil modification, on other than a hunter-gatherer scale, has been restricted to the last two hundred years. In these places there are experiments which, arguably, allow us to reflect on the changes that occur when virgin sites are first used. In Europe, there is also an established literature on agricultural management of land dating back to Hesiod (West 1988) in the eighth century BC (See Winiwarter in this volume); the role of these management techniques in mitigating deterioration and even in ameliorating the soil will be examined. Erosion and its corollary, deposition, have influenced the nutrient supply throughout the period but, as they are fundamentally dependent on the vegetation, which has been heavily altered by human action, their effect will be considered together with the management of the land rather than as a separate topic.

A BRIEF BACKGROUND ON THE ENVIRONMENT FOR FARMING IN EUROPE

Europe contains no desert areas and, apart from the highest mountains and the area north of about 60° latitude, the growing season is adequate for at least some cereal and vegetable crops. Many of the cooler and moister areas that are less suitable for cereal production provide seasonal animal grazing. The cereals, ignoring rice and maize, vary from wheat, requiring the longest growing season, through barley, which is the most drought and salt tolerant, to oats and rye, which tolerate moister, more acidic environments with short growing seasons (Leonard and Martin 1963). On the wetter western fringe, ripening and drying cereals can be difficult and here dairy products and the more recent introduction, potato, are preferred. The upper and northern boundaries between the areas that are more or less suitable for particular crops have varied with temperature – in the last millennium with the variation ranging from the Little Optimum through the Little Ice Age (Parry 1978) into the modern amelioration. The pattern of rainfall varies from a summer maximum between about 45° to 56° N to a very distinct late autumn maximum in the south (Table 8.1). The furthest north locations also have an autumn maximum but not the summer minimum of the southern sites. All have on average more than the total needed for wheat growth and as, in the driest areas, the rain falls predominantly in the cool months when wheat is grown, there will be relatively few sites at which wheat fails completely on a regular basis. The effect of the moisture balance on crop growth has, like the temperature regime, varied considerably over the last millennium and this variation is most likely, as with the temperature, to affect growth near the extremes.

The parent materials from which soils have formed are very variable. The continent has been affected by a number of orogenic episodes, the most recent being the Alpine, all of which have led to intense denudation of the uplands and deposition of alluvium in valleys on mountain fringes. In the Mesozoic however much of the continent was flooded and, being on the edge of the Tethys Ocean, is underlain by younger marine rocks. The main areas of ancient metamorphosed rocks lie along the western edge. These rocks weather and release minerals for plant growth slowly, whereas the younger rocks provide a more on-going supply. Although, in earlier time, the continent lay at low latitudes and suffered intense chemical weathering, its more recent northern location has led to much of the weathered rock being eroded or capped with recent alluvial or glacial deposits rich in plant nutrients. The glaciation affected the northwest and mountains directly (Flint 1971), where it left areas of glacial till, and other regions were indirectly affected by the deposition of large areas of loess and outwash deposits. The least fertile of these deposits are the fluvio-glacial sands, which contain few weatherable minerals. In the areas that were not glaciated there has been severe episodic erosion of soil, especially in the drier areas where vegetation cover is poor. Although this erosion did not always move soil far, it frequently reduced the total area of soil without producing a compensatory increase in soil quality in the areas on which eroded material was deposited (van Andel 1987). A very active period of erosion occurred at the end of the glacial when plant communities were destabilised and there was an increase in intense rainfall events.

Table 8.1. Monthly and annual rainfall averages (mm) for European cities between 37° and 60° latitude. The three wettest months are marked

Latitude	37°	41°	42°	43°	45°	48°	48°	52°	56°	59°	60°
	Athens	Madrid	Rome	Toulon	Florence	Vienna	Paris	Berlin	Moscow	Stockholm	Helsinki
January	48	45.1	80	48.7	64.1	38	45.8	43.2	34.4	31.4	46
February	41	43.2	70.9	39.5	61.5	37.2	38.9	37.6	29	25.3	36.6
March	41.2	36.8	68.6	44.5	69.4	44.2	41	37.8	32.7	26.3	34.9
April	23.4	45.4	66.8	46.2	70.5	51.3	44.4	41.1	38.2	28.6	37.4
May	17.9	39.7	51.5	46.1	73.3	68	55.7	49.4	51	33.9	42.1
June	7.4	25.2	34.1	26	56.4	70.9	56.6	64.1	65.6	44	45.9
July	5	9.4	16.3	14.6	34.2	74.6	57	71.1	81.5	64.4	61.5
August	7.6	10	24.4	24.3	46.9	68	55.2	62.1	71.8	66.1	74.7
September	9.8	29.3	69.2	63.4	83.4	47.5	52.8	44.1	57.7	48.7	66.5
October	53	46.4	113.3	93.6	99.1	49.1	56.5	44.3	50.4	50.6	68.5
November	55.3	63.9	110.7	76.3	103.4	47.7	53.7	45.5	44.1	44	65.7
December	61.8	47	97.1	58.6	79.4	45.8	49.3	47.9	42.4	39	55
Total	371.4	438.9	802.9	582	842.1	642.9	607.4	588.7	600.6	502.8	635.4

Source: http://www.worldclimate.com/

SOILS AND VEGETATION IN THE EARLY HOLOCENE

The most recent convenient collection of data on the current soils of Europe is that for the European Union (CEC 1985) but this excludes Norway and the strip of East European former Soviet Bloc countries and Switzerland. In the 1985 EU (Benelux, France, Germany, Italy, UK, Ireland, Denmark, Greece), cambisols – brown earths – cover 44 per cent and provide a good environment for all the cereals and forages. Luvisols – including terra rossae – (16 per cent) are more weathered and are commonest south of the glaciated area; they may have poor reserves of potassium. Podzols (9 per cent) occur most commonly on sandy materials and occur along the Baltic fringe on outwash sands and on slowly weathering metamorphic rocks. They have poor reserves of nutrients and are acid; they also retain little water and are prone to drought. Fluvisols (6 per cent) occur particularly in the great river valleys and in flood plains and depressions. Without drainage and flood control measures, use of their rich nutrient supply is very risky for other than animal grazing. Lithosols (5 per cent) occur on steep slopes where there is much rock exposed and the environment is usually only suitable for low-intensity grazing. Rendzinas (5 per cent) are found on the extensive chalk and limestone outcrops where these are not covered by non-calcareous deposits. The soils are often thin and can be droughty if the underlying rock is impenetrable to roots. They provide grazing and cereal production but, in addition to drought, crops can suffer from micronutrient deficiencies. Including the Eastern European countries would increase the small area of phaeozems – chernozems – which give good cereal growth and are fertile, but otherwise would have little effect on the overall balance of soils. Locally, the resource can deviate a long way from these proportions. Thus whole modern countries (eg Finland) may be dominated by a small, and not always agriculturally attractive, subset of the soils found across Europe. The natural vegetation over most of the area would be forest (Eyre 1968); deciduous except in the north and at high elevations, where conifers are more common, and in parts of the Mediterranean basin, where in addition to coniferous species there are evergreen broadleaf woodlands. Tundra is only found in the far north and at high elevation. Natural grassland is restricted and mostly lies to the east of the Carpathians.

The evolution of the soils has of course continued under both natural processes and under the modifications to vegetation and soil induced by human activity. After the initial postglacial increase in soil organic matter and profile development in the areas more affected by the climatic changes, one of the main changes influencing soil fertility has been the on-going leaching of nutrients from the topsoil. The amount of leaching that occurs depends partly on the balance between rainfall and evaporation, and also on the vegetation and soil physical properties. It should be noted from Table 8.1 above that in average years at all sites although there will be an excess of rainfall over evaporation, usually in the winter, the amount and frequency of that excess will vary strongly from year to year and place to place. In general, the effect on the soil's fertility will be proportionally greatest on soils with a low natural nutrient content in areas with a lower temperature and a larger winter rainfall. In addition, the amount of leaching will be greater if there is short vegetation, such as grassland, because it intercepts less of the rainfall before it reaches the soil (Lockwood and Sellers 1983). Woodlands therefore not only reduce the amount of leaching but tend to have a greater ability to recycle nutrients from depth to the surface. Life is not however simple and some trees – oak and pine for example – produce polyphenolic compounds which lead to surface acidification of the soil and even to podzol formation. Thus, under woodlands a spectrum of soils from brown earth to podzol can be found. Once the natural vegetation cover was complete there would have been little opportunity for soil erosion, and soil profile development would have continued unhindered. Evidence in major river valleys indicates a fining of the sediments upwards which supports the view that the landscape became relatively stable at an early date. Thus once the climax vegetation was established the rate of soil change slowed. Soils tended to have a large organic matter content with a large store of nitrogen, a good soil structure of vertical pores down which water could drain and roots grow and, with active recycling of nutrients from depth to the surface, the loss of nutrients was smaller than under shorter vegetation. Nutrients consumed by animals tended to be recycled locally. The effect of the postglacial climatic amelioration and the soil and vegetation succession was therefore to provide conditions that were attractive for the growth of a wide range of plants, which could provide food for a variety of animals. The areas that were too cold or where the soils were too wet or 'heavy' were the main locations that would have been unattractive to early farmers.

MESOLITHIC COMMUNITIES AND THE SOIL

Although the first Agricultural Revolution is believed to be linked to the Neolithic Revolution, it does seem that vegetation may have been heavily altered as a result of Mesolithic activity; there is even a suggestion from pollen data that in the previous interglacial an increase in grass pollen may be the result of human action (Flint 1971). Human population density would have been small though the people may have been very adept at amending the vegetation to provide themselves with more food. Although woodland provides high-energy seeds and some fruits, the clearing of areas to encourage grazing by herbivores would provide a convenient and less seasonal food supply. The impact that this change from woodland to grassland had on the soil's properties has probably been underestimated – it went on for a long time in North West Europe, even though in the south arable agriculture originates early in the Holocene and has been linked by Renfrew (1987) to the spread of the Indo-European language group. The removal of woodland, either for grazing or for arable crops, increases the rate of leaching and hence acidification. It also can lead to the soil becoming wetter; if this is widespread it may even appear to alter the climate. Woodland removal has increased the area of organic soils and gleysols in wetter areas though there has also been ongoing natural soil and vegetation change (Askew et al. 1985). It is unlikely that these changes had a large effect on the nitrogen content of the soil, though a change to grassland can be associated with some reduction of soil organic matter content from that in deciduous woodland; it also leads to a change in the carbon to nitrogen ratio (Jenkinson 1988). Generally, woodland soils have more organic matter but with a wider C:N than grasslands; organic soils have an even wider C:N. When first cultivated the C:N narrows and, if it is over about 20:1, then there will be little mineral nitrogen available to increase plant growth. The major benefit to cultivators of conversion of woodland to grassland by graziers is that grasslands are much easier to clear for cultivation (Askew et al. 1985). The warm countries, Greece, the Balkans and Italy, where agriculture first took hold in Europe, tend to have the oldest soils, the lowest natural soil organic matter contents and the most rapid rate of organic matter degradation. They are also the part of Europe in which there is currently a small risk (~1:20) (Arnon 1972) of summer drought severe enough to prevent wheat growth. Floodplain soils, though the most fertile in these and other areas, have a relatively large risk of damaging spring or autumn floods, which would be exacerbated by clearance of vegetation from surrounding slopes (Shiel 1997).

The clearance of land by fire can also expose the soil to water erosion in the period between burning and the regrowth of plant cover. This problem and the risk of loss of the nutrient-rich ash by leaching or wind erosion (Sillitoe and Shiel 1999) will be worst in dry areas with episodic intense rainfall. Plato lamented, as a folk memory in the fourth century BC (Lee 1971), the loss of soil from the hillslopes in Attica, in a region with a mean annual rainfall of about 450 mm but where there is currently a one year in 10 risk of a month with ≥200 mm rain and a similar risk of three consecutive months having in total ≥400 mm rain (Shiel and Stewart 2003). In the areas which do not have a carbonate-rich subsoil, removal of the woodland will lead to increased leaching of bases so that the soil becomes progressively impoverished in plant nutrients and also becomes acidic. It seems therefore that although the properties of the soils over a large part of Europe ameliorated greatly in the early postglacial, even the low intensity use of land for grazing and opportunist collecting of fruits, associated with the removal of competitor species may have started a process of soil deterioration, with increased soil wetness, loss of organic matter, leaching of nutrients and reduction of soil structure and moisture retention.

GRAZING ANIMALS AND SOIL FERTILITY

'Grass' as a forage for animals contains a wide range of species including herbs and legumes. The deeper rooting species recycle nutrients from depth and the legumes fix atmospheric nitrogen (Whitehead 2000). Stock will also eat shrub and tree foliage and thin twigs, though they are very selective over which species are preferred (Rackham 1982). Sheep are unable to eat coarse vegetation, whereas cattle and goats are more adaptable. Where stock has a range of vegetation they will usually graze the mineral-rich herbs most intensively (Shiel and Batten 1988). The major nutritional needs of plants are phosphorus, nitrogen, potassium and a moderate pH. A wide range of other nutrients are also required although under low intensity management these are only likely to become deficient in extreme situations – for example very sandy soils, where summer drought is a greater constraint to growth. Leguminous plants, which are ultimately responsible for most of the nitrogen in the soil, need phosphorus and most but not all need a high pH. Under acidic soil conditions the legumes and herbs tend to die out and less palatable grass species become more common.

The animals void over three quarters of the plant nutrients ingested, though some of the nitrogen is lost in forms that are not available to plants. The nutrients are not, however, always returned where they were consumed and hence the animals may alter the fertility of the soil with some areas being enriched while others are depleted. The most important other requirement of the shallow rooting species is a continuing supply of water, particularly during the growing season. Variation in this has been shown to have a major effect on forage yield. Without an adequate water supply, growth stops or at best slows during periods when otherwise temperature and light are excellent for growth. During summer and early autumn in the drier areas of south and east Europe growth stops for all but the deepest rooting species or where there is a high water table – at this time stock will eat any remaining standing vegetation or will browse trees and shrubs. In many areas it has become normal for stock to be moved during the summer to wet floodplains or to mountains where there is a better supply of water – transhumance. Over all Europe bar the very western fringe and extreme south it is too cold during at least part of the winter and forage growth ceases. The length of this season determines the extent to which the farmer must make alternative arrangements for animal feed, usually by conserving some of the spring or autumn excess as hay. Farmers may even decide to feed some hay during the summer in areas with a prolonged drought. The amount of nutrients removed from the soil in conserved forages is much larger than that removed by grazing animals; it incidentally is also greater than the amount removed in the majority of arable crops, and so could result in rapid depletion of soil fertility. Conversely, the animals will leave areas where they are fed and housed in winter enriched with nutrients. If manures can be collected, these are of particular benefit in maintaining fertility of cultivated soils, but most are not applied to the soil from which they originate. We can therefore conceive of zones of enrichment and depletion; the enrichment is often close to the habitation area with the impoverishment occurring around the fringe (Shiel 1991). In seasonal grazing areas which are distant from the permanent cultivated area there is commonly little nutrient relocation, other than that which occurs locally into areas where animals are penned at night or during the heat of the day.

If the first spring growth is cut for forage on part of the land then, if unmanured, it will probably yield about 2 t/ha of dry matter, as it has at the long term British experiments (Hall 1905; Shiel 2002) – traditionally as hay. The addition of minerals from say wood ash will increase the hay yield to about 4 t/ha dry matter. Although manure would increase hay growth further, to over 5 t/ha, it is doubtful if this would be the most advantageous use of the manure; it would enable more people to be fed if applied to cereals. The spring growth of grass will be less affected by local rainfall than is the later growth, as for all but the western edge of Europe there is a large water deficit which becomes severe from the end of May (Table 8.1). Where water is available during the summer, growth continues and will give about another 2 t/ha but in the drier areas there will be less than an extra tonne. If the conserved forage is fed to animals which are kept indoors then about twice as much fresh weight of manure will result compared to the weight of the dry material fed (Shiel 1991). Growth of grass on land which is grazed in situ and is not cut regularly for hay will tend to be substantially larger than these figures because of the nutrients returned in the urine and manure; there will of course be a correspondingly smaller amount of manure available to spread on arable land.

THE REQUIREMENTS OF THE VARIOUS MAJOR TRADITIONAL CROPS

Cereals have been and remain a major food source. A wide range are grown and have been thoroughly reviewed by Renfrew (1973) and Leonard and Martin (1963). They have a high protein content and need a large nitrogen supply from the soil; if the nitrogen supply is too generous, however, the crop will lodge (fall over) and will then be difficult to harvest. They suffer from a range of foliar diseases, which are controlled by crop rotation or fallowing. All give grain with comparable nutritional qualities, but the output per unit area of land, if there is sufficient water, is greatest from wheat followed by barley, rye and oats (Table 8.2). Wheat can be either autumn or spring sown though the former gives a substantially larger crop, which is harvested earlier with easier drying. In the Mediterranean area, wheat needs a minimum of ~300 mm moisture from a combination of water retained in the soil and rainfall during the growing season. The crop yield is reduced by excessive water partly due to leaching of nitrogen (Hall 1905). It tolerates a wide range of soil conditions including some waterlogging and frost; it can not be grown over winter in the continental interior north of about 45° unless there is reliable snow cover.

The other cereals tend to be spring sown, though there are autumn sown forms. A combination of spring and autumn sowing spreads the annual workload and reduces risk of crop failure. Barley has a shorter growing season than wheat and is more drought-tolerant; it will crop with about 200 mm total water available. It is more sensitive to waterlogging but has forms (bere) which are very tolerant of soil acidity. Modern barleys require soil with a high pH. Oats are very tolerant of acid soil conditions and also of a cool wet environment; they have been the traditional crop in the north and west fringe of Europe. Rye grows at lower temperatures than the other cereals and matures relatively early. It tolerates soil acidity very well and has been widely grown on sandy soils on the Baltic fringe. Under moist conditions, it suffers particularly badly from ergot (*Claviceps purpurea*) which can poison stock and humans; if fermented by storage of damp grain it produces LSD (Bryson and Murray 1977).

The amount of energy produced from one hectare by wheat and barley is sufficient to feed 3.5 people for one year at 2,500 Cal/day while oats feed just over

Table 8.2. The energy and protein content of four major temperate cereals and their yields according to sources in Norfolk and Hampshire, England during the last millennium.

Energy and protein in grain*				
	energy kcal/100g	protein %	energy	protein
			wheat =1	
wheat	348.2	14.1	1	1
barley	334.2	11.8	0.96	0.84
rye	366.8	16.2	1.05	1.15
oats	319.8	13.9	0.92	0.99

	Norfolk 1250–1710**			Hampshire 1600–1700***		
	yield kg/ha	energy Mcal/ha	protein kg/ha	yield kg/ha	energy Mcal/ha	protein kg/ha
wheat	916	3190	128.9	852	2967	119.9
barley	899	3003	105.9	970	3242	114.3
rye	636	2333	102.8			
oats	602	1925	83.6	796	2546	110.5

*USDA 2003

**Campbell 1991

***Glennie 1991.

two people. This however ignores losses due to vermin, waste and rotting as well as that for seed; equally it ignores other sources of food. As these values are from inventories they may of course be tactical underestimates of the crops but they are seen later (Table 8.3) to be comparable, in the case of wheat, with yields from the unmanured plots at Broadbalk in the nineteenth century.

LAND USE STRATEGIES

The combination of cereals and forages in a farming system can be traced back to the earliest neolithic communities in Europe (Chapman et al. 1996). The forms of the cereals have evolved to suit the local conditions and some of those formerly used – Emmer, Einkorn and Spelt wheats for instance – are no longer widely cultivated (Renfrew 1973). The integration of animal and human refuse into the farming system also occurs from an early date judging by the distribution of detritus spread around settlements and the presence of residual phosphorus in the soil (Russell 1915). One important aspect of the detritus was almost certainly wood ash; this is high in pH and several nutrients (Sillitoe and Shiel 1999; Bear 1931); the vigorous growth of plants near fires, on burnt land and where ash had been fortuitously scattered would have been as obvious to the 'street wise' cultivators of the past as was the observation that plants grew better near manure. Use of these materials would allow a monoculture to persist for longer but it would eventually fail because the soil became difficult to manage, weeds became rife or pests and diseases destroyed the crop. The land would then be abandoned, perhaps to be reused at a later date – shifting cultivation. Additional sophistication involves not growing the same species every year on the same land or in not using the land at all for one year – fallowing. This reduces the pest, weed and disease problems and may also affect the rate at which nutrient reserves in the soil are depleted. Certainly such a rotation of crops was not news to Virgil (Wilkinson 1978); his rotations describe the benefits from wild forage legumes such as vetches and Columella (Col.r.r. II, 7, ed. Ash 1941) indicates that a wide range of these were intentionally sown for forage for animals. Surprisingly, however, this knowledge appears to have been subsequently mislaid, and in England it was not until the eighteenth century AD that farmers were encouraged by the popularisers of the new Agricultural Revolution to sow particular forage species within a rotation, rather than sowing the mixture of grass and other seeds swept up from the floor of the hay loft (Moore 1944).

The use of non-waste soil ameliorants also originates from an early date; Pliny the Elder reports the use by the native inhabitants of Britain and Gaul of marl to improve soil (raise the pH) (see Chapter 7). The benefits of irrigation were also understood, though the role of the suspended silt in floodwaters may have been confused with the impact of the water itself. Silt from rivers was also used intentionally until recently in warping (Ferro 1949) – and may have been exploited unwittingly in the case of the management of water meadows. These remained the main external ameliorants until the use of ground bones in the eighteenth century. Other external materials include marine resources – seaweed, shell sand and surplus fish, but the use of these seems to have been limited geographically.

How much of the classical knowledge survived the Dark Ages after the fall of the Western Roman Empire is as yet unclear, although the Venerable Bede, writing in Northumbria in the seventh century AD, was certainly quoting widely from Roman sources. As his writings penetrated through Europe fairly efficiently and the Germanic tribes who settled the north and west of Europe had a well established farming system using the same range of crops and animals, diet and land management may have continued with less change than might be assumed. The origin of the mediaeval 3-field rotational system in this period is uncertain but it was in place with the fully fledged feudal system at the beginning of the second millennium AD. The knowledge necessary to modify the 3-field system into the rotations which remain in use today was in place from the middle of the second millennium AD but it was the eighteenth century before the intentional use of root crops (e.g.

turnips) and forage legumes became widespread; the reasons for the delay seem to have been as much social as technological. Even when the sophisticated rotations were introduced, fallowing, which was replaced by the root crops, did not completely disappear. It must be noted that although these systems were used in such a way that fertility appeared to be managed satisfactorily, the only external sources of ameliorant involved the use of marl and biological nitrogen fixation. All of the other methods of maintaining fertility of the arable land operated by the invidious procedure of robbing Peter to pay Paul! Thus animal manure spread on arable land decreased the nutrients in the surrounding grasslands and woodlands, wood ash use similarly debilitated the forests from which the wood was taken, river silt settling on irrigated land degraded the soils from whence it was eroded and even human wastes were rarely applied to the soils from which the nutrients had been taken. The effect of these practices was that fertility diversified within the landscape with those areas which received a net surplus of nutrients being ameliorated and the others being degraded; this can have a long-term effect on future fertility. Russell (1915) notes that the phosphate in manures applied to fertilise fields in Egypt in the pharaonic period could still be measured. The extent to which nutrient leaching and erosion of soil occurs also varies with crop grown and system of production. Compared with grassland, all arable crops leave the soil bare at some season, facilitating erosion, and even the longest-season cereal – wheat – evaporates a lot less water than grassland, increasing the leaching of nutrients (Lockwood and Sellers 1983). A rotation of crops can reduce the risk of both leaching and soil erosion by ensuring that the soil has a cover of vegetation for more of the year than is possible with a monoculture. The worst option of all is fallow, which leaves the soil intentionally exposed and where it is subject to the effects of intense rainfall and strong winds, and where the amount of water available to leach out nutrients is much larger than when a crop is grown.

INPUTS TO THE SOIL FROM CROP, ANIMAL AND URBAN WASTES

Manure was essential to maintain yields on the arable but the amount that could be extracted from the other land types was the limiting factor. Techniques such as night-folding or housing of stock would increase the proportion of nutrients transferred during the half of the year it was common for the animals to be outside grazing; much of the nutrients they consumed fell where they rested or grazed. Crop wastes if burned lost nitrogen though the soil pH was raised and there was an immediate flush of nutrients; animal wastes if poorly managed removed much nutrient from the grazing land but may not have enriched the arable on which they were spread. Home produced wood ash would be available from cooking and heating and would also result from land clearance and occur in urban waste. The amount of ash produced from burning wood is small – about 2 per cent of the dry weight of the wood. As the wood consumed annually might be in the region of 1 t/person in a warm area to 3 t/person in a cold area (Lefevre et al. 1997), a group of 10 people burning 10–30 tonnes of wood would only produce 200–600 kg ash containing proportionately more calcium and potassium than manure but no nitrogen and less phosphorus; ash from five tonnes of wood would give a comparable amount of mineral elements to one tonne of manure. As the manure application on the experiments described later (Table 8.3) was 35 t/ha the ash from 10 people would at best only manure 0.16 ha. In contrast, the manure from animals housed for half the year and fed conserved feed from 2 ha at the average yield for unmanured grass would produce 8 tonnes manure, and more if night folding were used – sufficient for 0.23 ha. The development of urban societies created a traffic in nutrients; the night soil they produced could be used to enrich peri-urban land but transport difficulty and expense restricted the distance they could be moved (Anderson, 1794). People however produce relatively little manure in comparison to ruminant animals, so from urban societies the main source of manure was their domestic animals and wastes from food preparation and the ash from cooking and heating. Some at least of the human waste is likely to have been 'informally' recycled onto garden plants.

MONOCULTURAL PRODUCTION OF CROPS

There are in England two long-term grass monocultures and two cereal monocultures at Rothamsted Research Station (Broadbalk wheat and Hoos field barley); others, as in Australia, have been running for a long time in Mediterranean environments. These and other long-term experiments, but by no means a comprehensive list, are reviewed in Leigh and Johnston (1994). There are also long-term experiments on wheat in America such as the Sanborn experiment in Missouri. One problem with many sites is that the nutrient input levels are high and do not include a manure treatment; this makes it rather difficult to relate them to conditions in the past. There are also problems in working with these long-term datasets because of variation in weather, which may influence the outcome more than the treatments do. One way to deal with this is to relate the result from each treatment to the mean of all the treatments. Unless there is a temporal shift in the long-term mean of all treatments this method will help at least to minimise the effect of weather variation and will in any case show the relative change in output due to the treatments.

The yield trend on the Broadbalk tratment receiving farmyard manure was strongly upward and the yield good (Table 8.3). If nitrogen was absent but the other mineral elements were present, as when ashes are applied, the yield was low as the only source of nitrogen was the soil organic matter and atmospheric inputs. Yield remained low and tended downwards; with no nutrients added the situation was even worse. The causes of the yield levels and trends are immediately clear. A

Table 8.3. Yield of Broadbalk continuous wheat plots between 1852 and 1901 and the percentage of the mean yield in each decade for treatments receiving nothing, minerals and farmyard manure (FYM) (Hall 1905).

Treatment	1852–1861	1862–1871	1872–1881	1882–1891	1892–1901	Mean kg/ha
FYM	118.3	126.5	128.5	140.9	154.6	2427
Untreated	55.0	48.9	46.6	46.5	48.5	897
Minerals*	63.7	52.3	54.2	50.9	58.4	1018

*Potassium, Magnesium and Phosphorus equivalent to the nutrients in ashes

balanced combination of nutrients maintains the yield but at a level which depends on that in least supply, especially if the nutrient in question is N; this is a classic example of Liebig's (1843) Law of the Minimum. Judging by the yield from the plot receiving no nutrients the soil fertility at Broadbalk for a range of nutrients was poor when the experiment began; it had been in continuous arable use before the experiment began in 1843 and had a low soil organic matter content. Where the supply of added nutrients is large then the yield increases, possibly due to accumulation of surpluses in the soil and to the increased return of organic matter from the larger crop residues. If the mixture of nutrients is imbalanced then yield trends downwards, the trend rate depending on the magnitude of the store in the soil of the inadequately provided nutrients. The nearby Hoos Field continuous barley experiment showed a much greater decrease in yield with time for the unmanured and mineral treatments than was found at Broadbalk; the yield at the outset was higher than at Broadbalk but decreased much more rapidly reaching 600 kg/ha within 50 years. The change in yield with time on the differently fertilised barley plots is very similar to that for the wheat; large balanced inputs of nutrients from manure maintained yield at a high level throughout the period.

The yield over the first 50 years on the untreated wheat plots at Broadbalk decreased to just below 900 kg/ha and remained about there. This compares favourably with the yields of wheat on the Sanborn untreated plots in Missouri (Brown 1994) which, even averaged over the first 30 years of the experiment, were only 600 kg/ha. This figure is similar to that at the Australian Mediterranean climate zone wheat experiments – the yield at Urrbrae in the second decade was 62 per cent of that in the first (Grace and Oades 1994). It has been shown here that continuous cultivation of a single crop cannot continue unless there is a large area of land available for manure production and even then it is difficult. If the grass was unfertilised and all of the manure from hay on half the grassland was collected then at Broadbalk 17.5 ha of grass would be needed to produce sufficient manure to fertilise 1 ha of wheat.

The classical Rothamsted experiments began on fields that have for long been in use for wheat growing so it is difficult to measure changes in soil nutrient content. Where the leaching of nutrients was measured the amounts were very much larger from the plots that were being fertilised (Hall 1905). At Sanborn field (Brown 1994) there were no original measurements made on the soil and hence considerable care must be taken in comparing the current properties of the soils receiving different treatments. It seems that the topsoil thinned considerably and the organic matter has decreased, especially on the unmanured plots. This loss of soil on cultivated land is confirmed by many field experiments. For example, over 14 years at a nearby experimental station in Missouri, permanent grass lost 760 kg soil/ha while under continuous wheat the loss was 30 times greater and rainfall runoff increased from 12 per cent to 29.4 per cent (Miller and Krusekopf 1932). This would tend to make the soil droughtier and reduce crop growth. Cultivation appears to have multiplicative effect on loss of nutrients by removing them physically in the crop, by increasing leaching of water soluble nutrients, by reducing the soil's ability to retain water and to transport it to depth and by weakening the structure, as well as by exposing the surface to raindrop impact. Catchment studies today seem to be more attractive than the chore of the traditional field experiment and they tend to confirm all that has been said about the risk of soil loss and the high content of nutrients in the eroded material (Wicherek 1993).

Wind erosion has not been mentioned as it only becomes a problem when large areas of bare soil are exposed. Where there is a mixture of land uses then soil eroded from a bare area may be retained in nearby vegetated areas – this idea is exploited today as strip cropping (Gustafson 1941). Although the bare soil may be degraded the soil remains in the catchment and offsets part of the effect of the loss from the eroded area. Tillage does however also cause soil movement on a smaller scale than in classical erosion, where soil is redeposited several kilometres away. On slopes there is a tendency for loosened soil to move downhill under gravity resulting in a 'creep' whereby the slopes have a thinner soil than level areas or footslopes. There is often also localised erosion which contributes to this same process; over a period of centuries the differences in surface level at either side of a fence line can be over one metre (Shiel and Chapman 1988). All of these result in the redistribution of the most fertile fractions of the soil.

WET AND DRY YEARS

In the 10 wettest years in 50 the overwinter rainfall at Rothamsted, England was 330 mm compared with the 147 mm of the 10 driest years; the yield of wheat in the dry years was 33 per cent greater than in the wet years on plots receiving fertiliser. Even when the 19 years with above average water percolation through the soil are compared with the 19 driest the yield difference is 18 per cent (Hall

1905). Measurements made on the timing and composition of the drainage water confirm a peak in nitrate content in the water in November and December which corresponds closely with the months in which the drains flow most commonly. The result is a substantial peak of nitrate loss from the soil in the final months of the year which is greatest in wet years and which can restrict future growth if the added fertiliser is inadequate. In dry areas in contrast spring wheat yield increases linearly with growing season rain and soil-stored water up to around 400 mm (McConkey et al. 1996). Barley at Rothamsted also yielded slightly more in bushel terms in dry years but as the bushel weights were smaller there was no clear benefit except on the minerals (ash) plot where increases of up to 50 per cent were obtained (Hall 1905). The benefit on these plots of the dry year is that there is less leaching of the nitrogen mineralised from the soil over the winter and spring. This is confirmed by a more detailed analysis which showed that the benefits were largest when the period March to May was drier than average but June and July were wet. Although the most apparent effect to the cultivator is on the crop yield, the loss of nutrients by leaching means that less are recycled and the long-term impact on the soil's fertility is clear. Low yields will also tend to encourage an expansion of cultivated area to maintain total output; this will encourage even more leaching of nutrients and will reduce the area of uncultivated land from which manure or ash can be obtained (Shiel 1991). The driest areas will suffer the least loss of nutrients by leaching, but may be even more susceptible to the more rapid loss of nutrients by erosion!

THE EFFECT ON SOIL OF SHIFTING AGRICULTURE

Although some shifting agriculture is equivalent to a long rotation, land with a small content of basic nutrients may only be used for a single episode of several years; the land is then abandoned and never reused; an example for this comes from the North York Moors National Park in north east England (Dimbleby, 1962). The initial clearance, often with fire, produced a brief bonanza of all plant nutrients except nitrogen, which was lost to the atmosphere. However, much of the nutrients in the ash are lost by leaching from sandy soil while nitrogen only increased relatively slowly after abandonment. The rate of nitrogen increase depends not only on the management of the land but is also affected by the climate and the chemical conditions in the soil. Cold, lack of available soil phosphorus and acidity militate against legume activity. The exploitation of nutrients and their non-return coupled with soil structural deterioration and increasing wetness, either due to vegetation or climatic change, have led to the spread of organic soils in areas (Askew et al. 1985) and extensive podzolisation elsewhere. Without appropriate technology these soils become suitable for only a limited range of uses. The range of cereals that can be grown is restricted to the acid tolerant forms of barley (bere), rye and oats. All of these offer perfectly satisfactory food sources though oats has lower energy density and barley lower protein content than wheat (Table 8.2). In terms of total yield there are large variations in estimated yield between the crops and it may be that the difference in total output of food value per hectare is relatively small.

For other soil types, changes in soil properties are smaller though organic matter can decrease rapidly under cultivation (Jenkinson 1988), restricting the amount of nitrogen available and the ability of the soil to retain water and nutrients, and erosion can occur. It would therefore be facile to imagine that the soil had not been altered before neolithic farming began. Archaeological evidence indicates that though sites may be in use at several periods there may be times when they are abandoned, often for several hundred years, so even the neolithic cultivator may have shifted his location if soil fertility decreased. Apart from the narrowing of the C:N ratio and the short-term benefit from burning even the shifting cultivator is likely to have left soil more acidic, nutrient depleted, more difficult to cultivate and more droughty, due to the reduced organic matter content, and possibly wetter and less well structured than they had been under natural conditions. Shifting cultivation can be seen as a transition to rotation but by creating a diverse landscape and breaking up the area of cultivated land and, during the recovery phase, increasing plant cover it acts to minimise wind erosion and reduce the severity of and localise water erosion. The extent to which soil recovery occurs during the uncultivated recovery phase will depend on its length and on the properties of the remaining soil and underlying parent material. Thus on loess soils, provided some loess remains, erosion has a small effect on future output while if the underlying material is hard rock or unweathered till then the recovery of fertility will be very slow. At Rothamsted on a weathered paleosol the organic matter has more than doubled in 81 years since the arable land was abandoned to natural woodland regeneration; when old grassland was ploughed out however the organic matter content fell to half in 18 years with a bare fallow. Even with a 6-course rotation including grass, the organic matter decreased by 30 per cent in 26 years (Jenkinson 1988).

FIXED FARMING AND THE ROTATION OF CROPS

The decrease in yield on all but the most generously manured treatments, together with the problems with pests, weeds and diseases, which have made monocultures difficult to maintain even on modern research stations, means that sooner or later any monoculture is likely to be broken. There are a variety of ways of avoiding the problems associated with monocultures; these range from regularly shifting the areas of land cultivated through fallowing to crop rotation. As a result of clearance of land for grazing and tillage the landscape consisted in many cases of separate arable, grassland and woodland components. If there is a surplus of land then the areas which were formerly used for grazing or woodland can be used for arable – at a large expense of labour. The woodland and grassland however were commonly on sites which were less easy to manage than the arable land. As more of the virgin land is exploited there comes a point when land that has been abandoned already and which has 'recovered' becomes attractive for cultivation, as the other land is becoming less fertile. This can be viewed as being equivalent to a long rotation; as it may be physically easier to recultivate formerly cultivated land then this system has its attraction and is used today in Papua New Guinea (Sillitoe and Shiel 1999). The alternative to shifting the area cultivated is to alter its use temporarily. The simplest form of such a rotation is the introduction of a fallow during which no crop is grown on the land. The next system is to combine this with a change in crop, although it is feasible to alternate the crops without a fallow. In the absence of a fallow, even if there is no nutrient advantage to the crop from being grown in rotation there may be a difference in susceptibility to pests, weeds and diseases between the crops which results in more of the available nutrients being harvested in the crops. Finally, one reaches the more sophisticated rotations which include botanically diverse plants including legumes and the use of the cultivated land temporarily for animal grazing. Rotation in the Mediterranean seems to have been in place by early in the first millennium BC and to have continued in this form for some 2,500 years at least.

Fallows

Because of the problems with pests, weeds and diseases it has not always been possible to grow the crop on all of the plots in every year in the monocultural research station experiments even under the high standards of management and with the tools and technology available. The result of this is that there have been unintended fallows or changes of cropping; one can be certain that this would have happened in the past, and the benefits to the crop of growing after a fallow would have been noted by farmers. When the yield from the Broadbalk wheat grown as a monoculture is compared with the adjacent Hoos Field plots, where the wheat is grown after fallow without manure, there is an increase over the 47-year period from 900 to 1210 kg/ha grain and a comparable increase in straw yield due to the use of the fallow – but there is only one crop every alternate year (Hall 1905). If the amount sown was about 200 kg/ha then the net yield on a monoculture is about 650 kg/ha and with fallow it is 480 kg/ha, after allowing for the uncropped area. Clearly there have to be other benefits than the larger crop to make fallowing an attractive proposition. These advantages are that work on the fallow is less critical than on the crop, the crop is thicker after the fallow and therefore weeds are less of a problem and, most importantly, the gap allows for weed control of perennials and also breaks the cycle of pests and diseases. Most of this benefit due to fallowing occurs in the drier years – in the 16 years with below average rainfall between 1872 and 1902 the yield increase due to fallow was 51.5 per cent while in the wet years it was only 7.9 per cent. In Australia, fallow reduced the rate of yield decrease with time in the Mediterranean zone – in the second decade yield was 94 per cent of the first and in the third 66 per cent of the first compared with 60 and 56 per cent respectively in the monoculture (Grace and Oades 1994). Here the benefits of fallowing were very much greater than in England, and even allowing for only half the area being cropped the fallowed plots yielded 12 per cent more than the continuous monoculture wheat. It seems that in the fallow year, as in the temperate region, organic matter breaks down and releases nitrogen which benefits the following crop. The long-term sustainability of this system is, however, suspect due to rapid rates of organic matter loss. In the short term (~100 years) fallowing will increase yields on the cropped area – at the expense of cultivating twice the area cropped – but after a century of use both systems will result in a very low soil organic matter content and the fallowed soil may be even less fertile than the annually cropped land! For a fallow to be effective the next crop must be established before there is a surplus of water to leach out the accumulated available nitrogen. There can also be problems as fallow soils are exposed for a considerable time to wind and water erosion.

In very dry areas fallow also gives an advantage to growth in alternate years by conserving water in the soil. However, the net effect is the same; in wet years there is little advantage of fallow because both systems have plenty of water and in dry years the extra water, and/or nitrogen stored in the soil, benefits the plants grown on the plots which were fallowed.

All-Arable Rotations

The separation of land used for crops and stock seems to have been widespread, other than for the temporary use of stock to clean up the 'waste' in fields after harvest. Many of the rotation experiments carried out include such a compartmentalisation even though manure may be applied – indicating that stock were held elsewhere. The simplest of these rotations is the 'Three field system' with the rotation of wheat, barley, fallow. The closest experiments to these are the Australian rotations of wheat, fallow, oats at Dooen and Urrbrae in the Mediterranean zone (Grace and Oades 1994). At Urrbrae the wheat and oat yields on the rotation were 10 per cent better than on the wheat–fallow rotation; this is excellent as the ratio of cropped land to unproductive fallow is larger. Allowing for the area uncultivated, the grain yield averaged over 60 years was 47 per cent greater on the rotation than on the wheat–fallow and was 64 per cent greater than on the wheat monoculture. Although the traditional strip farming approach used for such rotations reduced the risk of wind erosion the presence of a fallow and of long periods between crops being sown and harvested meant that all of the processes which led to the loss of organic matter and nutrients from the soil, and of the soil itself by erosion, had their opportunity to work. These soil losses may explain the lack of increase in crop yield during the Middle Ages. Seen in this light, episodes such as the Black Death may have had the effect of reducing pressure on the soil and giving it the opportunity to recover whereas during periods of population growth the pressure to increase total output would have decreased the area of local grassland to the point that insufficient manure was available to adequately fertilise the arable land. The reduced crop yield has a vicious cycle effect by exposing the soil to greater leaching, erosion and organic matter reduction while forcing the farmers to further increase the area of land ploughed to meet market demands.

Improvements to the Rotations in the Seventeenth and Eighteenth Centuries

The introduction of new crops, such as brassicas (usually turnips or swedes) and clover, the intentional sowing of short-term species-rich grass mixtures, which included legumes, in place of the fallow and the integration of grassland and arable areas led to the improvements in total output and crop and animal feed quality from the sixteenth century AD. Provided the deficiencies in bases and phosphorus were corrected these systems were very effective as producers of human food and remain in widespread use. In the 12 years on which comparison is possible between Broadbalk, Hoos Field and the Agdell rotation then on comparably manured plots the yield of wheat at Agdell rotation exceeds the 880 kg/ha at Broadbalk by 1145 kg/ha and the Hoos Field fallowed plots by 740 kg/ha (Hall 1905). Clearly rotation has an advantage that exceeds the simple effect of leaving a fallow gap in the cropping. The Agdell plots in question do not have legumes grown on them so the benefit can not be attributed to nitrogen fixation from this source.

When no fertiliser or only ashes (no nitrogen) is used rotation doubles cereal yields compared with a monoculture. However, when a complete manure is applied the monocultural plots give a similar, large (>2500 kg/ha) yield for wheat and a better yield for barley compared to the rotation; it must however be remembered that the animal manure is applied annually to the monoculture but only once per rotation, to the swedes, which responded spectacularly to it (a 25-fold increase), removing the nutrients which would benefit the following barley while the wheat followed either a legume or a fallow. The increase in animal feed provided by the rotation obviously creates more animal products but also more manure. The beneficial effect of ashes and of the minerals in a complete manure were considerable but the additional nitrogen in the complete manure had little effect. Unmanured permanent grass grew much better than did new sown clover but otherwise the clover gave similar or better yields. There are no monoculture swede experiments in the same years with which to compare, but in other years the effect of a comparable range of fertilisers was similar in trend though the yields on the fertilised monocultures remained less than half of those on the rotation.

The effect of the rotation is also to increase the overall level of soil organic matter above that which could be produced by only manuring the soil. Partly this occurs because of the increased yields and better returns of soil organic matter and partly because of the presence of a forage crop in the rotation. There is a substantial reduction in erosion (Table 8.4), more than can be explained by averaging out the erosion on the various crops, and also the length of time that the soil is left bare and exposed is reduced. Also there is more time over which plant roots are present to take up nutrients and prevent them being lost; the greater evaporation from the plants will also reduce the leaching volume. The rotation may not be as good as the permanent grass cover in preserving soil fertility but provides a larger amount of food than any other system and reduces soil erosion and nutrient leaching below anything but a permanent grassland cover.

Table 8.4. Effect of cropping system on erosion and runoff of a silt loam soil

	Runoff %	**Soil loss t/ha**	**Relative loss grass=1**
Fallow	30.7	93.2	122
Maize	29.4	44.1	58
Wheat	23.3	22.6	30
Grass	12.0	0.76	1
Maize/wheat/clover rotation	13.8	6.22	8
Average of maize/wheat/grass	21.6	22.6	30

Source: Miller and Krusekopf 1932

A LACK OF KNOWLEDGE?

When the experimenters began in the nineteenth century their findings helped to explain some of the existing practices, but also indicated serious shortfalls in the availability of materials with which to fertilise the soil. In many cases relatively simple changes made great improvements to output. This is however not to demean the accumulated experience of land managers, who in many cases were operating remarkably efficiently. The extent to which the soil had been altered has only become clear to us in the last century when it was realised that many of the arable soils probably had less than one-third of the organic matter that they had contained before woodland clearance. This also demonstrates the extent to which soils can continue to produce valuable outputs with severely altered properties.

AN OVERALL SYSTEM TO PRODUCE FOOD AND ITS EFFECT ON SOIL PROPERTIES

Although it has been feasible to demonstrate the likely effect of changes in land use on the properties of the soil in Europe and also to show the levels of food output that might have been achieved with various inputs the only estimate made so far for those inputs was for the production of manure from grass and of ashes from wood. The farmer's decision on what to do will be some combination of providing himself with food and sufficient economic gain from his business at least to pay for his expenses. If he is a subsistence farmer it is food for his immediate dependents that will be foremost, together with any tithes or other payments to 'superiors'. If he is operating in a cash economy his economic maximisation may involve quite a different group of products from that of the subsistence peasant. These two strategies may have very different impacts on the soil's properties. Thus if the land is used for wool production the loss of nutrients will be much smaller than if it is used for monocultural wheat, and the final soil organic matter content will be much smaller if wheat is grown.

As any use of land without external input of lime or fertiliser tends to lead to a reduction in the content of plant nutrients, basic cations and soil organic matter, relative to undisturbed natural conditions, then it can be argued that the most sustainable system will use these inputs where available and otherwise will minimise the area and intensity of land use. This latter constraint is a problem, as until the mid-nineteenth century AD use of land tended to lead to it becoming less productive of human food over time. In extreme conditions, when the soil was sandy this could occur rapidly and lead to a situation which was difficult to reverse with the technology available; the soils became podzols and at best the range of products was restricted both in diversity and quantity. In wetter areas the reduction in soil structural quality due to clearance led to gleying of the soil and in some situations to the spread of peat, which could then engulf drier surrounding areas. The use of drainage and lime from Roman times did lead to the 'reclamation' of large areas of such land and, interestingly, may be the reason for the use of a word which implies that the land had formerly been in use. Restrictions on the use of these methods include the need for large labour forces to effect land drainage – such as that along the Tisza in Hungary – and the proximity of a source of carbonate-rich deposits, together with fuel to burn them if they were in the form of hard limestones; this last does not seem to have been common for land amelioration until the eighteenth century AD, though softer limestones, chalk, marl and shell sand were in common use from an early date. Generally, however, the output per unit area of land decreased with duration of use until the land was periodically taken out of production and left for a prolonged period in a semi-natural vegetation; the more intensively the land was used the more rapidly would its properties deteriorate. Fallowing gave better crops but caused even more rapid soil deterioration. As productivity could only be maintained by manures and ash from non-arable land the deterioration was transferred to these other soils with the arable land struggling along with lower inherent fertility until the inclusive rotations and use of external fertilising materials and lime from the eighteenth century AD. The use of terms to describe land as 'in-by', which received animal manures, and 'out-by', which did not, and the different uses of the soils, indicates the flow of fertility within the farming unit with the land near the homestead being enriched relative to the outlying areas. The differences between the in-by and the out-by soils would become considerable and many years after the nutrients had been applied the phosphorus, which tends to be best retained by the soil, could still have a beneficial effect on crop growth which could not be achieved by the use of fresh fertilisers on formerly unfertilised soils (Roscoe 1960). There is also a trade-off aspect for the land user, with repeated reuse of land, and therefore soil deterioration, being easier in the short term than using a larger area less intensively. At a larger scale of distance the growing urban areas were centres towards which nutrients flowed and where the wastes created a ring of more fertile soils up to several kilometres wide around the area. Later urban sprawl has engulfed many of these fertile locations which retain only the place name – Covent and Hatton Gardens in London for example – to indicate the former use. All crop rotations involve both an exploitation phase and a recovery phase. The ultimate 'owner' of the land must also be considered, for landlords may constrain the activities of their tenants so as to prevent long-term deterioration of the soil – this is seen in the constraining clauses in many tenancy agreements which prevented the sale of hay and straw and even prescribed a particular rotation of crops. The ploughing up of permanent grassland was commonly banned without payment of a large fine.

The outsourcing of grain by Athens and Rome demonstrates the effect of decreasing soil productivity in these areas. That grain was then imported from Egypt, where the soil fertility was maintained by erosion in Ethiopia and redeposition in the lower Nile valley, and elsewhere, demonstrates the scale of the reduction in soil productivity in Greece and Italy and the need for their urbanised societies to have access to products from a large area where the soils had not been so severely degraded. Although Athens retained a substantial area of fertile soil in the valleys, where much of it had been redeposited after erosion from the hillsides, this rejuvenation was not, as in Egypt, an on-going process so that the fertility of these areas decreased. To what extent major population movements were caused by the need to seek new soil can be debated but it is clear that the productivity of considerable areas of Europe was seriously reduced long before the new Agricultural Revolution of the seventeenth and eighteenth centuries AD brought more sustainable systems of management. The benefits to the properties of soils provided by these new rotations – the soil organic matter content and structure were improved and, where necessary, drainage was carried out, lime was applied to raise the pH, urban wastes were recycled and bones and basic slag were used to raise the phosphate status – were to enable more people to be fed better off a smaller area of land. Better and faster plant growth also dried out the soils to greater depth, provided better cracking of the subsoil and returned larger amounts of roots, increasing soil organic matter and biological activity in the soil while reducing the risk of soil erosion. These changes raised the general fertility base, reducing between-soil differences and giving greater flexibility of use of a wider range of soils. To some extent these changes were remedying the adverse changes of the previous land use, although the now-limed sandy podzols would probably have been too acidic for agricultural use before farming became widespread in north-western Europe. Care must be taken not to over-interpret the extent of deterioration through exploitation, for some of the earlier land abandonment had been as a result of climatic change, though even the nineteenth-century amelioration of the climate did not lead to some of the degraded areas, for example the North Yorkshire Moors, being reused, even though this was now technically feasible. In areas of Europe which were less economically developed traditional practices remained, and the only redeeming feature in these areas was that the management units were small and tended to integrate livestock grazed on surrounding hills, so that a supply of nutrients was available in manure. Much of the more intensively managed land in such areas, for example Dalmatia, was in basin sites, which received run-on of water and eroded soil from the degraded uplands (Chapman et al. 1996). There was therefore a continuation of the decrease in cultivable area, due to soil loss from the slopes, which had begun before Plato commented on it (Lee 1971) and it continues today. The more-sustainable rotational farming system was to be swept aside over large areas by the expansion of industrial agriculture, particularly after the Second World War. Manufactured fertilisers, often subsidised, and with food prices buoyed up to maintain farm output led to maximisation of yield and intensification of use of the land. The larger units of management that resulted to fit with the enlarged machinery led to soil erosion (Wicherek 1993) and to widespread concerns over ecology and environment. The increasing interest, more recently, in organic farming has encouraged a return to the rotational systems of land use, at a cost to total output, and is probably leading to an amelioration of soils similar to that seen after the introduction of rotations in the seventeenth and eighteenth centuries.

REFERENCES

Anderson, J. 1794. *General View of the Agriculture of the County of Aberdeen.* Edinburgh: Blackwood.

Arnon, I. 1972. *Crop Production in Dry Regions*. London: Leonard Hill.

Ash, H.B. 1941. *Lucius Junius Moderatus Columella on Agriculture*, vol. 1. London: Heinemann

Askew, G.P., R.W. Payton and R.S. Shiel. 1985. 'Upland Soils and Land Clearance in Britain During the Second Millennium BC'. In D.Spratt and C. Burgess (eds), *Upland Settlement in Britain: The Second Millennium BC and After*, BAR British Series 143, 5–33.

Bear, F.E. 1931. *Soil Management*, 2nd edn. New York: John Wiley.

Brown, J.R. 1994. 'The Sanborn Field Experiment'. In R.A. Leigh and A.E. Johnston (eds), *Long-Term Experiments in Agricultural and Ecological Sciences* (Wallingford: CAB International), pp. 53–82.

Bryson, R.A. and T.J. Murray. 1977. *Climates of Hunger: Mankind and the World's Changing Weather.* Madison: University of Wisconsin Press.

Campbell, B.M.S. 1991. 'Land, Labour, Livestock and Productivity Trends in English Seignorial Agriculture, 1208–1450'. In B.M.S. Campbell and M. Overton (eds), *Land, Labour and Livestock* (Manchester: Manchester University Press), pp. 144–82.

C.E.C. 1985. *Soil Map of the European Communities 1:1000000.* Luxembourg: Commission of the European Communities.

Chapman, J.C., R.S. Shiel and S. Batovic. 1996. *The Changing Face of Dalmatia: Archaeological and Ecological Studies in a Mediterranean Landscape,* Reports of the Research Committee of the Society of Antiquaries of London No. 54. Leicester: Leicester University Press.

Dimbleby, G.W. 1962. *The Development of British Heathlands and Their Soils,* Oxford Forestry Memoirs, 23. Oxford: Clarendon Press.

Eyre, S.R. 1968. *Vegetation and Soils* 2nd edn. London: Edward Arnold.

Ferro, R.D. 1949. 'Warping at Blacktoft, 1948'. *Journal of the Yorkshire Agricultural Society*, 56–65.

Flint, R.F. 1971. *Glacial and Quaternary Geology.* London: Wiley.

Glennie, P. 1991. 'Measuring Crop Yields in Early Modern England'. In B.M.S. Campbell and M. Overton (eds), *Land, Labour and Livestock* (Manchester: Manchester University Press), pp. 51–77.

Grace, P. and J.M. Oades. 1994. 'Long Term Field Trials in Australia'. In R.A. Leigh and A.E. Johnston (eds), *Long-Term Experiments in Agricultural and Ecological Sciences* (Wallingford: CAB International), pp. 53–82.

Gustafson, A.F. 1941. *Soils and Soil Management.* New York: McGraw Hill.

Hall, A.D. 1905. *The Book of the Rothamsted Experiments.* London: John Murray.

Halstead, P. and J. Tierney. 1998. 'Leafy Hay: An Ethnoarchaeological Study in NW Greece'. *Environmental Archaeology*, **1**: 71–80.

Jenkinson 1988. 'Soil Organic Matter and its Dynamics'. In A. Wild, *Russell's Soil Conditions and Plant Growth* (Harlow: Longman), pp. 564–607.

Lee, H.D.P. 1971. *Plato's Timaeus and Critias.* Harmondsworth: Penguin Books

Lefevre, T., J.L. Todoc and G.R. Timilsina. 1997. *The Role of Wood Energy in Asia.* http: //www.fao.org/docrep/w7519e/w7519e00.htm

Leigh, R.A. and A.E. Johnston (eds). 1994. *Long-Term Experiments in Agricultural and Ecological Sciences.* Wallingford: CAB International.

Leonard, W.H. and J.H. Martin. 1963. *Cereal Crops.* London: MacMillan.

Lockwood, J.G. and J.G. Sellers. 1983 'Some Simulation Models Results of the Effect of Vegetation Change on the Near-Surface Hydroclimate'. In A. Street-Perrott, M. Beran and R. Ratcliffe (eds), *Variations in the Global Water Budget* (Dordrecht: Reidel), pp. 463–77.

McConkey, B.G., C.A. Campbell, R.P. Zentner, F.B. Dyck and F. Selles. 1996. 'Long-Term Tillage Effects on Spring Wheat Production on Three Soil Textures in the Brown Soil Zone'. *Canadian Journal of Plant Science*, **76**: 747–56.

Miller, M.F. and H.H. Krusekopf. 1932. *The Influence of Systems of Cropping and Methods of Culture on Surface Runoff and Soil Erosion*, Research Bulletin 177. Missouri Agricultural Experimental Station.

Moore, H.I. 1944. *Grassland Husbandry*. London: George Allen & Unwin.

Parry, M.L. 1978. *Climatic Change, Agriculture and Settlement*, Studies in historical geography. Connecticut: Archon Books.

Rackham, O. 1982. 'Land Use and the Native Vegetation of Greece'. In M. Bell and S. Limbrey (eds), *Archaeological Aspects of Woodland Ecology*, BAR (Int Ser) **146**: 177–98.

Renfrew, C. 1987. *Archaeology and Language: The Puzzle of Indo-European Origins*. London: Jonathan Cape.

Renfrew, J.M. 1973. *Palaeoethnobotany*. London: Methuen.

Roscoe, B. 1960. 'The Distribution and Condition of Soil Phosphate under Old Permanent Pasture'. *Plant and Soil*, **12**: 17–29.

Russell, E.J. 1915. *The Fertility of the Soil*. Cambridge: Cambridge University Press.

Shiel, R.S. 1991. 'Improving Soil Fertility in the Pre-Fertiliser Era'. In B.M.S. Campbell and M. Overton (eds), *Land, Labour and Livestock* (Manchester: Manchester University Press), pp. 51–77.

Shiel, R.S. 1997. 'Surviving in the Tisza Valley; Plants' and People's Perspectives on the Environment'. In J.C. Chapman and P. Dolukhanov (eds), *Colloquia Pontica 3, Landscapes in Flux: Central and Eastern Europe in Antiquity* (Oxford: Oxbow books), pp. 181–92.

Shiel, R.S. 2002. 'The Future of Farm Grassland as a Diverse Productive Environment'. *Transactions of the Natural History Society of Northumbria*, **62**: 115–22.

Shiel, R.S. and J.C. Batten. 1988. 'Redistribution of Nitrogen and Phosphorus on Palace Leas Meadow Hay Plots as a Result of Aftermath Grazing'. *Grass and Forage Science*, **43**: 105–10.

Shiel, R.S. and Chapman, J.C. 1988. 'The Extent of Change in the Agricultural Landscape of Dalmatia, Yugoslavia, as a Result of 8,000 Years of Land Management'. In J.C. Chapman, J. Bintliff, V. Gaffney and B. Slapsak (eds), *Recent Developments in Yugoslav Archaeology*, BAR Int Ser **431**: 31–44.

Shiel, R.S. and Stewart, A. 2003. 'The Soils and Agricultural Potential of the Thespiai Area'. Unpublished report of the Cambridge–Durham Boeotia Project, available from R.S.Shiel@ncl.ac.uk

Sillitoe, P. and Shiel, R.S. 1999. 'Soil Fertility under Shifting and Semi-continuous Cultivation in the Southern Highlands of Papua New Guinea'. *Soil Use and Management*, **15**: 49–55.

Smith, L.P. 1976. *The Agricultural Climate of England and Wales*, MAFF Technical Bulletin 35. London: HMSO.

USDA, 2003, *Food and Nutrition Information Center*. http://www.nal.usda.gov/fnic/

van Andel, T.H. 1987. 'The Landscape'. In T.H. van Andel, and S.B. Sutton (eds), *Landscape and People of the Franchthi Region* (Bloomington: Indiana University Press), pp. 3–64.

West, M.L. 1988, *Theogony and Works and Days*. Oxford: Oxford University Press.

Whitehead, D.C. 2000. *Nutrient Elements in Grassland: Soil–Plant–Animal Relationships*. New York: CABI.

Wicherek, S. (ed.). 1993. *Farm Land Erosion in Temperate Plains Environment and Hills*. Amsterdam: Elsevier.

Wilkinson L.P. 1978. *The Georgics of Virgil: A Critical Survey*. London: Cambridge University Press.

9

Human Interaction with Soil-Sediment Systems in Australia

R.J. Wasson

INTRODUCTION

Most environmental consequences of economic development are unanticipated, ignored because they are not believed to be significant, or responded to either by attempts at repair or by adaptation well after they have become plain to see. Many environmental changes occur well after their human or non-human causes, making the link between cause and effect difficult to identify. There is therefore a delay between cause and effect, response, and either improvement or adaptation. History is essential to understand such a system of interconnected processes and phenomena (Newell and Wasson 2001) so that changes and delays resulting from present and future economic development can be anticipated; the causes of current changes identified; and more timely responses developed.

Environmental history therefore can be of great practical value for present day natural resource and environmental management (Wasson and Clark 1985, Wasson 1994, Dovers 2000). River catchment (or watershed) response to land cover and land use change, and the ways in which people have perceived and responded to the changes in rivers in particular, are the subject of this chapter. The erosion and transport of soil and sediments through river catchments provides a useful case for analysis because, in Australia, European settlement in catchments produced a shock (or perturbation) that can be identified and is still working its way downstream. Much of the European history of Australia is recorded, and in some places former farmers are still able to remember river changes set off by removal of native vegetation. These people, and others, can also tell us how and why they responded to environmental change. All of the difficulties of oral history cannot be overcome, but the method of Starr (1989) has been used at different study sites to at least provide consistency of approach.

THE SOIL-SEDIMENT SYSTEM

The idea of a soil-sediment system needs some explanation. Soils are the result of weathering both of rocks and of mineral sediment that has been transported from erosion of rock, weathered rock, and soils. Soils developed in rock are not only the product of weathering and new mineral formation, but also involve transport of materials as illustrated by an important conclusion from pedological research in Australia. Duplex soils consist of a coarse-grained topsoil (A-horizon) and a fine-grained subsoil (B-horizon). The topsoil has been washed over the B-horizon (Paton et al. 1995) and the B-horizon is largely the product of ***in-situ*** weathering, with some downslope creep and a small input of clay by eluviation from the topsoil. So duplex, or texture-contrast, soils are a product of both weathering and slopewash.

All soils on hill slopes must be a result of such a combination of processes but it is easiest to demonstrate in the case of duplex soils.

Topsoil formation in duplex soils is aided by animals. Burrowing ants, worms, termites, wombats, and animals that disturb the soil surface to build nests, such as mound-building fowl and lyre-birds, make soil available for erosion by rain splash and slopewash. Fine particles are moved downslope leaving coarse particles behind to form the A-horizon. The topsoil becomes finer downslope, grading into alluvium and colluvium in valley floors. Mobilisation of this material by concentrated flows produces the alluvial deposits of rivers even further downstream. Once deposited, alluvium is also subject to weathering, new mineral formation and horizonation to form different soil types. The rate and type of soil formation in actively accumulating alluvium is dependent upon all of the usual factors of soil formation, modulated by the rate of sediment accumulation. When a floodplain is abandoned by a river, either by incision of the river or by uplift, a river terrace is formed. Soil formation in such a landform proceeds without the effect of continuing sediment accumulation; except for the occasional very large flood in the case of low-lying terraces, and by dust accumulation.

Soils are therefore intimately connected to sediment transport and deposition. The continuum of the soil-sediment system can be thought of as a cascade of energy and material that begins with photosynthesis and the breakdown of rock by biological agencies, and ends with the transport of sediment and organic matter by rivers into the ocean.

Such a cascade is never in equilibrium with causative factors and perturbations. This is because of the time lags between perturbing influences, such as climate change or land use change, and the time required for the effects of a perturbation to make its way through the soil-sediment system. A change in sediment transport rate at the mouth of a river, for example, may reflect a perturbation in the headwaters decades or centuries before. This idea is extremely important for analysis of human response to change where delays are large, and the cause of change distant in space.

A CONCEPTUAL FRAMEWORK

A conceptual framework is needed for analysis of the biophysical changes triggered by land use and climate change, for the human response to such changes, and for the biophysical changes caused by human responses.

A Spatial Framework

The first part of this framework is a conceptualisation of the sources, transport and sinks of sediment in a river catchment (Figure 9.1). The catchment is subdivided into sub-catchments according to the Strahler method, as follows. Each division is given an order number, beginning with zero for the sub-catchment upstream of a channel head, and then subsequently numbered downstream. Channels are divided into links, so a first-order link is that which begins at the downstream margin of a zero-order sub-catchment and ends at the first stream junction downstream. A second-order link is produced by the junction of two first-order links. A third-order link is produced where two second-order links join. A first-order sub-catchment is the total area from which water drains to a first-order link. A second-order sub-catchment is the total area from which water drains to a second-order link; and so on for higher orders.

Soil is eroded from hill slopes by sheet erosion, rill erosion, gullying, and landslides. Some of this material reaches a river channel link, and the rest remains on colluvial footslopes perhaps to be moved during subsequent runoff events. Therefore, some eroded material goes into a sink (Figure 9.1), at least temporarily. Over long periods, all such material reaches the channel network. Within the channels, some sediment is temporarily stored (the in-channel deposits in Figure 9.1), to be moved later. Second–order links commonly are flanked by floodplains, where sediment is stored and rudimentary soil formation occurs. Floodplains are also sources of sediment as channels migrate laterally, very large floods enlarge channels, or land use reduces the resistance of stream banks to erosion. Rivers also incise, that is they cut downward into their own deposits to produce a channel that contains all flows. This process releases very large amounts of sediment to downstream reaches. Further downstream, for links greater than third-order, floodplains are wide enough to prevent soil eroded from adjacent hill slopes reaching the channel. Still further downstream, sediment reaches a reservoir, lake, estuary or ocean.

Ideally each of the inputs and outputs to the components in Figure 9.1 is quantified, and a history of change provided in the form of a time-series of fluxes.

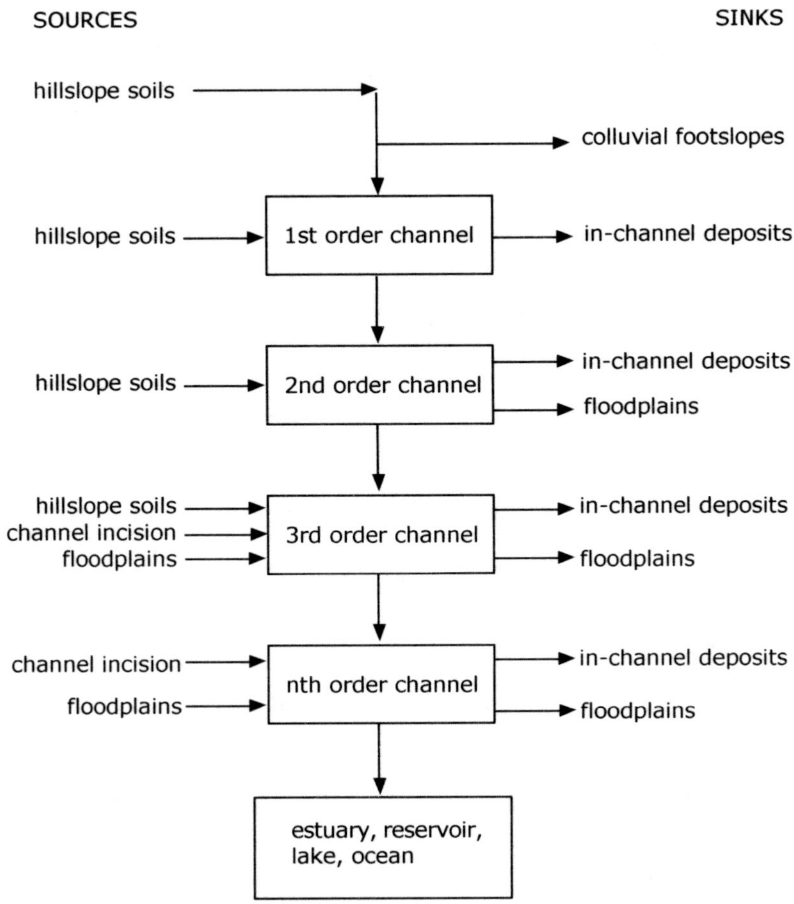

Figure 9.1. River catchment soil–sediment spatial framework

That is usually only possible over periods of decades to centuries by documenting changes in stores. The result is a time-resolved sediment budget that not only allows analysis of perturbations and delays, but also provides vitally important information about ways to reduce the impact of sediment transport. There are several cases in Australia and elsewhere where the identification of sediment sources has dramatically altered catchment management and led to environmental improvement (e.g. NSW Department of Land and Water Conservation 2000). Sediment budgets also provide a straightforward way of linking to other disciplines because many use a budget approach (Wasson 2003).

Biophysical Change

The second part of the conceptual framework requires a way of including perturbations and biophysical responses. Each of the components of Figure 9.1 is susceptible to change by external forces, such as climate change, large magnitude rainfall events, and/or land use change. They also change by internal processes such as by passing a gradient threshold so that an external perturbation that previously had no effect now has an effect.

One way of envisaging these changes is by means of the diagrams of Figure 9.2. Perturbations can be thought of as ramps, steps and spikes. The ramp (A) may be the gradual change of land cover as hunting and gathering gives way to patches of settled agriculture and then large areas of agriculture. The response to this ramped change may be instantaneous (solid line) in sensitive systems where soils are highly erodible, or delayed in low sensitivity systems where soils are resistant to erosion. The delay is equivalent to a threshold. Where soil organic matter and tree roots persist after clearing, erodibility is low and a threshold (t) caused by organic matter decay must be passed before erodibility is increased. Another threshold might be the connection of patches of cleared land sufficient to allow eroded soil to reach a stream channel, so that the dashed line represents stream sediment load.

The step change perturbation (Figure 9.2B) mostly occurs in places where industrial agriculture has been introduced very quickly, such as in Australia and North America. Here the tools were available for rapid clearing of forest, for example, producing a near-instantaneous increase in runoff and soil erosion in the case of sensitive soil. Where organic matter helps the soil to resist erosion, erosion may increase slowly soon after clearing (Ba), or a threshold (t) of soil organic matter decay has to be passed before erosion increases markedly (Bb).

Both ramps and steps are representations of changes that do not recover quickly, such as land cover change. The responses, on the right-hand side of Figure 9.2, are represented as smooth (average) conditions in A and B. In reality, there will be infrequent large magnitude events superimposed upon the smooth curves. Such 'spikes' are shown separately in Figure 9.2 C and D for ease of representation. A single spike (C), such as a large rainfall event, is responded to quickly by soils on steep slopes that fail by for example rotational slump (Ca). On gentler slopes, soils may fail by earthflow, and so the response to a single spike of rainfall will be slower and last longer as a period of movement (Cb).

Multiple spikes produce rapid recovery in highly sensitive systems such as small catchments where there are large numbers of channels that drain quickly (Da). A spike of runoff and sediment transport in larger catchments will wane more slowly because of the larger travel time. If the interval between spikes is shorter than the recovery time then sediment transport will continue to increase, as depicted in Figure 9.2(Db), making it appear that there is an increasing cause.

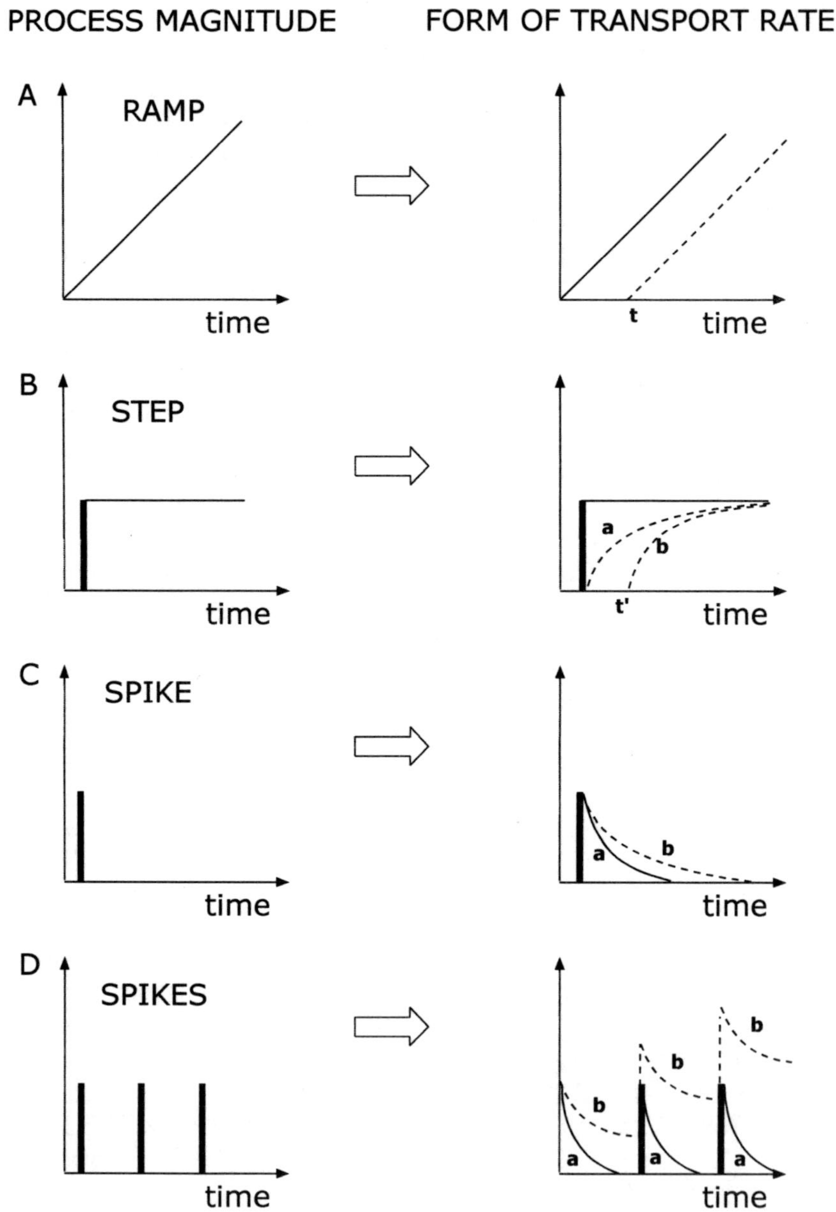

Figure 9.2. The temporal framework

Causation and Responses

The third part of the conceptual framework involves causal loops that run from a perturbation (either human or non-human in origin) to a response in the biophysical system, and then a response in the human system involving either amelioration or adaptation to a perturbation.

We can imagine situations in which biophysical change occurs but it is not observed and so no action results (Figure 9.3); for example a change to soil formation processes following land cover change. Other biophysical changes, such as enhanced sheetwash or fresh gully erosion, are observed but the impact on people is slight or non-existent at least in the short term, and no action is taken. If human impact occurs, but the cause is not understood, there is no action. In this case, the cause might not be known because it is distant in time and space.

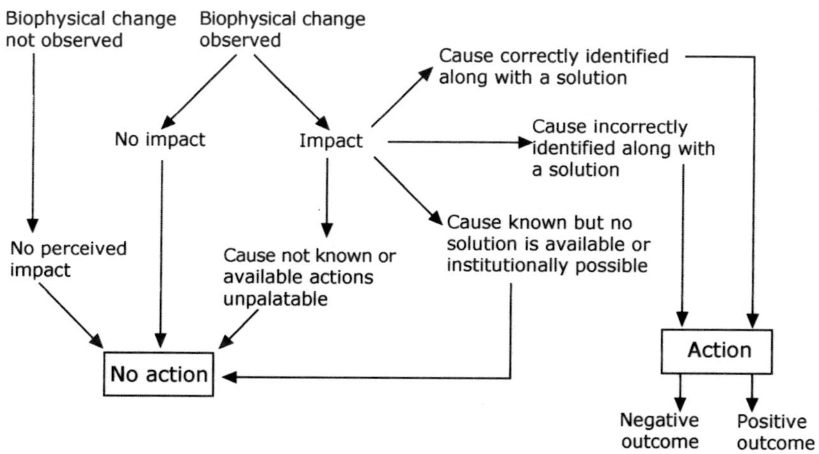

Figure 9.3. Links between biophysical change and human responses

Other examples contained in Figure 9.3 include the case of human impact where the cause of the biophysical change is known but there is no solution that is economic, technically feasible, politically tractable or socially acceptable. Therefore, there is no action.

All of the cases described so far from Figure 9.3 explain why no action occurs for very long periods of time following biophysical change. Particularly in the last few decades, impacts on non-human organisms have attracted attention. So along with impacts on people in Figure 9.3, impacts on other organisms could also be considered. 'Impacts' in Figure 9.3 are therefore diverse.

The remainder of Figure 9.3 consists of cases that result in action, for better or worse. The cause of a biophysical change may be incorrectly identified and a solution is found that is believed to be viable, leading to either no amelioration of the biophysical change or a worsened condition. If the cause is correctly identified, and a solution found that is tractable, then action will lead to a positive outcome. There may still be a delay between observing a biophysical change and taking effective action because the cause may be discovered well before a viable solution is found; or other matters have a higher priority for expenditure and effort; or institutional arrangements do not permit action.

To Figure 9.3 can be added a fourth component, namely feedback from action to the biophysical system. If the cause of the initial biophysical change is correctly identified, and an effective solution adopted, then the damaging biophysical change may be slowed, stopped, or even reversed. Depending upon the effectiveness of actions, there may or may not be further investigation and action. If action is adaptation, then the only continuing biophysical change will be according to the response trajectory that is determined by the processes and feedbacks within the biophysical system. In some cases, a biophysical system once perturbed will relax to a new state or take a new form which may be less problematic for people (and other organisms). An example is the triggering of gullying by land cover reduction (Wasson and Sidorchuk 2000). A gully network grows until headcutting stops as the catchment area to each gully head is reduced to a critical size. Sediment yield in such a system therefore rapidly increases as the network grows, and then falls as the network stops growing. Downstream sediment transport also slows, not as a result of human intervention but as a result of a negative feedback that stabilises the system.

In summary, the conceptual framework that is adopted in this chapter has the following components:

- a spatial framework (Figure 9.1) based on the hierarchy of the stream network and its sub-catchments, taking explicit account of the sources, sinks and yields of, in this case, sediment

- a temporal framework (Figure 9.2) in which the magnitude of processes inducing change are represented by a few simple conceptualisations, and the responses, measured as changes in form (of rivers particularly) and transport rate of sediment, are conceived as averaged trajectories. If individual events are known, the curves can be represented more accurately.

- the causal links between observations, impact, and action (or no action) in which delays are implicit (Figure 9.3). Impacts on both people and other organisms can be treated within this component.

- The feedbacks from action, either as intervention or adaptation, to the biophysical system and then to impacts.

While this chapter is devoted to the soil-sediment system, the conceptual framework can be used to analyse nutrient movement, pollutants, and even organisms.

This conceptual framework is a product of geomorphology and system dynamics developed principally in the latter part of the twentieth century. It will not be superimposed on the ideas and perceptions of people in the past, except where they have used a similar scheme. Rather it will be used as an analytical framework, for identifying which part of the soil-sediment system is being discussed, and for conceptualising the human-environment relationship. There are no studies in Australia that allow all of the components to be documented completely. The framework is therefore an ideal, and can be used in future studies.

Three river catchments in Australia will now be examined, and the framework used as much as possible. The three catchments come from the southeast, southwest, and northwest of the continent (Figure 9.4)

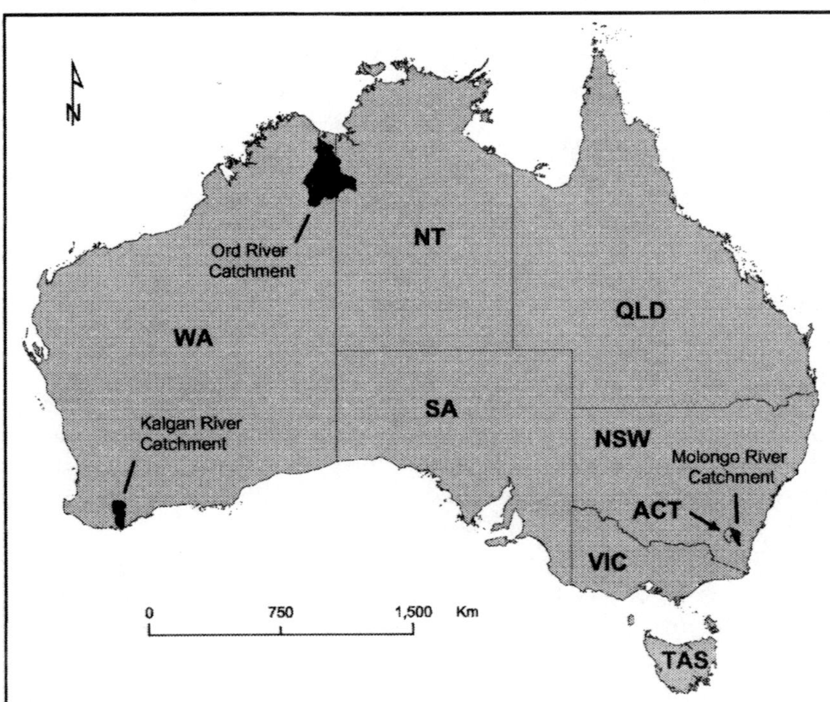

Figure 9.4. Location map of the three river catchments

Molonglo River Catchment

The Molonglo River Catchment (Figure 9.4) has an area of 717 km^2 on the Southern Tablelands of New South Wales (NSW) and the Australian Capital Territory (ACT). The river flows into the ACT from NSW, through Canberra and joins the Murrumbidgee River. The Molonglo has been dammed in the ACT to create Lake Burley Griffin, for recreational use and beautification of the nation's capital, and named after the principal designer of Canberra, Walter Burley Griffin. The catchment is largely used for grazing sheep and cattle, and housing in small areas, while the upper reaches produced small quantities of low-grade timber during the early years of settlement by Europeans.

Sheep and shepherds entered the region during the 1820s and soon after subsistence gardening and cropping was established near homesteads. There was little removal of trees as the valley floors were naturally treeless, providing ample grazing land (Starr et al. 1999). Grazing of grassy hill slopes and woodlands, along with the end of burning by Aboriginal people, radically altered the terrestrial ecosystem. On steeper hillsides, the understorey thickened in woodlands, but on grassy slopes grass cover was thinned. Introduced rabbits increased grazing pressure, and the establishment of permanent dams as watering points for sheep and cattle almost certainly increased kangaroo numbers by providing water, and grazing pressure therefore increased.

The Molonglo catchment is susceptible to droughts of several years duration. Cold winters and hot summers reduce the length of the growing season as well, and recovery of grass cover from grazing can be slow and often incomplete. Constant grazing pressure from domestic stock, rabbits and kangaroos has almost certainly increased hill-slope erosion by sheet and rill processes. But grazing was not restricted to hill slopes. The valley floors in this catchment at the time of European settlement were often swampy with poorly defined channels. They were sources of water and food for stock, and apparently were preferentially grazed.

Diaries of early settlers, survey maps, and stratigraphic evidence have been used to document changes to the soils and erosion rates of the Molonglo catchment (Eyles 1977a, b; Wasson et al. 1998). Particular attention has been paid to the Jerrabomberra Creek catchment, an area of 130 km^2 that is a sub-catchment draining into lake Burley Griffin (Figure 9.5).

The swampy valley floors contained oval ponds linked by short lengths of channel, known locally as chains-of-ponds. They appear on survey plans from the nineteenth century in second and third order sub-catchments, and continuous channels are shown on plans of higher order sub-catchments. First order sub-catchments do not appear to have had chains-of-ponds, but were drained by shallow continuous channels.

By the 1850s there is diary evidence of incision of the swampy valley floors, and by 1912 the geographer Griffith Taylor described the lower reaches of Jer-

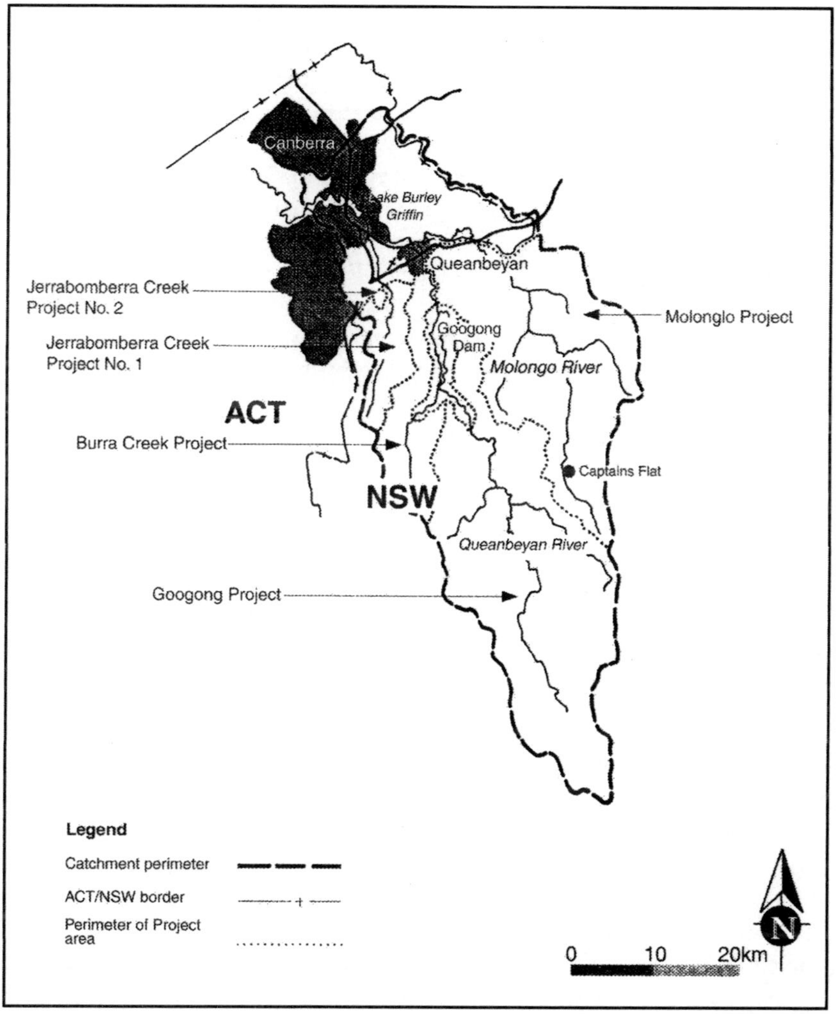

Figure 9.5. Location of the soil conservation projects in the Molonglo Catchment

rabomberra Creek much as they are today; that is, vertical walled and incised. The diary account is by John William Buckle Bunn, a remarkable teenager living on Woden Station. His sketches and descriptions show that incision of Woden Creek, a tributary of Jerrabomberra Creek, had occurred a little before his diary entry. This conclusion is evidenced by his sketch of tree roots protruding near the top of the gully well. A well had been sunk in the bed of the gully, the remains of which can be seen today (B Starr, pers. comm.). There is no evidence however of any response to this incision of Woden Creek. It is not until the 1950s that attempts were made to control stream-bank erosion by planting poplars and willows, on the middle reaches of Jerrabomberra Creek. At one particularly active bend, car bodies were piled at the base of a channel wall and poplars planted. The car bodies are now stranded in mid-stream, showing that they had no effect on stream-bank erosion.

For most of the history of farming in the Molonglo catchment, farmers concentrated their efforts on productive land. They introduced non-native grasses, a process known as 'improving pasture', managed stocking rates, fenced land, established dams, controlled feral animals, reduced weeds as best they could, and benefited from the reduction of rabbit numbers by the spread of the myxoma virus in the 1950s (Starr et al. 1999).

It is likely that farmers noticed the incision of valley floors and the destruction of valley floor swamps, because they went to some trouble in some places to build bridges where once crossings of valley floors were simple matters (Starr, et al. 1999). But apart from planting a few trees they either did not have the resources or the motivation to do more to stabilise the channels. It seems that the farming community adapted to the changed valley floors.

The Lake Burley Griffin Catchment Protection Scheme

As early as 1909, the idea of a lake in Canberra was suggested by the NSW district surveyor, Charles Robert Scrivener (NSW Department Land and Water Commission 2000). Walter Burley Griffin's plan for the city included a lake, an idea adopted in 1925 (Lyne 1978). The lake formed behind Scrivener Dam in 1964.

Before construction of the dam, concern was expressed about sedimentation in the lake, and turbidity. The visual prominence of the lake, its symbolic significance as the Water Axis in Griffin's plan for the city, and its recreational potential were all reasons for concern about sedimentation and turbidity. In 1958 the National Capital Development Commission asked the NSW Soil Conservation for advice. L.J. Burham of the Soil Conservation Commission reported in the same year:

> Present rates of silt and sand movement and stream bank erosion will cause considerable deposition in a relatively short time. It must be considered also, that unless land use in the catchment is greatly improved within the next few years, the erosion rate and deposition will probably be much higher 20 or 30 years hence.

Discolouration of the water, sediment shoals and reed beds were of concern, and commentary on erosion continued:

> If allowed to go unchecked, this could lead to the silting up of large sections of the lake (National Capital Development Commission 1962).

This concern appears to have grown out of experience of reservoir sedimentation in the United States, and concerns about storage loss in the reservoirs being constructed in the Snowy Mountains of NSW for an ambitious hydroelectric scheme (NSW Department of Land and Water Conservation 2000).

Analysis in the Molonglo Catchment was restricted to survey of erosion phenomena (Durham 1958), and an estimate by Munro et al. (1967) of the amount of sediment entering the lake. This estimate was based upon very few measurements of sediment concentration and was an over-estimate. Crucially, there was no estimate of the proportionate contribution of sediment from different sources, so there was no clear idea of how to reduce erosion.

In 1964 the Commonwealth and NSW Governments established the Lake Burley Griffin Catchment Protection Scheme. The agreement had three components (Smith 1968):

- A soil conservation program for 10 years in the NSW part of the catchment, to be carried out by the Soil Conservation Service of NSW;
- One-third of the cost would be met by the Commonwealth for structural soil conservation works.
- The remaining two-thirds of the cost would be met by the NSW Government and landholders in equal amounts.

This is an early engagement of landholders in what was essentially a top-down government conservation programme. There is little record of the views of the landholders, but they seem to have accepted the need for the scheme and contributed as best they could, mainly in-kind. That is, they provided labour, adjusted their land management practices, improved pastures, and fenced land to control stocking rates.

The first stage of the Protection Scheme was in the Jerrabomberra Creek catchment where erosion was recorded by survey to be particularly severe. Emphasis was given to building absorption banks on hill-slopes to slow runoff and therefore reduce sheet erosion, and gully control structures. Diversion banks were built to catch runoff before it could enter gullies, and dams sited to provide water for stock and stop gully growth. Deep ripping and chisel ploughing of hill-slopes was designed to encourage rainfall infiltration, thereby increasing vegetation and reducing sheet erosion.

The Jerrabomberra works were complete in 1970, but stream-bank erosion continued. Squires (1977) claims that one of the intentions of the structural works was to reduce stream-bank erosion by reducing runoff and peak flows. But it is also clear that the NSW Soil Conservation Act 1938 did not empower the Soil Conservation Service to become directly involved with stream-bank erosion; leaving aside the difficult issue of distinguishing a gully from a stream in many cases. The NSW Water Conservation and Irrigation Commission had authority over streams, but attempts to involve them in the Molonglo catchment broke down in 1973 (Starr 1991).

While the significance of stream-bank erosion had been recognised as early as 1958, its control had thus far been indirect; and it is not at all clear that peak flows in large erosive events were reduced by structural works because there are no flow data for most of the catchment. The views of local landholders on these matters are not recorded, except for a report in the *Canberra Times* (10 May 1966) of the concerns of Mr Bernard Morrison, a local grazier, which have been interpreted to refer to stream-bank erosion not being under control (NSW Department of Land and Water Conservation 2000).

The Protection Scheme continued in stages (Figure 9.5) with most emphasis on sheet erosion and gully stabilisation until quantitative sediment sourcing studies began to alter the thinking of the NSW Soil Conservation Service. By estimating the erosion rate by sheet processes from different land cover types, from gullies,

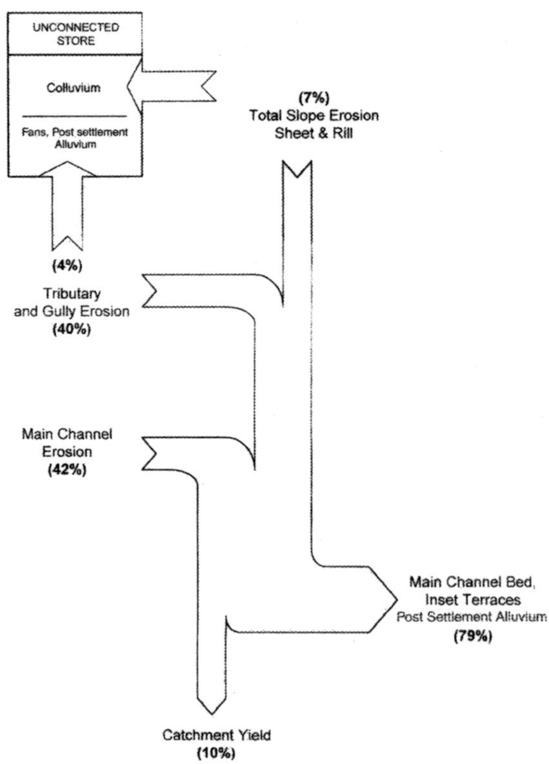

Figure 9.6. Sediment budget for the Jerrabomberra catchment

and by stream-bank erosion, it was possible to construct a sediment budget for the Jerrabomberra catchment (Figure 9.6). This showed that only 5–10 per cent of all sediment removal from the catchment was derived from sheet and rill erosion. The remainder came from gullies and stream incision, with most from stream incision (Wasson et al. 1998). This result was consistent with results from the use of radionuclide tracers of topsoil erosion in a study of the sediments trapped in Lake Burley Griffin (Wasson et al. 1989).

By applying geomorphic, stratigraphic, and historical techniques, it was also possible to reconstruct the history of sediment yield from the Jerrabomberra catchment (Figure 9.7) (Wasson et al. 1998). This showed that most erosion had occurred in the nineteenth century and that gullies had mostly not cut headward since 1944 when the first aerial photographs were taken. Also, while stream-bank erosion remained the single largest contributor to stream sediment loads, it was now at a rate much lower than 100 years before. So most of the erosion was over before the Protection Scheme began.

In 1992/93 the NSW Soil Conservation Service re-oriented the Scheme to focus on the main sediment sources. Streams were fenced and de-stocked, and protective devices such as rocks in wire cages (so-called 'rock sausages') were placed at the base of eroding stream banks and planted with native species. In a few cases channels were re-aligned and in-stream bars removed to stop flow being directed into banks.

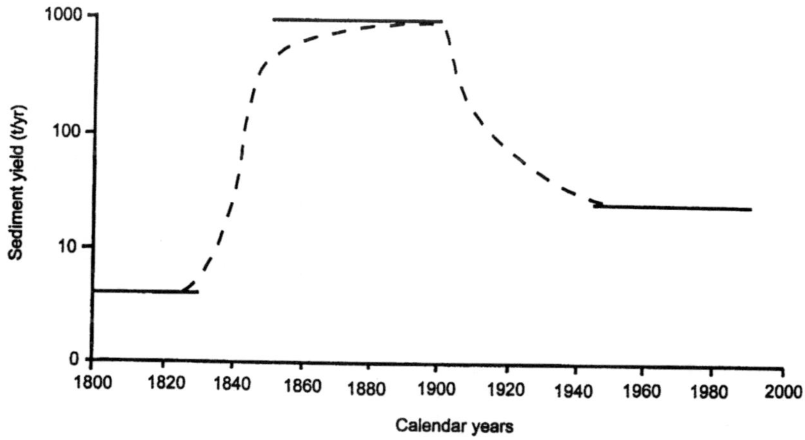

Figure 9.7. History of sediment yield from the Jerrabomberra catchment

Lessons had also been learned about broad-acre soil conservation (Fogarty et al. 1989). Absorption banks reduced access across paddocks. Grass flumes often failed because grass cover was difficult to maintain in a climate with a short growing season. Dams on larger gullied catchments could now be built by using PVC pipes as primary spillways, because earth spillways, where flows are large, often fail. Finally, dams were being located more strategically to catch water and sediment from ploughed paddocks, roads and gullies. It is clear that the NSW Soil Conservation Service had been pushed very hard by the technical difficulties of the Protection Scheme, and innovation had occurred throughout the Scheme. Over 154 km of banks had been built and over 226 km of gullies shaped or filled. Nearly 1000 gully control structures or dams had been built, often at the expense of landholders. Not only had erosion been slowed but land was now more profitable (NSW Department Land and Water Conservation 2000).

The Commonwealth's interest in the Scheme came to an end in 1998 as it was clear that most continuing effort would mostly benefit landholders, and all that could be done to save Lake Burley Griffin from sedimentation had been done. Also, the Scheme was out of step with total catchment management objectives of solving multiple problems rather than just erosion and sedimentation. Revegetation has become a priority, often planting trees where they did not exist naturally!

The environmental history of soil erosion and conservation in the Molonglo Catchment is summarised in Table 9.1 within the conceptual framework constructed earlier. The spatial framework used by farmers appears to have been their own properties, and soil conservators thought of the catchment as a collection of properties rather than as a hierarchical nesting of catchments as in Figure 9.1. The hierarchical view, with connectivity between hill slopes and channels, and between channels, only became clear in the 1980s and 1990s as scientists interacted with soil conservators. Treatment under the Protection Scheme of important sediment sources only became focused when a sediment budget was available.

There is little detail of Perturbation and Biophysical Responses, but it can be inferred that grazing of valley floors led quickly to incision and export of sediment. While there is no documentation, it is likely that observant farmers would have seen sheet erosion once stocking rates became high in the late nineteenth century. It is not clear that they did anything about this situation until rabbit numbers were reduced by the myxoma virus and improved pasture became possible in the 1960s with the aid of fertilisers.

People responded to incision by adaptation, although there were a few attempts at channel stabilisation before the Protection Scheme. But most farmers focused on their properties, and it was not until a top-down government program of catchment protection was developed that landholders became interested in larger areas. While this conclusion is reached largely by inference, the few remaining farmers of sufficient age support this conclusion. Soil conservators altered their views of

Table 9.1. Summary of soil erosion and conservation in the Molonglo catchment

Spatial framework	Perturbations and biophysical responses	Causation and responses, and feedback
Priority given to sub-catchments badly eroded by sheet processes. Other processes noted. No explicit attention paid to connectivity. Late in the Scheme, quantification of sediment sources led to more explicit treatment of particular locations of high sediment yield.	Grazing of valley floors led to rapid incision (as a step change, Figure 2). Delay between grazing and incision probably less than 10 years. Sheet erosion probably became obvious soon after heavy stocking in the late nineteenth century.	Early recognition of incision but no effective control until 1990s; a delay of at least 30 years. Most responses were to sheet and gully erosion for both technical and institutional reasons. Soil conservators adapted their methods both as they learned from their traditional techniques, and as they focussed on stream-bank erosion.

appropriate techniques, and their priorities shifted as quantification of sediment sources became available. The institutional framework also changed, allowing them to move off the hill slopes and into the streams when the Soil Conservation Service was merged with NSW Government agencies responsible for water resources. Total catchment management has broken down many of the institutional and conceptual barriers that once plagued natural resource management in Australia.

Ord River Catchment

The Ord River catchment in the East Kimberley region of north-western Australia (Figure 9.4) has an area of 46,000km^2 upstream of Lake Argyle, a reservoir that began to fill in 1972 to supply water for irrigation and that has also been used for recreation, aquaculture and most recently hydroelectric power generation. The catchment lies in the semi-arid monsoonal tropics of Australia, with a long hot dry season and a short wet season with periods of intense rain.

Alexander Forrest reported in 1879 that the area is '… so well watered, and the soil so good …' and is well grassed (Forrest 1880). He was referring particularly to what is now known as the Hardman Syncline (Figure 9.8), a geological feature consisting of Cambrian age sedimentary rocks (Wasson et al. 2002). His glowing report of a healthy grassland enthused cattlemen in eastern Australia who moved cattle into the area by 1884.

By 1905, there were substantial numbers of cattle in the area (Bolton 1953). Weir, a pastoral inspector, reported in 1906 that river frontages had been largely de-vegetated and Forrest's 'fine grassy plains' were bare. The prevailing leasehold policy of subdivisions based on river frontages actively discouraged development of permanent watering points away from the rivers (Bolton 1953). Yet cattle had managed to de-vegetate large areas away from the rivers, as well as along them; albeit in a time of drought.

Teakle (1944) and Medcalfe (1945), two agricultural scientists, reported massive erosion. They saw exposed tree roots, soil pedestals, and an absence of A-horizon (topsoil) over large areas. Sheet and rill erosion of between 15 and 30 cm was estimated. This amounts to 3,500 to 7,000 t/km^2/year, a very high rate of erosion.

Forrest noted gullies along the Behn River frontage because he nearly fell into them, but he did not record other gullies either along or away from rivers. Today gullies are common, particularly across the Hardman Syncline and also along the frontages of most rivers. Teakle and Medcalfe focused on sheet and rill erosion presumably because of their interest in the pastoral enterprise. Medcalfe, however, concluded that the serious erosion he observed in the Hardman Syncline must be producing 'excessive' sedimentation downstream.

In 1917, the Ord River Station, a cattle property first occupied by N. Buchanan in 1884, was subdivided into seven stations (Pratchett 1990). Impetus was given to the cattle industry in 1918 by the opening of a meat works at Wyndham and the beginning of live cattle export. Grazing pressure probably increased as a result of both subdivision and increased access to markets.

The irrigation potential of the alluvial soils in the vicinity of the modern town of Kununurra was early recognised. Agricultural experiments between 1941 and 1945 by a member of the pioneering Durack family (Durack 1945) were promising, and initial exploration for dam sites by the Director of Works Western Australia occurred in 1941 (Dumas 1944).

Durack also recognised that uncontrolled grazing by cattle leads to pasture destruction and erosion. But it was not until the 1950s that a systematic survey of erosion occurred (Stewart et al. 1970). Meanwhile, the idea of further development of north-western Australia, either for pastoral use or irrigated agriculture, would serve no useful purpose (Ord River Irrigation Area Review 1978). But factors other than economics were at play. Decisions taken after the Second World War were made when the threat of 'Asian invasion' had just been shown to be possible. The defeat of the Japanese Imperial Army in New Guinea had been a 'close-shave', and, for many, demonstrated the correctness of the earlier policy of populating the north because if 'we' did not the Asians would. In addition, the developmental and political ambitions of the Western Australian government drove plans for irrigation on the Ord River.

As the bureaucratic and political arguments about a dam and the development of irrigation continued, there is no evidence that stocking rates declined. Fencing

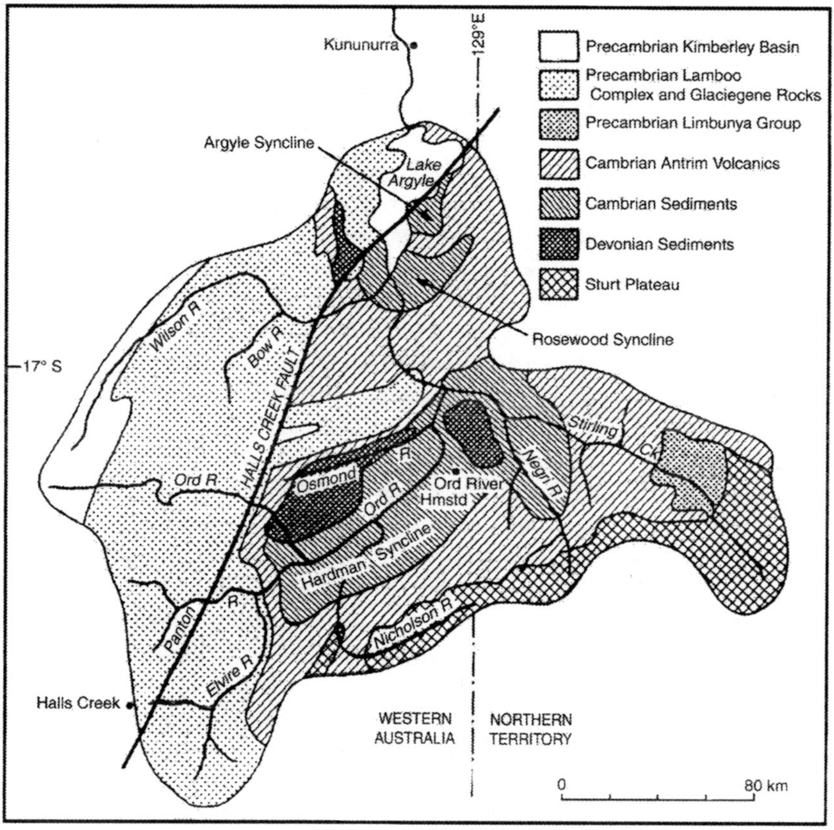

Figure 9.8. Geological map of the Lake Argyle catchment

occurred, allowing the control of stock locations, and bores were sunk to provide water away from the river frontages. This may have reduced stocking pressure near the rivers, but the damage in the form of gullying had already occurred before the 1960s when the first high quality photographs of erosion phenonoma were taken by officers of the WA Department of Agriculture (A. Payne, pers.comm.). Written descriptions from a decade before (Stewart et al. 1970) suggest that gullying had not markedly changed from that photographed in the 1960s.

In 1959 the decision was taken to build a Diversion Dam near Kununurra, prior to construction of the main Ord River Dam (Patterson 1965). The alarming reports in 1944 of soil erosion upstream led to revegetation of part of the visibly worst degraded area in the Hardman Syncline.

This response was thought to be essential to save the Diversion Dam from sedimentation. Large areas of the soil of the Hardman Syncline (Figure 9.8) had been de-vegetated by cattle. The soil surface rapidly sealed as pores infilled with particles washed across the surface and rain splash 'puddled' the surface. Infiltration by rainfall into the soil was reduced and revegetation by natural means was almost impossible. This sequence of events is common throughout the world's semi-arid tropics and sub-tropics (Walker et al. 1981). There is generally a ten-fold reduction of infiltration rate when a grass-litter cover is removed to create bare soil. This change is supported by Fitzgerald's (1968) infiltration measurements in the area that became known as the Ord River Regeneration Reserve (ORRR) where most of the revegetation occurred.

Mechanical pitting of the soil surface by means of tools pulled by tractors was necessary to break the surface seal, along with direct seeding into the pits. Exotic species were used because they produced more seed than the native species. Absorption banks were also built to slow runoff and thereby reduce sheet erosion, and to trap seeds and speed regeneration in areas away from lines of pitting. Re-seeding was restricted to areas that could be reached by tractors, limiting it to relatively flat areas between gullies. Some attempts were made to re-seed gully walls, where they were gentle, and some seed reached gully floors resulting in limited revegetation. But there is no evidence of an explicit attack on gully erosion. Once again, the agricultural scientists involved in this superhuman revegetation program believed that reducing runoff from areas between gullies by revegetation would slow gully erosion. (A. Payne, pers. comm.).

In 1964 the Western Australian government continued to lobby the Commonwealth government for financial assistance with the Ord Irrigation Scheme. By 1965 there was a new argument for building the Main Dam; to slow sedimentation of the Diversion Dam (Patterson 1965). Revegetation of the most severely eroded land was assisted in 1968 by resumption of the land by the Western Australian Department of Agriculture, and the creation of the ORRR. While the active reseeding program was largely finished by this date, the resumption allowed the Department to control stocking rates and thereby maintain vegetation cover.

Funding for the Main Dam was approved and the reservoir began to fill in 1972. Measurement of sediment concentrations and discharges in the Ord River by the Western Australian Public Works Department were used to estimate the annual average sediment transport rate to the reservoir at $29.76 \pm 8.60 \times 10^6$ t/yr (Wasson et al. 2002). The large uncertainty of 8.60×10^6 t/yr, which is 29 per cent of the average, is entirely the result of interannual variability of flow rates and therefore sediment transport rate. Errors due to sampling are unknown.

In 1985 a survey of the total amount of sediment in Lake Argyle was carried out by the Western Australian Water Authority (Wark 1987), using coring, re-survey of cross sections established when the dam was built, and echo-sounding. On this occasion measurement errors were estimated, giving a result of $23.5\pm4.7\text{x}10^6$ t/yr for the average sedimentation rate. The two figures are statistically identical, and R.J. Wark, an insightful engineer, calculated that revegetation of the Hardman Syncline had not been successful.

This conclusion led to a sediment tracing study that attempted to determine if revegetation of the soft rocks and erodible soils of the Hardman Syncline was the correct strategy. Mineral particle magnetic properties, uranium series nuclides, and rare earth isotopes, were used to determine which sub-catchments produced most sediment. In addition, the proportion of topsoil in river sediments was determined measuring the amount of ^{137}Cs in Lake Argyle sediments, a radioactive nuclide that occurs in topsoils across the planet. Two major conclusions were reached (Wasson et al. 2002). First, the Hardman Syncline, and similar geomorphic zones nearby, produced most of the sediment in Lake Argyle. Second, most of the sediment was produced by gullying and channel erosion, not by sheet and rill processes. These conclusions meant that while the revegetation project had targeted the right part of the catchment, the major erosion process had not been substantially affected.

While the tracer study was going on, officers of the Western Australian Department of Agriculture had independently recognised that gullies were an important source of sediment. They had built small brushwood and bamboo fences in some gullies, but they were washed away during the next wet season. Revegetation of the gullies by an expansion of the 1960s scheme was not seriously considered, and so the ORRR was destocked of both domestic and feral cattle. The low value of the land, and a shift to a provider–providee method of internal allocation of funds (i.e. the need for any research project to have an internal user other than the proponent of the research) within the Western Australian bureaucracy made any other course of action impossible. However the Department plans to re-survey the sediment in Lake Argyle and examine the gullies in the ORRR to determine if any change has occurred.

In summary (Table 9.2), the problem of sedimentation only came to prominence once plans were underway for a large irrigation reservoir. Soil erosion had been commented upon earlier by agricultural officers but there is little evidence that pastoralists did much to ameliorate it. Fencing and provision of waters away from rivers probably reduced grazing pressure and therefore erosion along river frontages, but these measures were almost certainly designed to improve stock management rather than control soil erosion.

With a reservoir being planned, scientists focused attention on the area most badly eroded. This approach is essentially the same as that used in the catchment of Lake Burley Griffin. There was no attempt to quantity sediment sources, rather they

Table 9.2. Summary of soil erosion and conservation in the Ord River catchment

Spatial framework	Perturbations and biophysical responses	Causation and responses, and feedback
Priority for conservation given to the Hardman Syncline, but emphasis given to sheet erosion. Quantification of sediment sources confirmed for location but not the assumed dominant erosion process.	Heavy grazing by cattle of grasses with low seed setting rate in a semi-arid monsoonal environment produced massive erosion. Decreased infiltration produced a de-vegetated condition. Increased downstream sediment transport	Early recognition of erosion, but no effort at control until a reservoir was envisaged. Large scale revegetation of the most eroded area. Later assessment showed that the scheme was not focussed on the main erosion process, gullying. No further attempts at active erosion control; rather, the most eroded area has been de-stocked.

relied simply upon field survey and description. An enormous effort at revegetation was based upon a very limited understanding of erosional processes and rates. The role of local pastoralists in this venture appears to have been slight, all of the effort and cost being borne by the Western Australian government.

By the 1980s R.J. Wark decided that the effectiveness of the revegetation of the ORRR needed to be determined. This is a rare, and early, case of a government agency testing the effectiveness of an intervention. The conclusion from re-survey of the amount of sediment in Lake Argyle led to an attempt to quantify sediment sources because it was possible that revegetation had occurred in the wrong place. The tracer study showed that revegetation was spatially appropriate but had not paid sufficient attention to the main erosional process, gullying. Once this was known, the response was swift and appropriate to the institutional setting of the time. The ORRR was destocked and left to revegetate by natural means.

The Kalgan River Catchment

On the south coast of Western Australia, the Kalgan River flows into Oyster Harbour at the town of Albany (Figure 9.9). The catchment is 3,041 km^2 in area, and rises in the Stirling Ranges 70 km to the west of Albany. Much of the catchment is cut in Tertiary age Pallinup Siltstone with a cover of sands both residual and aeolian. Outcrops also occur of Proterozoic granite and adamellite. Summer is mild and dry, and winter is cool and wet; a typical Mediterranean climate. Mean annual rainfall ranges from 400 mm in the northeast of the catchment to 900 mm near the coast. Natural vegetation was tall open (and some closed) eucalypt forest in the

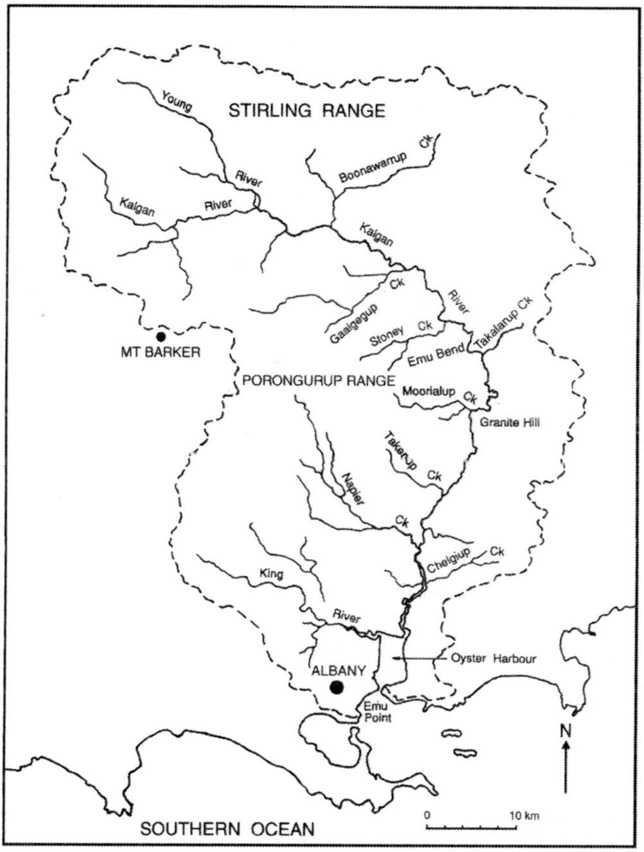

Figure 9.9. Map of the Kalgan River catchment

west and south, grassland south of the Stirling Ranges, and mallee (multi-stemmed eucalypts) woodland and tall scrubland in the eastern part of the catchment. About 72 per cent of the catchment has been cleared, mainly for sheep and cattle grazing, cereal cropping, and limited viticulture (Seal 1995).

European exploration of the area began seriously in the 1820s and 1830s. Most early settlement was along the coast, but by 1840 settlers had spread from Albany along the Kalgan River (Blight 1996). Major land clearing occurred after the Second World War, largely at the instigation of the government to settle returning soldiers and to open up the country for economic development (Gaynor 2002).

Phosphatic fertilisers were an important part of the agricultural system, given the low natural nutrient status of the soils of the catchment.

In 1961 the seagrass of Oyster Harbour was surveyed and found to be in good condition (Seal 1995). Subsequently, macro algae began to appear in the estuary, and seagrass was smothered. When resurveyed in 1981, the seagrass biomass was only 23 per cent of that in 1961. By 1992 it was only 10 per cent of that in 1996, and macro algal biomass had increased 14-fold. This fundamental ecosystem shift was attributed to excessive quantities of phosphorus reaching the estuary from eroded paddocks where fertilisers had been applied.

The Environmental Protection Authority (of Western Australia) monitored nutrient loads starting in 1988, but it was not until the 1990s that sufficiently detailed measurements showed that about 70 per cent of the total phosphorus was particulate, and the rest dissolved (Weaver, et al. 1994). This may even be an underestimate of the particulate fraction. Therefore, erosion of phosphorus-enriched soil particles is clearly the most likely source of the nutrient that was damaging the Oyster Harbour ecosystem. Sedimentation of the harbour was also noted by local fishermen (the *West Australian* newspaper, 4 January 1994).

In 1990 the Environmental Protection Authority recommended that nutrient input to Oyster Harbour should not exceed its assimilation capacity. Targets for each source of phosphorus were established, and management plans constructed. Importantly, scientific research by the Western Australian Department of Agriculture began at the same time as, rather than decades after, a problem had been recognised. In this way, the Kalgan differs markedly from the Lake Burley Griffin and Ord cases.

The management response consisted of determining the phosphorus needs of crops and pastures and tailoring fertilisation to these needs, and planning the re-establishment of riparian vegetation both to intercept runoff of soil from paddocks and to slow stream bank erosion. These measures were targeted particularly to low order channels because they are by far the largest proportion of channels, and soil can reach them most easily. This is the only example in our three case studies where the stream order hierarchy has been used in management interventions.

The management planning has been aided by a total phosphorus budget for the catchment (Weaver et al. 1999). From the total channel length in the catchment, an estimated average riverbank erosion rate and bank height, and an assumed 100 per cent delivery ratio (that is, all mobilised phosphorus reaches Oyster Harbour), Weaver estimated that 23 per cent of the annual average phosphorus load reaching the estuary could be from riverbank erosion. Rivers in this case include all channels including gullies. This calculation probably overestimates the amount of phosphorus coming from stream banks because the delivery ratio is almost certainly less than 100 per cent and the assumed bank erosion rate too high. Nonetheless, to know that more than 77 per cent of particulate phosphorus in the estuary is coming from surface erosion, mostly off agricultural land, has been of enormous value to the catchment managers and farmers in the Kalgan area.

So far emphasis has been given mainly to the observations and responses of government agencies and officers. What of the farmers? Systematic interviewing of farmers in small sub catchments in the vicinity of Takalarup supply the only available information (Figure 9.9). The four sub catchments are different from most of the rest of the Kalgan catchment because here channel incision has been more pronounced since significant clearing along valley floors began in about 1907. Incision of Takalarup Creek began in 1939 as the result of an intense rainstorm on cleared land. Gullies were also formed in the area by the same event. None of the present or former landholders spoke of attempts to control incision or gullying. Rather they adapted by building bridges and by fencing. Salt scalds existed in 1949 in parts of these catchments, and attempts were made to control their spread and to slow the attendant erosion by using straw. This was of limited success.

Sediment budgets for Takalarup Creek show that between 1907 and 1955, 80 per cent of sediment yield came from channel incision. As clearing spread up slopes away from valley floors, the sheet erosion component of the budget increased so that channel erosion produced only 38 per cent of sediment yield between 1955 and 1997.

The farmers in this area, like those throughout the Kalgan catchment, are engaged in riparian vegetation reconstruction, minimum tillage and careful use of fallowing. While there is some evidence that the changed cultivation practices reflect a desire to maintain productivity in the face of both water and wind erosion, riparian zone reconstruction is a response to eutrophication of Oyster Harbour. Large-scale revegetation to reverse dryland salinity remains a largely uneconomic proposition. Interviews with farmers show that they are now living with saline scalds and seeps, and salinised streams. Some are growing saltbush (Atriplex spp) on areas of saline discharge to both reduce discharge and to feed sheep. These responses are the product of pragmatism and the nationwide Landcare movement, a government-assisted program involving local people in natural resource management.

It is too early to judge the effectiveness of all of these responses to land degradation. Reversal of eutrophication should be possible, although anoxic conditions in bottom waters of Oyster Harbour will continue to release phosphorus from sediments on the bed for decades to come. Reversing sedimentation is a much more difficult proposition.

Adaptation and local but minor attempts at amelioration of land degradation occurred before a catchment-wide solution was necessary to reverse eutrophication of Oyster Harbour (Table 9.3). Advice from Western Australian government agencies before the introduction of catchment management was largely focused on farm productivity, and certainly progress with controlling wind erosion in particular has been made.

Table 9.3. Summary of soil erosion and eutrophication in the Kalgan River catchment

Spatial framework	Perturbation and biophysical response	Causation and responses, and feedback
Until eutrophication was an issue, farmers focussed on their farms. Catchment-wide thinking introduced the idea of stream order for management priorities	Land clearing and fertilisation caused increased sedimentation and eutrophication of Oyster Harbour, and infilling of pools in major rivers	Local erosion was adapted to, while local salinisation received some local attention. Once eutrophication was recognised, farmers worked together to reduce the transport of soil and phosphorus from their farms, with both economic (on farm) and environmental benefit.

In summary (Table 9.3), until a catchment-scale problem arose, namely eutrophication, the spatial framework for land management was individual farms. Because scientific research began simultaneously with management of nutrient sources, stream order was used as the spatial framework early in the response phase. The response to eutrophication has involved scientists in identifying phosphorus sources and transport processes, optimising fertiliser applications to both reduce phosphorus use and to reduce farm input costs, and reconstructing of riparian zones to reduce stream-bank erosion and slow transport of phosphorus-enriched topsoil from paddocks into streams. Otherwise, farmers have adapted to incision of streamlines by fencing and building bridges.

CONCLUSIONS

In all three case studies, changes to the erosion–sedimentation system were observed by local farmers and inhabitants to varying degrees. Almost all responses were adaptive, involving bridges and fencing to keep stock safe and allow access on farms now divided by deep gullies. A few instances of stream-bank protection occurred in the Molonglo catchment, but in the Ord the size of the individual farms and the cost of such measures militated against their use. In the Kalgan catchment, stream banks of high order channels were protected by not clearing riparian vegetation, but this was probably because the agricultural benefit of clearing would have been minimal.

It was only when a problem was identified that required a catchment-wide management response that farmers adopted a wide spatial perspective. In the

Molonglo this was co-ordinated in a formal agreement between government and property owners. In the Kalgan it has been organised by voluntary co-operative arrangements, where government-organised information campaigns have convinced most farmers to become involved. How much of this co-operation is the result of genuine concern for the environment, the economic benefits of reduced fertiliser use, or peer pressure is yet to be discovered. In the Ord the response to a perceived threat of sedimentation in irrigation reservoirs has been almost solely driven and funded by government. Some private expenditure on fencing and watering points has helped, but most of this was probably driven by a desire to improve animal husbandry.

In the case of the Molonglo catchment, the prospect of a lake full of sediment triggered soil conservation. The cause of this potential problem was identified as sheet erosion of hill slopes, although other sources of sediment were noted. It seems that the emphasis on sheet erosion was institutionally driven. That is, the Soil Conservation Service of NSW was expert at treating sheet erosion, had a charter to improve farming, and sheet erosion among erosion types is usually the greatest threat to crop productivity, and the Service was not permitted by legislation to treat stream channel erosion. This response turned out to deal with only 5–10 per cent of the sediment sources. Also, the Service did not have a rigorous method of determining the proportional contribution from different sediment sources. Therefore, the efforts at soil conservation were not necessarily well targeted spatially. If these two limitations in their actions had not been corrected there would have been no effective outcome. Finally, they and other organisations and commentators had overestimated the magnitude of likely sedimentation in Lake Burley Griffin. It is not clear if this was a deliberate attempt to scare decision-makers into action, or a simple error based upon very limited information.

In the Ord River catchment the magnitude and extent of erosion in the Hardman Syncline could not have been overlooked by those planning irrigation reservoirs. But once again the experience and skills of those responsible for rehabilitation of the worst eroded land led them to focus on sheet erosion. By contrast with the Molonglo catchment, where lower energies of stream flow do allow for effective control of channel and gully erosion, in the Ord the high energies of runoff during the summer monsoon have so far defeated any attempt at control of channel erosion. This experience may have had some effect on the decision to control only sheet erosion. The effect, nonetheless, was that a massive revegetation effort has probably had little impact on sedimentation rates in Lake Argyle.

In the case of the Kalgan catchment scientific research, use of a stream-order spatial framework, and careful management planning to reduce phosphorus input to Oyster Harbour appears to be paying off. This set of responses and actions came at a time in Australia when catchment management was not only an accepted idea but was also maturing. By comparison, the management interventions in the Molonglo and Ord catchments were either before or during the early years of catchment management in Australia. Management of the Kalgan has therefore benefited from this development.

The conceptual framework set up at the beginning of this chapter has been used where possible to analyse the three case studies. But most of the data necessary to use the framework properly and fully do not exist. The framework nonetheless stands as a device for others with more data, and possibly for structuring new studies.

ACKNOWLEDGEMENTS

For collaboration, many fieldtrips, and stimulating discussions, I wish to thank Barry Starr, Jon Olley, David Weaver, Paul Novelly, and Graham Clifton.

REFERENCES

Blight, J. 1966. *South Coast Regional Initiative: Land and Water Care Strategy.* The Great Southern Development Commission and Agriculture Western Australia

Bolton, G. 1953. A Survey of the Kimberley Pastoral Industry from 1885 to Present. MSc Thesis, University of Western Australia.

Dovers, S. (ed.) 2000. *Australian Environmental History and Policy: Still Settling Australia.* Oxford: Oxford University Press.

Dumas, R.J. 1944. 'Development of the North-West of Western Australia'. *Journal of the Institution of Engineers of Australia*, **16**: 65–70.

Durack, K.M. 1945. 'The Monsoonal North of Western Australia'. *Journal of Agriculture Western Australia (2nd Series)*, **22**: 248–74.

Durham, L.J. 1958. *A Report on a Reconnaissance Survey of Erosion in the Molonglo River Catchment Area.* New South Wales Soil Conservation Service.

Eyles, R.J. 1977a. 'Changes in Drainage Networks since 1820, Southern Tablelands, NSW'. *Australian Geographer*, **13**: 377–86.

Eyles, R.J. 1977b. 'Erosion and Land Use in the Burra Catchment, Queanbeyan'. *Journal of Soil Conservation Service of New South Wales*, **33**(1): 47–59.

Fitzgerald, K. 1968. 'The Ord River Catchment Regeneration Project', *W.A. Dept. Agric. Bulletin 3599.*

Fogarty, P., G. Clifton, B. Griffiths and R. Hammond. 1989. 'Soil Erosion and Soil Conservation in the Lake Burley Griffin Catchment, NSW and ACT'. *Australian Journal of Soil and Water Conservation*, **2**(1): 28–32.

Forrest, A. 1880. *North West Exploration. Journal of the Expedition from De Gray to Port Darwin.* Perth: Western Australian Government.

Gaynor, A. 2002. 'Looking Forward, Looking Back: Towards an Environmental History of Salinity and Erosion in the Eastern Wheatbelt of Western Australia'. In A. Gaynor, M. Trince and A. Haebich (eds), *Country: Visions of Land and People in Western Australia* (Western Australian Museum), pp. 105–24.

Lyne, J. A. 1978. *Canberra: A Planned City.* Melbourne: Sorrett.

Medcalfe, F.G. 1945. 'Notes on the Ecological Aspect of Soil Erosion'. In *Report on the Vegetation and Climate of the Kimberley Division of Western Australia with special reference to the Ord River Region* (Government Botanist).

Munro, L.H., R.T. Hettersley and A.J. Buchan. 1967. *Studies of Inflow and Deposition of Sediment in Lake Burley Griffin.* University New South Wales Water Research.

National Capital Development Commission. 1962. *Lake Sedimentation: Report to Interdepartmental Committee on Lake Burley Griffin.*

Newell, B. and R.J. Wasson. 2001. 'Social System vs Solar System: Why Policy Makers Need History'. In S. Castelein and A. Otte (eds), *Conflict and Cooperation Related to International Water Resources: Historical Perspectives.* UNESCO Document SC. 2002/WS/53 (Grenoble: UNESCO), pp. 3–17.

New South Wales Department of Land and Water Conservation. 2000. *The Lake Burley Griffin Catchment Protection Scheme 1965–1998: A Short History.* Sydney: NSW Dept. Land and Water Conservation.

Ord River Irrigation Area Review. 1978. *Joint Commonwealth and Western Australian Review.* Canberra: Aust. Gov. Publishing Service.

Patterson, R.A. 1965. 'The Economic Justification of the Ord River Project'. 38th Congress ANZAAS.

Paton, T.R., G.S. Humphreys and P.B. Mitchell. 1995. *Soils: A New Global View.* London: UCL Press.

Pratchett, D. 1990. *The Ord Regeneration Area: A Short History.* WA Dept of Agriculture.

Seal, C. 1995. *Albany Waterways Management Programme.* Perth: Waterways Commission.

Smith, L.W. 1968. 'Soil Erosion Control in the Lake Burley Griffin Catchment'. *Journal Soil Conservation Service of New South Wales,* **34**(1): 27–34.

Squires, L.J. 1977. 'Soil Conservation Projects in New South Wales'. *Journal Soil Conservation Service of New South Wales,* **33**(1): 7–14.

Starr, B. 1991. *Soil Conservation in Lake Burley Griffin Catchment: Catchment Management or Manipulation?* B. Starr Pty Ltd.

Starr, B.J. 1989. 'Anecdotal and Relic Evidence of the History of Gully Erosion and Sediment Movement in the Michelago Creek Catchment'. *Journal of Soil and Water Conservation,* 2(3): 26.

Starr, B., R.J. Wasson et al. and G. Caitcheon. 1999. *Soil Erosion, Phosphorus and Dryland Salinity in the Upper Murrumbidgee: Past Change and Current Findings.* Upper Murrumbidgee Catchment Coordinating Committee.

Stewart, G.A., R.A. Perry, S.J. Paterson, D.M. Travers, R.O. Slatyer, P.R. Dunne, P.J. Jones, and J.R. Sleeman. 1970. *Lands of the Ord-Victoria Area, Western Australia and Northern Territory,* CSIRO Land Research Series 28.

Teakle, L.J.H. 1944. Soil Erosion and its Relationship to Soil Type and Geological Formations of the Upper Ord Basin. File 23/44 Vol. 1 Western Australian Department of Lands and Surveys.

Walker, B.H., D. Ludwig, C.S. Holling and R.M. Peterman. 1981. 'Stability and Semi-Arid Savanna Grazing Systems'. *Journal of Ecology*, **69**: 473–98.

Wark, R.J. 1987 *Deposition of Sediment in Lake Argyle*. Report Number WP47. Western Australian Water Authority

Wasson, R.J. 1994. Living with the Past: Uses of History for Understanding Landscape Change and Degradation. *Land Degradation and Rehabilitation*, **5**: 79–87.

Wasson, R.J. 2003. 'Effective Policy Interventions in Environmental Systems using Material Budgets'. In S.R. Dovers, D.I. Stern and M.D. Young (eds), *New Dimensions in Ecological Economics* (Cheltenham: Edward Elgar), pp. 129–92.

Wasson, R.J., G. Caitcheon, A.S. Murray, M. McCulloch and J. Quade. 2002. 'Sourcing Sediment Using Multiple Tracers in the Catchment of Lake Argyle, Northwestern Australia'. *Environmental Management*, **29**(5): 634–6.

Wasson, R.J. and R.L. Clark. 1985. 'Environmental History for Explanation and Prediction'. *Search*, **16**: 258–63.

Wasson, R.J., R.K. Mazari, B. Starr and G. Clifton. 1998. 'The Recent History of Erosion and Sedimentation on the Southern Tablelands of Southern-Eastern Australia: Implications for Soil Conservation'. *Geomorphology*, **24**: 291–308.

Wasson, R.J., A.S. Murray, H. Crockford and G.G. Caitcheon. 1989. *Sources of Sediment Reaching Lake Burley Griffin*. Progress report to ACT Parks and Conservation Service and NSW Soil Conservation Service.

Wasson, R.J. and A.Yu. Sidorchuk. 2000. 'History for Soil Conservation and Catchment Management'. In S. Dovers (ed.), *Australian Environmental History: Still Settling Australia* (Oxford), pp. 97–117.

Weaver, P.M., L.J. Pen and A.E.G. Reed.1994. 'Modifying the Phosphorus Cycle to Achieve Management Objectives in the Oyster Harbour Catchment'. *Water Journal*, Feb. 1994: 28–32.

Weaver, D.M., P.B. McCafferty and E.G. Reed. 1999. 'A Phosphorus Inventory for Albany Hinterland, South Coast of Western Australia'. In C. Barber, R. Humphreys and J. Dixon (eds), *Proceedings, International Conference on Diffuse Pollution, Solutions and Innovations* (Perth, May 1999), pp. 105–16.

10

The Dynamics of Soil, Landscape and Culture on Easter Island (Chile)

Andreas Mieth and Hans-Rudolf Bork

A. INTRODUCTION

1. The Special Role of Islands in Ecological and Land Use Research

Island ecosystems have repeatedly been the subject of research on the interaction between land use and ecosystem change. Island ecosystems are well delineated, making it easier to determine the balance of material import and export than for continental systems. In most cases a rather short settlement history allows for a more exact temporal and causal analysis of the effects of human impact than on the mainland. Reaction of island ecosystems is particularly sensitive to anthropogenic impact, not at least because of the high degree of adaptation of the indigenous flora and fauna to ecological niches and a generally low diversity of species. The quantitative and spatial limitation of resources lead to very sensitive reactions of the ecosystem to land use (Fosberg 1963a, p. 5; Fosberg 1963b, p. 559, quoted by Kirch 2000, p. 58). Even islands which became settled only a few centuries ago have changed much faster than comparable mainland systems with much longer periods of settlement and land use. Particularly degrading is the introduction of animal and plant species (Coblentz 1978). Ecosystem change comprises the displacement of animal and plant species characteristic of specific sites, the extinction of endemic species, and also the degradation of soils. For the Polynesian region, including Easter Island, there has been an ongoing discussion as to whether negative modifications of the systems actually date back to the land-use practices of the indigenous island peoples or only occurred after the introduction of European-style land-use systems. From the anthropological perspective, degradational land use in prehistoric times tends to be negated, whereas from the archaeological one there is much evidence of significant ecosystem changes in Polynesian times (Kirch 2000; Diamond 2005).

Frequently, ecosystem changes of islands have been studied for the purpose of deriving models for larger scales, including, in the case of Easter Island, planet Earth (Bahn and Flenley 1992).

Easter Island, because of its cultural uniqueness and its isolated location, may well be the archaeologically and ethnologically most intensively studied island of the world. Intensive archaeological research began with the studies by Routledge (1919 [1998]). In the 1950s William Mulloy and Thor Heyerdal triggered a considerable new wave of research (Gill 1993; Bahn 1993b).

Hardly any research has been devoted, however, to the ecological effects of prehistoric and historical land use on Easter Island. Bahn and Flenley (1992, p. 212) and Flenley (1993b, pp. 43–5) were the first to present a comprehensive theory of the interaction between prehistoric land use, resulting ecosystem changes, and cultural shifts on the island, but based their argument almost exclusively on the palynological analysis of the crater sediments of three volcanoes. They did not use landform analysis tied in with soil and sediment stratigraphies for understanding vegetation and landscape change, although – with appropriate methods – a much higher regional, temporal and causal resolution is possible (Bork et al. 1998).

The archaeologists have almost exclusively inferred the conditions and techniques of prehistoric cultivation from the soils as they are today, and not much is known about the prehistoric and historic land use of Easter Island beyond the level of conjecture. It is true that Bahn and Flenley (1992) and Flenley (1993b) postulated that forest clearing and resulting soil erosion dramatically worsened living and cultural conditions. There has been no proof, though, derived from the analysis of landforms, soils and sediments. No systematic use has been made of these geo-archives for identifying different techniques and phases of horticulture, or for reconstructing former vegetation patterns. Nothing is known of the effects various types of historic and prehistoric land use had on soils and soil erosion. The destructive effect of intensive cattle farming since the late nineteenth century on the vegetation was described by Skottsberg (1953, p. 494), but not the effects of overgrazing on the soils. The cause of the large-scale erosion features most prominent in the eastern part of the island is as yet unknown.

2. Settlement History of the South Pacific

The history of settlement and cultural development of the Pacific region is one of the most dramatic chapters in the recent history of mankind. Overcoming the enormous distances between the islands and atolls of the Pacific called for nautical and logistical skills never attained by other peoples and hardly comprehensible today. The conquest of the so-called 'Polynesian Triangle', the central Pacific region between New Zealand, Hawaii and Easter Island, was a matter of a mere

1,200 years. Environments were settled where humans had never set foot before. Exquisitely adapted land use strategies and techniques were indispensable for the permanent settlement of human populations on the isolated islands with their poor resources. Biodiversity of the islands was low because of their extreme isolation. Therefore domesticated plants were always part of the cargo of the large double-hull canoes of the Polynesians, as were animals such as poultry, pigs, rats and dogs for supplementing the fish from the ocean and the meat resources found on the islands (Orliac and Orliac 1995). The production systems set up on the islands had to be organised on the principle of subsistence and be sustainably adapted to their resources if they were to function for a long time.

There are several possible reasons for the unprecedented conquest of the Pacific region:

- curiosity in general and the human urge for conquest and innovation
- catastrophes affecting settled regions, e.g. volcanic eruptions, earthquakes, floods and typhoons
- overpopulation of the limited space of the settled regions
- shortage of resources due to catastrophes, overpopulation or natural and man-made changes of the environment.

Settlement took place in three cultural, temporal and spatial phases:

- the settlement of western Oceania by seafaring peoples which, coming from South-east Asia, reached the western part of the South Pacific via New Guinea and Australia about 35,000 to 40,000 years ago (Kirch 2000).
- expansion of the Lapita people, around 3000 BC, southward from the Indonesian-Malayan region, reaching the Papua region by 1500 BC and moving eastward to western Polynesia. Known by its highly developed pottery techniques, the culture group reached the archipelagos of Fiji, Tonga and Samoa between about 1200 and 1100 BC (Orliac and Orliac 1995). The Lapita culture was the precursor of the Polynesian culture (Esen-Baur 1989).
- advance of the Polynesians into the central Pacific region, most likely originating from the Tonga and Samoa region (Kirch 1984). Around 200 BC the centre of Polynesia was reached. The Hawaiian islands were discovered about AD 400, New Zealand probably around AD 1000 (Kirch 2000). Like all these datings, that of the settlement of Easter Island, the easternmost island of Polynesia, is still not fully established, with estimates ranging from AD 300 to AD 800 (e.g. Orliac and Orliac 1995; Kirch 2000; Lee 2001; Martinsson-Wallin 2002). The age determinations are based on radiocarbon

analysis of charcoal. The oldest datings are now thought to be unreliable (Bahn 1993a).

The pottery techniques of the Lapita were not adopted by the Polynesians. Aside from the many uses and ways of processing of organic materials, working in stone became their cultural hallmark. Not only houses, ornamental and household objects and weapons, but also spiritual objects were made of volcanic rocks, finding their strongest expression in stone buildings and statues. On Easter Island an extreme version of a megalithic culture evolved, expressed in the famous humanoid ***moai*** statues. The creation of these gigantic effigies, each weighing dozens of tons, is in stark contrast to a culture adjusted to high mobility, light-weight construction techniques for their boats, efficient long-distance transport of man and material over the seas, as well as sustainable and careful land use. This contrast alone is suggestive of specific conflicts between the extreme megalithic culture and land use on Easter Island.

Settlement of the island was subjected to yet another extreme factor: Easter Island may well be the most isolated place of human settlement on earth. In prehistoric times the island was isolated from other terrestrial environments and resources in an extreme way. It is still not known whether the isolation persisted after the first settlers had reached the island, whether other groups of Polynesians arrived in later times, whether there were contacts with other islands of Polynesia, or whether there may have even been contacts with the South American continent.

3. Settlement and Cultural History of Easter Island

The numbers of the first people reaching Easter Island between AD 300 and 700 were restricted to the crew of a few large canoes, comprising at best a few hundred people. It appears that the first settlements were set up on the north-east coast (Martinsson-Wallin 2004). Evidence of the intensive colonisation comes from numerous house foundations still visible in the countryside, stone-mantled ovens dug into the ground (***umu***), and platforms more than 150 m long in places and 3 m high (***ahu***), consisting of a rim of square basalt blocks with a mineral filling, erected along most of the coastline. In the beginning the ***ahu*** may have exclusively served ceremonial purposes. The oldest ***ahu*** are dated to the end of the seventh century (Bahn 1993a). Production of the megalithic figures did not begin until a few centuries later (van Tilburg 1994). Perhaps based on the technique of creating anthropomorphic stone figures developed on other Polynesian islands, the largest ***moai*** on Easter Island were being produced from about AD 1000 onward, unique and unequalled in size. Tools of hard basalt were used to chisel almost all known statues from the softer tuff of Rano Raraku volcano.

Ahu, moai and settlements formed spatial entities. Only recently archaeological excavations of the ***moai*** roads have started to find out how these heavy statues

Figure 10.1. Stone statues (moai) sticking deeply into the slope sediment of the quarry volcano Rano Raraku. These moai represent the last phase of the megalithic culture on Easter Island.

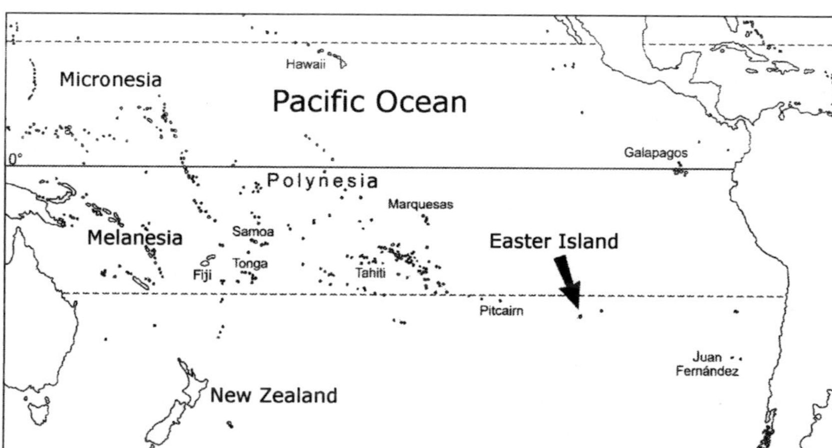

Figure 10.2. Location of Easter Island in the South Pacific Ocean

were transported. It is as yet unknown how the statues were lifted from the quarry, how they were transported downslope, how they were moved along the roads and, finally, how they could be erected on the platforms. Most archaeologists agree that tree trunks were indispensable for moving the statues (Esen-Baur 1989, Flenley and Bahn 2003, van Tilburg 1994, Grau 1998). Depending on the technology assumed, estimates on the number and quality of trees needed vary considerably. Answering the question precisely is of considerable ecological significance, though.

About 900 *moai* are known, of which about 300 were erected on *ahu*. The largest statue set up is about 10 m high and weighs about 80 t. The largest *moai* ever produced, however, have remained in the quarry of Rano Raraku, marking the climax and possibly also the abrupt termination of the *moai* culture. Impressive testimony to the end of this cultural phase are about 400 statues still standing or lying on the outer and inner slope of Rano Raraku, many of them buried in sediment up to their heads, and many never completed nor cut free from the tuff rock of the volcano.

Why the stone cult grew to such unprecedented dimensions and why it came to its sudden end are among the largely unanswered questions of the cultural history of the island. Ecological catastrophes are increasingly believed to have been the reason for the collapse, but opinions differ as to whether natural or man-made causes lie behind such a sudden change of environmental and thus also of living conditions. Flenley and Bahn (2003) created a model describing possible interrelationships between land use, ecosystem change and cultural effects.

No one knows how many people lived on Easter Island during the heyday of the *moai* culture. Some authors assume a maximum population of 8,000–10,000 (Bahn 1993a, van Tilburg 1994, p. 52), but there is no reliable basis for any well-founded estimates. All figures presented are extrapolations from an inferred number of the original population or have been estimated from the settlement structures found (Flenley and Bahn 2003).

Around the fifteenth or sixteenth century, a cultural shift occurred, characterised by – but not only by – the end of *moai* production. It possibly came together with food shortages, increasing social tensions, warfare and a population decrease to about 2,000 persons (Flenley and Bahn 2003; Bahn 1993a). Evidence of battles comes from oral traditions, and from weapons produced since the fifteenth or sixteenth century (Flenley and Bahn 2003). Thousands of obsidian weapons, such as spearheads, and the flakes from their production litter the surface and are buried in sediments. Fighting also led to the destruction of the once holy *moai*. The statues were toppled from their platforms and their faces, mainly the eyes, were deliberately destroyed (Flenley and Bahn 2003).

The time of war also saw the rise of new cultural and spiritual ideals expressed in the so-called birdman cult (Routledge 1919 [1998]; van Tilburg 1994, Martinsson-Wallin 2002). The spiritual centre of this cultural phase was the ceremonial village of Orongo on the narrow seaward crater rim of Rano Kau volcano. Ritualised competitions may have served the selection of high-ranking decision-makers and advisers of the society. Seabirds, new fertility symbols, new gods and new hierarchic ideals replaced the old peaceful society with its ancestral cult and a life strictly regulated by taboos. Living in hidden caves became important, probably as an expression of security needs in a conflict-ridden society.

Perhaps the birdman cult was an effort to transform deadly conflicts into ritualised fighting. Van Tilburg (1994, pp. 58–62) sees a close link between changing factors of the ecosystem, primarily the marine part of it, and the development of the birdman rituals. Other authors also attribute a shortage of resources and resulting tensions to the end of the *moai* culture and the rise of the birdman rites (Flenley and Bahn 2003; van Tilburg 1994; Martinsson-Wallin 2002). Various datings are given for the demise of the *moai* culture: Kirch (1984, quoted in Lee 1992, p. 8) thinks it was around AD 1500, for Bahn and Flenley (1992, p. 152) it was in the sixteenth century, for Lee (2001, p. 109) around AD 1550, and Martinsson-Wallin suggests about AD 1600. Some authors put the end of the *moai* culture, the beginning of the birdman culture and/or the beginning of the period of warfare as late as 1680, based on the seemingly good coincidence between radiocarbon datings from the western part of Poike Peninsula initiated by Thor Heyerdal and the oral tradition

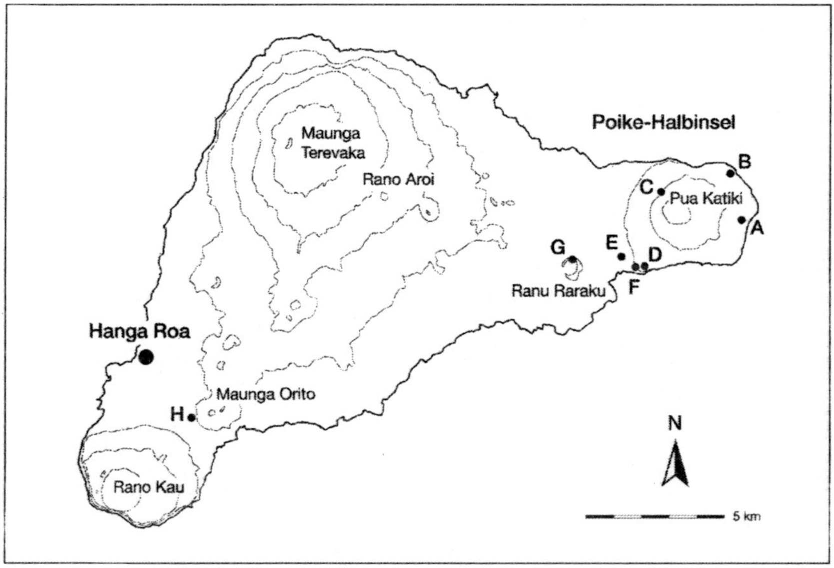

Figure 10.3. Easter Island with locations of recent investigations on soil history and landscape development

of a fight having taken place at the same time in that area (e.g. van Tilburg 1994, p. 51; Esen-Baur 1989, p. 107). It is assumed that food shortages and warlike conflicts drastically reduced the population from the seventeenth century onward (Flenley and Bahn 2003; Bahn 1993a).

On 5 April 1722 the Dutch captain Jacob Roggeveen was the first European to set foot on Easter Island. There followed a time of increasingly frequent visits by European seafarers (Gonzales 1770, Cook 1974, La Pérouse 1786, v. Kotzebue 1816, Beechey 1825, Palmer 1868, Geiseler 1882, and others). The first European visitors of the eighteenth century reported ***moai*** still standing upright on their ***ahu***, but by the beginning of the nineteenth century many, and by the end of it most likely all, of the stone statues had fallen over. This is taken as evidence that fighting on Easter Island continued well into the nineteenth century.

The seafarer's records contain the first population estimates. Gonzales speaks of 900 to 1,000 persons, Cook of 700 people, most of them men (quoted by Esen-Baur 1989, pp. 56, 59). La Pérouse estimated the island population to be close to 2,000 around 1786, a number Bahn and Flenley (1992, p. 179) think to be plausible. There are considerable uncertainties, though. Most of the European visitors stayed for a few days only. Roggeveen and Cook saw mostly men on the island, which may indicate that women and children hid inside their houses or caves during the ships' visits.

In 1866 the first Christian mission was set up on the island. Missionary activity and slavery in the 1860s brought a rapid end to the rituals of the birdman cult. The end of this cultural phase can therefore be dated almost by the year. In 1877 the population of Easter Island was reduced to its all-time low of 110 persons due to deportation of men and imported diseases (McCall 1994, p. 64). According to the 2002 census about 3,800 people live on the island today.

4. Prehistoric Land Use on Easter Island: State of Knowledge

Settlers of any Polynesian island had to bring cultigens with them and establish horticulture suitably adjusted to local conditions. For Easter Island numerous cultivated plants have been identified that were brought by the first settlers, among them the food staples taro (***Colocasia esculenta***), yam (***Dioscorea alata***), banana (***Musa sapientium***), sugar-cane (***Saccharum officinarum***) and Ti (***Cordyline fructicosa***) (Zizka 1991; Flenley 1993a).

It is still disputed whether the most important food plant of the Polynesian region, the sweet potato or ***kumara*** (***Ipomea batatas***) actually came with the first settlers. According to Yen (1974, p. 329) the sweet potato reached eastern Polynesia between AD 400 and 700; while Kirch (2000, pp. 241–3) dates its presence in the central Polynesian region to around AD 1000. It is confirmed that the sweet potato was common on Easter Island when the first Europeans discovered the island (Zizka 1991). Possibly the plant reached the island several centuries after its initial colonisation, either imported by islanders returning from far-reaching explorations or by Polynesian seafarers who came later (Wallin et al. 2005). Kirch (2000, p. 125) describes the sweet potato as a cultivated plant that produces high yields even in a somewhat cool climate and on nutrient-poor soils.

Chicken (*Gallus domesticus*) and Polynesian rats (*Rattus exulans*) were the main source of animal protein for the early settlers, supplemented by fish, turtles, dolphins and seals from the sea, and by the collection of molluscs, crayfish and sea urchins (van Tilburg 1994, Steadman et al. 1994).

Hierarchical structures as well as taboos, rules, tradition and technical experience regulating the way to deal with the natural foundations of their life were among the organising principles of the population for sharing the resources of the island (Martinsson-Wallin 2002). Land subdivisions were the basis of fair spatial access to resources (Métraux 1940 [1971]; Lee 1992).

In the beginning probably only small patches of vegetation were cleared for the construction of houses and *ahu*. Based on many years of archaeological research Stevenson and Christino (1986), Stevenson (1997), Stevenson and Haoa (1999), Stevenson et al. (2002) and Martinsson-Wallin (2002) describe the chronology of land use on Easter Island in relation to population growth. According to these studies, around AD 800 settlement was directly along the coast, later expanding to adjacent areas. With the construction of new settlements the population gradually moved inland, reaching the higher parts of the volcanoes around AD 1250. By AD 1400–1500 large tracts of the higher areas had been settled. Settlements in the higher inland areas had apparently been abandoned in the early 1700s.

The expansion of settlement went together with an increase in the area and intensity of horticulture. Stevenson et al. (2002), from archaeological evidence, describe phases of varying extent, intensity and techniques of horticulture. Extensive gardening started in the coastal areas around AD 800, moved into the lower inland areas by AD 1100 and was practised on the middle and upper slopes of the volcanoes by AD 1250. Stevenson et al. (2002) assume a general intensification of horticulture in the coastal areas as well as on the higher slopes by the early 1400s. They correlate intensified horticulture with the introduction of the sweet potato which they assume to have taken place at about that time (Stevenson et al. 2002, p. 21). Some authors take improved mulching techniques (using stones) as a sign of intensive horticulture (Wozniak 1998, 1999; Stevenson et al. 2002). Stones were also taken to erect walls 1–2 m high, mostly ring-shaped, for protecting sensitive plants, such as bananas, against the wind. Stone mulching and wall construction are likely to have been laborious. No calculations or estimates of the amount of work necessary for horticulture have been made so far.

According to Stevenson et al. (2002), intensive horticulture on the upper slopes was abandoned around AD 1700 and gardening became concentrated in the lower, near-coastal areas.

5. Land Use on Easter Island in Historical Times

The horticultural situation and the status of the soils can be inferred from the reports of European scientists and travellers from the eighteenth century onward. Thus Carl Friedrich Behrens, travelling with Roggenveen, describes for the year 1722 that the 'area is planted and well divided into fields', that they are 'well kept', and that 'fields and trees' carry 'abundant fruit' (quoted by Vogler 1998, p. 55). Contrarily Forster, a German botanist with the Cook expedition, described the island in 1774 as being 'so extremely infertile that not more than twenty different species grow on it'. According to his report, fields 'take up by far the least part of the otherwise waste-lying land'. He further describes 'traces of old plantations here and there on the tips of the mountains' (Forster 1784 [1983], pp. 506, 509). La Pérouse, for the year 1786, described intensive agriculture, e.g. in the west coast area (quoted by Wozniak 1999, p. 95). He estimated that 10 per cent of the island was cultivated land, and he also mentioned careful management techniques (quoted by Bahn and Flenley 1992, p. 94) Routledge (1919 [1998]), during an expedition of several months' duration estimated that about 50 per cent of the island was suitable for growing bananas and sweet potatoes.

Mainly in the second half of the nineteenth century the Europeans brought other cultivated plants to Easter Island, among them guava (*Psidium guajava*), pineapple (*Ananas comosus*), maize (*Zea mays*), potato (*Solanum tuberosum*), cotton (*Gossypium sp.*), coffee (*Coffea arabica*) and vines (*Vitis vinifera*) (Zizka 1989; Flenley 1993a). There are reports on planting experiments from the inner slopes of Rano Kau, where remnants are still found today (Zizka 1989, 1991). Commercial agriculture with these plants was not successful, mainly for climatic reasons. Since the end of the nineteenth century European land use centred on grazing. Reports of the first importation of sheep from Tahiti are from 1868. Beginning in 1888, sheep production was expanded under the management of a Scottish company. There are no detailed figures on the number of sheep. For the first half of the twentieth century, a maximum number of 70,000 animals has been reported (Lee 2001; Ramírez 2001). The grassland was burned regularly (Ramírez 2001), annually on Poike Peninsula (Velasco, pers. comm. 2002), to suppress unwanted species and to further the growth of the young grass. Sheep raising came to an end in the 1960s with the world-wide collapse of wool prices. Since then almost all of the island is open range for cattle and horses, about 6,000 of them (Ramírez 2001).

6. Flora and Vegetation of Easter Island

Because of its isolation, its rather young geological age, and its modest topographic and climatic diversity, Easter Island has a low diversity of species, compared to other Polynesian islands (Skottsberg 1956; Zizka 1991; Feeser 2003). Ziska (1991, p. 16) classifies a mere 30 flowering plant species as indigenous. Flenley (1993a,

p. 9) lists 48 indigenous flowering plants and ferns, eight of them endemic. Most likely also the autochthonous flora the first settlers met with on Easter Island was quite poor, but richer than today, judging from archaeological and palaeobotanical evidence. Due to differences in the long-term resistance of pollen to weathering, palynological analysis will permit only limited taxonomic differentiation and thus yield only incomplete information on the earlier diversity of species. From cores taken from the crater lakes of Easter Island Flenley et al. (1991) palynologically identified five woody plant species existing prior to the first settlement. They also identified six types of ***Compositae-tubiliflorae*** pollen, but the evidence did not allow further taxonomic differentiation. In more recent anthracological studies (analysis of charcoal macro-remains) in several archaeological contexts, nine more woody plant species could be identified, eight of them taxonomically identified (Orliac 2000; Orliac and Orliac 1998, 2001). Both from pollen and endocarp remains an indigenous but now extinct palm species was identified (Flenley 1993b). Comparison of the fossil with present material suggests that the palm was closely related to, if not even identical with ***Jubaea chilensis*** on the genus or species level (Dransfield et al. 1984; Grau 2001). Because of the uncertainty, Zizka (1991) classified this palm tree as a species of a new genus: ***Paschalococos disperta***. Following Grau (1998, 2001), for the present study the name ***Jubaea*** sp. respectively ***Jubaea***-palm was chosen. Another woody plant that had temporarily become extinct on Easter Island in the European period, endemic Fabacea ***Sophora toromiro***, now survives with only a few individuals in some European botanical gardens and in some house gardens on Easter Island.

In the Polynesian, and more so in the European period, the flora of the island became strongly modified by anthropochorous (introduced by man) species. Only 16 per cent of the present flora are indigenous species (Feeser 2003). Perhaps more than 100 plant species were introduced to the island in the historical period (Flenley 1993a). Dominant among them, because of the use of the island as grazing land since the end of the nineteenth century, are voluntarily imported grasses, together with those introduced inadvertently.

It is not easy to reconstruct the prehistoric vegetation from the palaeoecological evidence. Pollen results only allow limited conclusions as to the frequency and distribution of the plant species identified. In spite of the extreme isolation of Easter Island it cannot be fully excluded that pollen in the crater profiles may have originated from regions outside the island. Similarly charcoal from archaeological sites yields a selective picture only, reflecting, with a high degree of certainty, the presence of a species in the area (with the exception of driftwood), but not necessarily its distribution and frequency. Nevertheless some authors have attempted to describe the island vegetation prior to human settlement. Skottsberg (1953) assumed that an open park-like forest landscape existed at the time of the first settlement. Flenley and King (1984), Flenley et al. (1991) and Flenley (1993a,

1993b, 1996, 2001), from their pollen analysis, reconstructed a woodland of palm trees interspersed with ***Sophora*** and other shrubs that covered at least part of the island. Orliac (2000), from her anthracological studies, reconstructed a mesophytic forest for Easter Island as it still exists in the lower elevations of other East Polynesian islands. Until now there has been no direct evidence of the distribution and density of stands of the ***Jubaea*** palm. In his archaeological papers, Charles Love described root traces of this species, mainly from soils of Poike Peninsula (quoted by Bahn and Flenley 1992, p. 90), but did not study them in more detail. From root tubes, charred endocarp remains and the character of Ah- and B-horizons of soils Wozniak (1998, pp. 190–1, 2001, p. 93) inferred an open palm-tree vegetation in the north-west of the island. Our own recent studies on the reconstruction of a palm-tree dominated vegetation will be described below.

There is consensus among all researchers that there has been a dramatic change of the island vegetation since the time of settlement, but with many open questions as to its chronology. This is due to the problems of extracting spatial and temporal distribution of plant species from pollen and charcoal remains, but also to problems of dating the pollen stratigraphy obtained from drill cores (Flenley et al. 1999, pp. 92–3). According to Flenley (1993b, 1996, 1998) and Bahn and Flenley (1992) the decrease of tree and shrub pollen simultaneously with the in-

Figure 10.4. Grassland and surface covers of volcanic rocks characterise Easter Island's landscape today.

crease of grass pollen suggests a much progressed deforestation of Easter Island in the fifteenth century. The authors unequivocally attribute this change to man-made forest clearing. For the surroundings of the volcanoes Rano Kau, Rano Aroi and Rano Raraku they date the beginning of forest clearing to the ninth and eleventh centuries respectively. Orliac and Orliac (1998) as well as Orliac (2000) note a pronounced decrease of charcoal from trees in favour of charred remains of grasses in the fireplaces of several settlements from the seventeenth century onward and also take this as evidence of a reduced forest vegetation. The European seafarers of the eighteenth and nineteenth century unanimously describe the scarcity of trees on the island, though some stands of trees and shrubs probably still existed, among them ***Sophora toromiro*** which, as mentioned above, no longer grows in the wild. Forster (1784 [1983], pp. 494, 504) probably misnamed it as ***'Mimosa'***. In 1917 some small specimens of it were sighted on Rano Kau by Skottsberg (1956, p. 6). The last wild toromiro shrub was documented by photograph there in 1935 by Métraux (quoted by Zizka 1991, p. 47). Whether there were still some of the indigenous or endemic palms around in the eighteenth or even nineteenth century, cannot be determined from the European seafarers' reports with any certainty. From his visit to the island in 1868 J.L. Palmer reports wooden 'boles' which he held to be the trunks of 'large trees' like 'Edwardsia' (Toromiro), 'Hibiscus' and 'coco palm' (the last term is possibly for the ***Jubaea*** palm as the coco palm was not present on the island in this time) (Palmer 1870, p 168). The author saw these boles decaying in some places.

More recently the intensive use of the island for sheep grazing towards the end of the nineteenth century caused the extensive removal of any remaining tree or shrub vegetation, with the exception of a few hardly accessible areas like the inner crater of Rano Kau (Skottsberg 1956). Grassland now prevailed on almost all of the island. Reforestation with exotic trees (mainly ***Eucalyptus*** sp.) and shrubs began in the 1950s. Area-wide afforestation was continued from the 1960s to the 1980s for wood production and erosion control. Etienne et al. (1982) noted a tree and shrub cover of not more than 9 per cent at the beginning of the 1980s. The progressive anthropogenic transformation of the vegetation by imported species and intensive grazing (mainly by cattle and horses) has been described in detail in recent studies by Feeser (2003).

7. Climatic and Geo-Ecological Characteristics of Easter Island

Easter Island lies in the subtropical zone of the south-eastern Pacific, at 27°09' S and 109°26' W. Except for the small uninhabited island of Sala y Gómez 360 km to the north-east, Easter Island is the easternmost of the Polynesian islands. It is not known by which name the first settlers called it. The Polynesian name 'Rapa Nui' dates from the nineteenth century and was given to the island by seafarers from Tahiti (Reichardt and Reichardt 2000).

Distances to the nearest inhabited lands underline the extremely isolated location of Easter Island. The shortest distance to the South American coast is about 3,700 km. The closest inhabited island, Pitcairn, lies about 2,100 km to the west. The total area of the island, with its roughly triangular shape, is about 166 km^2, the longest side of the triangle measuring about 24 km. Up to a maximum distance of 1.5 km off the coast there are a few very small rocky islands called *motu*.

Climatically Easter Island lies within the drier part of the rainfall gradient which, in the subtropical Pacific, runs from West to East. Figures on mean annual precipitation and temperatures vary considerably among the authors. Mueller-Dombois and Fosberg (1998) suggest 1,365 mm and 20°C. All official records of the island are from a station near Mataveri Airport close to Hanga Roa, at 47 m a.s.l. (above sea-level). Annual rainfall is quite variable, for the period 1961–1999 ranging from 841 mm to 1,926 mm (Dirección Meteorológica de Chile Departamento Climatología y Meteorología Aplicada 2002). Spatial differences are also considerable. During our stays on the island we could often observe rainfall in the higher parts of Maunga Terevaka and Rano Kau while other areas did not receive any rain. Some authors reckon that the rainfall in the higher altitudes and on the windward side of the volcanoes is 1.5 to 2 times that of the coastal areas and the leeward side of the mountains (Honorato et al. 1991; Wozniak 2001). There is no reliable database for these estimates, however. The spatial differentiation of rainfall is important for interpretation of the prehistoric horticultural techniques.

The annual course of the climate alternates, between easterly to south-easterly trades from September to June and north-westerly trades in July and August. There are no pronounced rainy and dry seasons. Episodic droughts may occur at any season. Those recorded did not last longer than two months. Annual temperature variation is low. In the coldest season, from June to September, the average temperature at the Mataveri weather station from 1961 to 1990 was 18.3°. For the same years the average temperature of the warmest season was 23.3°.

Easter Island is of volcanic origin, owing its existence to a hot spot over which the oceanic Nasca plate has moved. The island is situated on the western rim of the Nasca plate and rises about 3,500 m above the ocean floor. Three volcanoes consecutively piled up the land mass of the island, today forming the three corners of the island: Maunga Terevaka (c. 510 m a.s.l.) in the north, Rano Kau (c. 310 m a.s.l.) in the south-west, and 370 m high Pua Katiki in the east. The oldest rocks dated so far are from the last-mentioned volcano and are from 2.5 to 3 million years old (Zizka 1989; Flenley and Bahn 2003; Fischer and Love 1993). Pua Katiki created the present Poike Peninsula, which owes its reduced size, almost rectangular shape and its more than 150 m high cliffs to marine abrasion. The landward continuation of an abandoned cliff is evidence of the former isolated position of the Poike volcano in the sea.

The next-youngest volcano of Easter Island, Rano Kau, is a caldera with a diameter of 1.6 km and a fresh-water lake with a mosaic of floating bog vegetation. Rocks of Rano Kau have been dated to about 1 million years (Zizka 1989; Fischer and Love 1993). On the southern flank of Rano Kau a steep cliff up to 300 m high has been created by abrasion. Most of the land of the island was created by the youngest of the three volcanoes, Maunga Terevaka, dated to about 240,000 years. In several steps it filled in the gap between the two older volcanoes. Its youngest volcanic activity may have taken place only 12,000 years ago (Flenley and Bahn 2003). Maunga Terevaka and its numerous parasitic volcanoes with their cone shapes and steep flanks determine the present landform character of the island.

Of high cultural significance is Rano Raraku, as it is the site of the former quarry where the ***moai*** were produced. The crater of this volcano also has a lake, which was drained in the second half of the twentieth century. In prehistoric times the three crater lakes were the only permanent surficial fresh-water reservoirs of Easter Island. Their peat deposits have been the object of palynological studies (Flenley and King 1984).

Except for the small amounts of marine sediment in the littoral zone, all rocks, soils and sediments of Easter Island are of volcanic origin. The rocks are basalts, hawaiites, trachytes and rhyolites (Baker 1997). Next in importance are yellowish and reddish tuffs. The volcanic bedrock is exposed in the cliffs and along parts of the immediate hinterland, on parts of the volcano flanks, on eroded surfaces above the cliffs, mainly so on Poike Peninsula, as well as on the floor of gullies, again mainly on Poike Peninsula.

About 70 per cent of the island surface is strewn with stones of all sizes, in places forming a complete blanket. Most of the latter is of human origin and came into being in the context of stone mulching (Stevenson et al. 2002; Bork et al. 2004).

The reddish-yellow, yellowish-brown to brown, mostly clayey soils of the island formed in the late Pleistocene and Holocene following deep-reaching weathering of the volcanic rocks. Evidence for the mostly Holocene development of the soil is based on dated humus taken from an undisturbed B-horizon on the south-western slope of Maunga Orito. The radiocarbon dating yielded a calibrated humus age of 6831±39 BP (2-σ cal 5790–5657 BC, 5652–5639 BC, KIA 17120).[1] For methodological reasons the younger humus is over-represented; therefore humus formation should have begun more than 10,000 years ago. The original soils are rich in kaolinite, typically have a very low density, a polyhedric structure and slickensides.

---

[1] KIA stands for radiocarbon datings carried out in the Leibniz Laboratory for Radiometric Dating and Isotope Research at the Christian-Albrechts University in Kiel, Germany.

8. Soil Fertility and Soil Erosion on Easter Island: The State of Knowledge

Until recently the soils of Easter Islands were hardly studied. Consequently reliable information on former soil fertility or on indications of soil erosion in the literature is scarce. Studies of the horticulture on Easter Islands have been able to partly document the prehistoric land-use techniques, but not the earlier soil conditions or soil qualities. No causal analysis of the relationship between soil quality and land use has therefore yet been possible. Instead, some authors have inferred former cultivation conditions from the present state of the soils. Presently exposed volcanic sediments and volcanic rocks, together with the stone-strewn surfaces, create the impression that large parts of Easter Island carry soils that are infertile or not suitable for horticulture.

The present high permeability of the volcanic rocks and the almost complete absence of deep humus-rich soils with a high usable field capacity in large parts of the island has been extrapolated by a number of archaeologists into prehistoric times, and has then been thought to have been the main reason for the high technical input in horticulture in those times. Thus Stevenson and Haoa (1999, p. 6), in their analysis of prehistoric horticultural techniques, assume that soil conditions in those days would have hardly been different from today. Louwagie and Langohr (2002), in contrast, emphasise our scant knowledge of the earlier suitability of the soils for horticulture. The authors took samples from two soil profiles of the island and applied standard tests to determine their capacity for crop production.

Wozniak (1999) describes the high permeability of the rocks and the reduced availability of certain plant nutrients such as nitrogen and potassium as the limiting factors explaining why prehistoric garden soils were mulched with charred plant remains. From only a few studies of the present state of the island soils, conclusions were drawn as to the prehistoric environmental conditions, and these were described as an 'uncertain environment' for a secure food supply. Spatial and temporal variability of climate as well as generally less favourable climatic conditions than on other islands of the Polynesian region are regarded as additional risk factors for prehistoric agriculture (Flenley and Bahn 2003; Hunt and Lipo 2001; Stevenson et al. 1999, 2002).

As yet there are also no reliable studies on soil erosion in prehistoric times. The first identification of erosion in prehistoric times was interpreted from cores taken from the crater lake of Rano Raraku (Bahn and Flenley 1992; Flenley 1993b). At a depth of 1–1.2 m the authors found a sediment that had been eroded from the inner slope of the crater. They could only very roughly date it as having been deposited at some time within the last 1,000 years, and attributed it to quarrying during the principal phase of ***moai*** production. From archaeological evidence Wozniak (1998, pp. 189, 191) described a sediment 1 m thick from the northwestern coast, at the foot of Maunga Terevaka, which had been deposited before and during the construction of an ***ahu***. The author attributed it to soil erosion following forest clearing in the area and dated the deposition to the fourteenth and fifteenth century. Because of temporal inconsistencies in the distribution pattern of archaeological surface finds on the upper volcano slopes, she also inferred prehistoric erosion events, but did not bolster her interpretation with geomorphological and pedological evidence.

It was not until the middle of the twentieth century that the nature conservation agencies began to pay attention to young erosion features and the resulting change of the landscape in large parts of Easter Islands (Diaz Vial 1949). Allochthonous trees and shrubs were planted, mainly on Poike Peninsula, in order to protect slopes above areas with severe soil erosion against any further loss of soil. Later studies showed, however, that many plants had not taken root permanently and that soil erosion had not been stopped (Sudzuki 1979). Although soil erosion could easily lead to severe ecological and economical problems on the whole island because of its isolated position and limited size, there has been practically no scientific analysis of the interrelationship between land use and soil erosion. Neither temporal and regional differentiation of soil erosion has been undertaken, nor has there been any calculation of erosion balances.

9. Poike Peninsula: A New Case Study of Soil and Landscape History

The 14 km^2 large Poike Peninsula is presented here as a new case study of the interaction between land use, landscape history and cultural development, based on a recent detailed analysis. Prior to this study hardly anything was known about prehistoric land use and landscape history in the area. The peninsula is in an isolated position, cannot be reached by road or path and is closed by a cattle fence, so that tourists and even natives hardly ever go there. The difficult access may be one of the reasons why only a few studies have been conducted there.

A remarkable trait of the peninsula is areas in the east and south-west with severe present soil erosion of the micropedimentation type as described by Rohdenburg (1977). The exposed surface of weathered volcanic rocks gives a reddish colour to the almost unvegetated surfaces. On the east, north and north-west slope of Pua Katiki are deeply incised gully systems. Away from the eroded areas, which comprise about 3 km^2 or 20 per cent of the peninsula, grassland prevails, with ***Sporobulus africanus*** the leading species (Feeser 2003).

During the first half of the twentieth century the peninsula was intensively grazed by sheep. Velasco (pers. comm. 2002) spoke of 7,000 to 10,000 animals on c. 900 ha between 1930 and 1960. Since the end of sheep production, grassland burning has been abandoned, though it is still widely applied on the rest of the island.

The research results presented below are based on detailed stratigraphic studies following the methodological approach of a four-dimensional landscape analysis (Bork et al. 1998; Mieth and Bork 2005).

Figure 10.5. Poike Peninsula is the easternmost and oldest part of Easter Island. Large areas of severe soil erosion are visible at the eastern tip of the peninsula.

Figure 10.6. The investigation of large soil profiles, here a 100.5 m long exposure above the south-western cliff of Poike Peninsula, is of significance in the four-dimensional landscape analysis.

B. RECONSTRUCTION OF THE LANDSCAPE AND LAND-USE DEVELOPMENT ON EASTER ISLAND BASED ON STRATIGRAPHIC AND PALAEOECOLOGICAL METHODS

10. The Palm Phase: Reconstruction of the Prehistoric Forest Vegetation

Stratigraphic analysis of pedogenic relics of *Jubaea* palms at several sites allowed a very detailed reconstruction of the prehistoric palm-tree vegetation for the first time. The pedogenic relics are

- root tubes of the palms in original soils, sediments and weathered volcanic rocks
- the arrangement of root tubes in the shape of root cylinders or cones that can be attributed to individual palms.

The root tubes of palms are mainly found in the clayey brown Holocene soils of Easter Island. They also occur in the underlying C-horizons of sediments and chemically much-weathered volcanics. On gully walls and micropediment scarps root depths of more than 6 m have been measured. The tubes end at lesser depths in hardly weathered volcanics because of their high resistance to penetration. The hollow root tubes have more or less round cross sections with diameters between 5 and 7 mm. There is only a little downward decrease in tube diameter of individual roots. Close to the surface of the former Holocene soil the tubes converge, but are almost parallel and vertical at greater soil depth. The root systems are never branched. Their shape and pattern leave no doubt that they are the root relics of monocotylodonous trees. In places where they have been completely preserved in the soil the root tubes are arranged in a cone-shaped pattern, circular to slightly oval

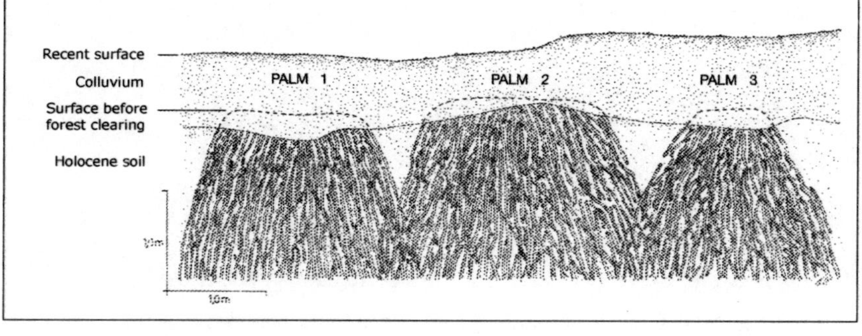

Figure 10.7. Cone-like formation of Jubaea palm root casts in the Holocene soil. The root casts represent the last palm generations which existed in certain areas.

in cross section. Where the upper part of the root system has been truncated, only the deeper-seated, more cylindrical part of the tubular pattern has survived. The diameter of the root discs at the former soil surface corresponds to the diameter of the palm trunks at their base. Diameters of up to 1.5 m were measured.

Generally the root biomass has been completely decomposed, but in some of the tubes organic relics can be found in the shape of very thin charcoal linings. More frequently the roots are coated by black to grey sesquioxide coatings (oxides of iron and manganese). The matrix surrounding the root tubes is very stable. The soil particles there have been glued together. The polyhedric pattern of the B-horizons in the root-tube areas is so stable that major aggregates containing root tubes can be taken from the soil without destroying the structure.

From the distance and diameter of the clearly identifiable root cones of individual trees within the soil profiles the distribution pattern of the former palm-tree vegetation could be reconstructed. Along a 100.5 m long section in the south-western part of Poike Peninsula, 29 individual palms of varying size and at varying distances could be identified (Mieth and Bork 2003a). Based on the morphology of the present *Jubaea chilensis* growth heights of the prehistoric palms from 5 to 18 m were reconstructed. The horizontal distances and the inferred height of the palms allow the reconstruction of a partly dense, partly open palm forest with a storeyed

Figure 10.8. Palm root casts of the extinct Jubaea palm traverse the soil vertically forming unbranched channels. The channels are frequently coated by iron and manganese oxides of dark grey colour.

Figure 10.9. True to scale reconstruction of the prehistoric palm vegetation for a slope segment in the south-west of Poike Peninsula

age structure. The presence, distribution and density of shrubs in this system are still open to conjecture, as these plants have left no autochthonous traces. The existence of a prehistoric palm vegetation could be confirmed from several other sites on Easter Island as well (Mieth et al. 2002; Mieth et al. 2003a; Bork and Mieth 2003). Based on several island-wide test sites we assume that at least 70 per cent of the island (equal to about 116 km^2) was once covered by the palm forest.

Based on an average growth distance between palms of 2.6 m – derived from several sites – a former island-wide palm-forest of more than 16 million trees has been calculated. Assuming that there were no efforts to increase the number of trees after settlement, this would be about the palm population at the time of initial settlement. A prerequisite for this estimate is that the root traces described always portray a *single*, namely the *last* palm generation at a given site. Had earlier palms also left their traces, the multiple interference of root systems would no longer have allowed the identification of individual palms. The situation is the same in all areas studied. It remains unclear why in each case only the roots of the last palm generation were preserved.

11. Horticulture in the Palm Forest: The Strategy of Sustainable Land Use

Several stratigraphic traits in the soil profiles of Poike Peninsula confirm that the palm forest was used for crop production (Mieth and Bork 2003a, Mieth et al. 2003a). Working the soil with a digging stick or appropriate stone tools created planting pits of different dimensions, depending on the kind of crop planted. Digging loosened the soil to an average depth of 30 cm. Some planting pits were as much as 60 cm deep. Ah- and B-horizons were fully homogenised by the digging. Because of its mostly crumbly texture, the cultivated soil is clearly different from the autochthonous relics of the Holocene soil. The base of the cultivated layer is weakly undulating. Less frequently, at high soil humidity, individual planting pits can be identified within the cultivated soil. Some of them are deep-reaching, also affecting the base of the cultivated layer. Occasionally small aggregates of the less fertile C-horizon have been worked into the lowermost part of the cultivated layer. In most cases any incorporation of the C-horizon seems to have been avoided.

Most likely the spatial density of planting pits existing at the same time was low, or planting may have been restricted to smaller areas. Some hundred years of crop production in the palm forest were sufficient, though, to completely loosen and homogenise the material of the cultivated layer to a rather uniform depth between the palms. The available space between the trees was used very efficiently, but the palm roots were protected, as can be seen from particularly deep planting pits carefully dug right next to the trunks. The palm roots there were ***not*** destroyed by the digging. The lateral boundaries as well as the (formerly) near-surface parts of the root cones have been fully preserved in the profiles. The reconstruction of the vegetation as described above would not have been possible without such a high degree of root preservation.

The palm forest had important protective functions for the crops. It provided shade and protected them against transpiration and wind. The forest vegetation also protected the soil, as the canopy of the trees weakened the impact of rain and wind and thus prevented soil erosion. The garden soils were sustainably enriched with organic substance by mulching of plant remains. Therefore the humus content of the cultivated horizon is higher than that of the original soil. In the oldest cultivated layers traces of burning are very rare.

In the eastern part of the island the phase of forest-related horticulture lasted about 600 years, beginning with the first settlement assumed for the seventh century, and ending in the second half of the thirteenth century with the clearing of the palm forest.

12. The Phase of Forest Clearing

In most soil profiles of Easter Island root relics are accompanied by traces of widespread burning:

- remains of burned palm trunks, appearing as pockets of charcoal, several centimetres thick, sharply delimited along vertical and lateral lines and overlying the palm root cones
- layers of ash and charcoal on the former surface of the Holocene soil overlying the root horizons and cultivated layers
- charred nutshells of the palms, often with their shape fully preserved, within the burned layer, some of them with tooth marks indicating gnawing by Polynesian rats
- thin charcoal linings in near-surface root tubes

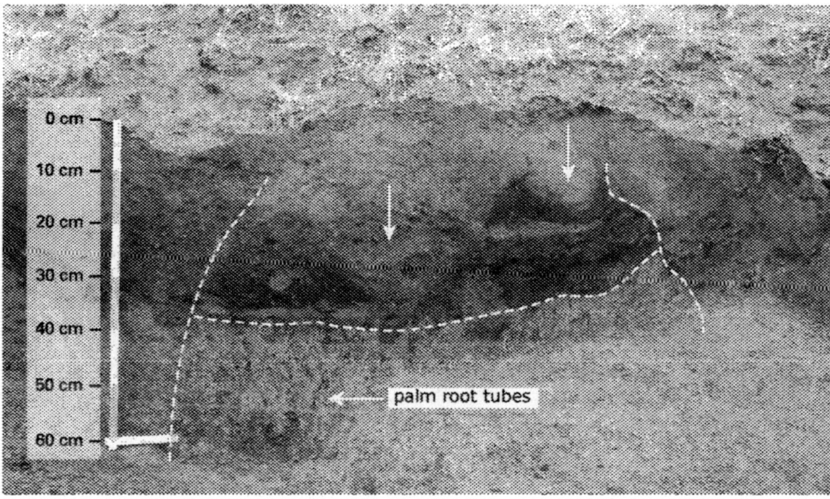

Figure 10.10. Burned stump of the indigenous Jubaea palm with indications of its use as umu and palm root channels below the fire site.

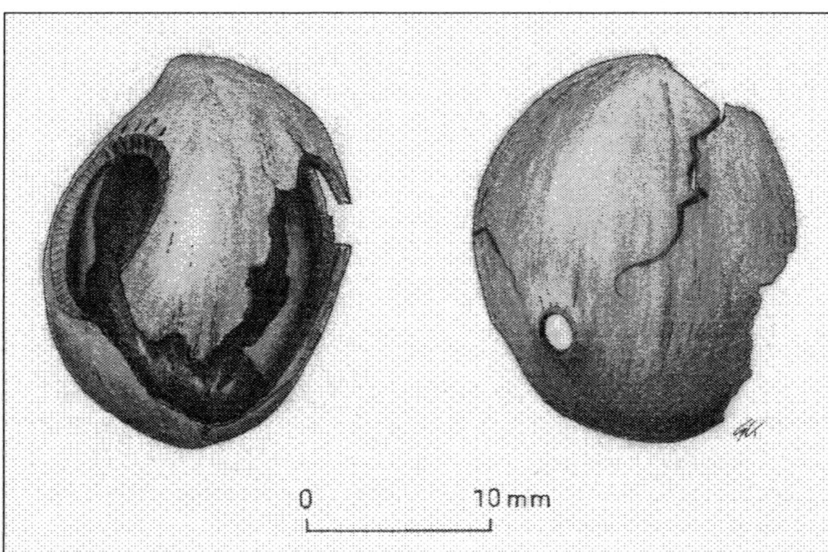

Figure 10.11. Burned nuts of Easter Island's extinct palm are found in charcoal layers, fire sites and umu. These nuts allow the dating of slash and burn events in certain areas with great exactness.

- fireplaces (*umu*) dug below the burned layer or placed above burned palm stumps
- stacked layers of fine-grained (transported) charcoal particles within fine-grained sediment immediately above the burned layer
- soil baked red at the base of charcoal veins and pockets.

All these findings clearly indicate human forest clearing and burning. From the soil profiles the systematic and efficient application of the slash-and-burn method can be reconstructed. The palms were chopped off only a few centimetres above the soil surface and the leaves and other useless parts were probably removed at the place of felling. The trunks were then taken out and put to various uses. The remains of the palms, their stumps, the litter – including some fallen-down nuts – as well as the waste from other non-usable plants were burned. The traces of gnawing by rats on the nuts (Mieth et al. 2002) are evidence that they had already lain at the surface before the fire. Sometimes dry grass, twigs and leaf litter were heaped directly on the palm stumps, perhaps to speed up incineration. Burned layers extending over hundreds of square metres, e.g. at several places on Poike Peninsula, confirm that large tracts of land were burned at a time. The preserved thickness of the surviving autochthonous charcoal layers – up to several millimetres – as well as the frequently stacked deposition of finely distributed charcoal at the base of bedded sediments suggest a considerable local thickness of the burned layer right after the fire, in view of the low stability of charcoal dust and ashes under the influence of rain and wind. Similarly the local, heat-induced chemical alteration of the clay soils – turned red by fire – speaks of a considerable intensity of the fires.

Radiocarbon datings confirm that burning the palm forest was not a single event, but consisted of a sequence of clearings, each limited in space. Datings from Poike Peninsula and a site on the Rano Raraku volcano allow the reconstruction of the consecutive phases of forest clearing by fire:

- The radiocarbon dating of a palm nut from a burned layer in the easternmost part of Poike Peninsula of 731±25 BP (2-σ cal AD 1244–1254, 1256–1299, KIA 17107) puts the beginning of fire clearing into the second half of the thirteenth century (Mieth et al. 2002). Two datings of charcoal from a large fireplace in the same area correlate quite well (631±22 BP, 2-σ cal AD 1297–1329, 1343–1396, KIA 18837; 588±22 BP, 2-σ cal AD 1304–1367, 1384–1408, KIA 18838).
- The youngest dating of a palm relict originates from the east of Poike Peninsula, only a few hundred metres away from the site just referred to. A burned nutshell from the fireplace there yielded a radiocarbon dating of 317±20 BP (2-σ cal AD 1492–1505, 1507–1600, 1614–1642, KIA 20383). This does not stand for a specific clearing at a specific place, though, but

only tells that palms existed somewhere in the area.

- About two centuries after the beginning of fire clearing in the east of the peninsula palms were also cleared in the south-west. The radiocarbon dating of charcoal from a burned palm stump is 573±20 BP (2-σ cal AD 1309–1356, 1358–1365, 1368–1416, KIA 18836), and for burned grass from the same place the calibrated age is 525±26 BP (2-σ cal AD 1330–1341, 1396–1438, 1396–1438, KIA 19369). The dating obtained from a root tube lining of another palm yielded a clearly higher age: 951±32 BP (2-σ cal AD 1018–1162, KIA 18835). It indicates the high age of the tree, 200–300 years.
- Five other datings of charcoal from planting pits, from a grave filling as well as from colluvium overlying the burned layer in the south-west of Poike Peninsula equally indicate intensive fire activity in the fourteenth to fifteenth century which, with a high degree of probability, can be attributed to the clearing of the palm forest in this area (Mieth and Bork 2003a, Mieth et al. 2003a). Datings indicate that clearing above and below the cliff was about simultaneous.
- The dating of a fireplace at the former site of a palm-tree, located at the northern inner slope of Rano Raraku yielded a radiocarbon age of 518±18 BP (2-σ cal AD 1402–1436, KIA 1719), equivalent to the clearing phase in the south-western Poike Peninsula.
- Palm root relics from the northern inner slope of Rano Raraku were found in a colluvium that had been deposited after the erosion of the original Ah/B-horizon. This is in contrast to the results obtained from all other sites, telling us that after forest clearing, erosion of the local soil and deposition of a colluvium there was a short new phase of palm-tree growth. A piece of charcoal embedded in the colluvium yielded a radiocarbon age of 659±33 BP (2-σ cal D 1283–1328, 1344–1394, KIA 17459), possibly indicating the time of the first fire clearing, only about 150 years before the second and final clearing.

13. Land Use after the Clearing of the Forest

The cleared and burned areas were immediately used again. Stratigraphic analysis confirms that in the east and north of Poike Peninsula stone platforms and tombs were placed directly on the burned surfaces.

At least for the sites studied in the south-west and east of Poike Peninsula any intensive or continuous horticultural use after the burning can be excluded. Mechanically working the soil would have destroyed the palm root relics, the burned palm stumps and the burned layer there, but the soil profiles attest to the opposite.

In the east of Poike occasional planting pits dug down from the burned layer suggest occasional planting right after the clearing (Mieth et al. 2002).

In the south-west of Poike Peninsula development of a humus horizon indicates a longer phase of calm after the clearing of the forest. Within some decades or even a few centuries a pronounced humic horizon developed underneath a grassland vegetation. Evidence of horticultural activity at this site is 100 to 200 years younger than the end of the first land-use phase, respectively the clearing of the forest. This can be read from the following characteristics found in a 110.5 m long soil section above the cliff (Mieth and Bork 2003a, Mieth et al. 2003a):

- Starting from several centimetres of a dark, grey-brown humus horizon there are clearly delineated planting pits, 10 to 50 cm deep. Their U-shaped cross sections are up to 40 cm wide and 10 to 30 cm apart.
- Between the planting pits there are undisturbed relics of the Ah-horizon which also cut through the upper parts of palm root relics.
- In contrast to the older cultivated layer the base of the younger one is quite indented because of the varying depth of individual planting pits.
- The root relics of the palms have only occasionally been affected by the digging.
- The planting pits contain loose fillings of the dark grey-brown material of the humus horizon, frequently mixed with charcoal and ash.
- The dating of charcoal from the younger cultivated layer yielded a radiocarbon age of 319$\pm$21 BP (2-σ cal AD 1492–1600, 1614–1641, KIA 18834)

The well-defined planting pits, the relics of the Ah-horizon between them as well as the good preservation of the palm root relics all indicate that also in the south-western part of the peninsula there was a short episode of at best a few years or decades' duration of low-intensity horticulture after the palm-forest clearing.

14. Sheet Erosion after Forest Clearing

In large tracts of Easter Island the volcano slopes are blanketed with well-bedded sediments. They result from sheet erosion that in each case had taken place a few metres to decametres above their site of deposition. On concave lower slopes, where the original soils have been preserved, these colluvia overlie the prehistoric land surface with its humus, fire, soil cultivation and palm relics. On the upper parts of the lower slopes, in mid-slope position and on the gently convex upper slopes relics of the Holocene soil or –where the soil was completely removed – barely weathered volcanics are buried by the colluvia.

Figure 10.12. Layered fine-grained sediment: result of rising sheet erosion after woodland clearance and intensification of land use on Poike Peninsula. Single layers of this colluvium represent single erosion and deposition events.

Only the uppermost parts of the volcano slopes, close to the divide, did not receive any of the finely bedded deposits. On the young micropediments and in the gullies of Poike Peninsula they have been removed by later erosion. The profiles studied show an average deposition of 60, in certain cases of up to 400 thin layers (Mieth et al. 2003). Generally two layers, in most cases just a few millimetres thick, form a unit that is representative of a single rainfall and runoff event. The upper layer of each unit consists of clayey micro-aggregates of silt to fine sand size, the lower layer of microaggregates and grains of mostly sand and grus size. The dominant clay minerals are kaolinite and iron oxides. Flow structures, grain-size distribution, bedding directions and the general structure of the deposits are evidence of fluvial erosion and deposition.

Analysis of downslope catenas revealed a kind of imbricated bedding of sequences about 10 m long and 70 to 150 cm thick, each with the fine-layer structure described above. From their position it is clear that the sediment packages deposited on the lower slope are older than those in mid-slope and upper-slope position, the oldest colluvia being those at the bottom of the lowermost sequence. The vertical profiles of the older and younger deposits have clearly distinct sedimentary and pedogenic characteristics. The clayey microaggregates of the older downslope deposits are rather fine-grained, mostly in the silt and fine-sand fraction, their grey-brown colour indicating elevated levels of humic material. The younger sequences on the middle and upper slopes mostly consist of reddish coarse sand and grus. Their pedogenic properties indicate that the older beds are reworked Ah/Ap- and B-horizons, whereas the younger deposits mainly consist of reworked C-horizons.

Radiocarbon dating of the colluvia allows the absolute age determination of these depositions. A piece of charcoal embedded 64 cm above the burned layer near the cliff in the east of Poike Peninsula was dated to 654±22 BP (2-σ cal AD 1289–1325, 1348–1391, KIA 17110), and thus to a maximum of 100 years after fire clearing (Mieth et al. 2002). Charcoal from mid-slope in the south-west of Poike was dated to an age between 570±22 BP (2-σ cal AD 1309–1356, 1358–1365, 1386–1419, KIA 18832) and 252±26 BP (2σ cal AD 1527–1553, 1632–1671, 1779–1798, 1944–1955, KAI 20381). Radiocarbon datings of the bedded colluvium of the upper slope of Pua Katiki indicate an even younger age of erosion and deposition, extending over a period of 600 years up to the beginning of the twentieth century (Mieth and Bork 2005). The age determinations thus confirm progressive upslope soil erosion and sedimentation. Imbrication-like sedimentation, beds continuing for a few metres only, relief and sediment characteristics all indicate that the sediment was transported for a few metres only. The short distances suggest that grassland vegetation was present in the areas of deposition, reducing the flow velocity and almost completely filtering the suspended load from the water. Obstacles to runoff such as major stones, rills or planting pits were not found in the profiles. There is no evidence from the fine bedding that attempts were made at checking the runoff. Except for microrills there are also no signs of linear erosion. Absence of dissection and the way the sediments are spread out suggest a high sediment load of the surficial runoff, largely unvegetated areas of erosion and sheetflood-like runoff.

Linking of the stratigraphic results from the various parts of Poike Peninsula allows a model reconstruction of prehistoric erosion:

The Holocene, up to the forest clearing towards the end of the thirteenth century, was a time of soil formation and surface stability. Also the land-use management during the first 600 years of human settlement was such that soil erosion was prevented. Conditions only changed with the clearing of the palm forest. Sheet erosion in the east of Poike Peninsula set in with the destruction of the forest vegetation, in the south-west of the peninsula with the beginning of horticulture after a few centuries of a grassland phase following forest clearing. A flat erosion scarp a few decimetres high ate its way up from the lower to the upper slopes over the centuries during frequent rainstorms. Along this flat scarp first the Ah-horizon with traces of digging, then the original brown Holocene soil and, in the end, the reddish material of the C-horizon were eroded. Immediately downslope from a flat scarp the eroded material was deposited in the inverse sequence as hundreds of fine

Figure 10.13. Prehistoric phases of land use and soil degradation, reconstructed for a downslope segment above the cliff in south-west Poike.

A. Agroforestry in a palm tree forest until about AD 1450
B. Slash and burn of the palm forest around AD 1450
C. Development of a humic horizon from about AD 1450
D. Few planting pits result from extensive agriculture from about AD 1550
E. Deposition of a layered fine-grained sediment after about AD 1675

layers. Some of the sediment will have been washed over the cliff; thus the erosion rates mentioned below are minimum estimates. Because of the gently concave lower slopes above the cliff, where the fines could easily be trapped, sediment loss into the sea will have been well below 20 per cent. Approximately 640 years after its beginning the erosion reached the near-summit slope areas of Pua Katiki volcano. Agriculture on the eroded land became impossible.

In contrast, intensive land use well into the last three centuries is reflected in the colluvium on the slopes of Maunga Orito in the south-west of Easter Island. There numerous embedded obsidian tools, weapons and charcoal as well as planting pits and *umu* are evidence of an intensive recent phase of land use (2σ cal AD 1665–1683, 1733–1784, 1789–1808, 1928–1955, KIA 17116; 2σ cal AD 1622–1699, 1723–1779, 1798–1814, 1832–1879, 1915–1944, KIA 17117; 2σ cal AD 1682–1734, 1807–1903, 1905–1930, KIA 17118).

15. The Intensity of Prehistoric Soil Erosion

Based on the volume of the dated sediment sheets, soil erosion dynamics on the eastern slope of Pua Katiki could be quantified. Assuming an average deposition of two layers per runoff event, 6,900 of them should have occurred over 640 years. The average of 10 events per year seems plausible in view of the present rainfall regime. The average retreat of the micropediment scarp would have been 0.3 m per event or 3 m upslope per year.

Volume and mass calculations were made for the finely bedded colluvium deposited on the slope (Mieth and Bork 2005). Per hectare and year an average of 8.6 tons was reworked by soil erosion, i.e. less than 1 ton per hectare and erosion event. The average erosion rate is thus comparatively low. Nevertheless, in the east of Poike Peninsula 80 to 100 weak soil erosion and deposition events within a few decades around AD 1300 decisively lowered the soil quality of the site. The fact that the basal layers in the east as well as in the south and north-west of Poike are undisturbed is evidence that horticulture was abandoned there with the onset of soil erosion.

16. Soil Erosion in the Nineteenth and Twentieth Centuries

Recent micropedimentation and gullying have partly eroded or at least dissected the finely bedded colluvia. Erosion reaching several metres down has not only removed the colluvia, but also the prehistoric cultivation layers, relics of palms and fire clearing as well as the brown Holocene soils and the underlying severely weathered volcanics that had survived in the depositional environment of concave slope segments. There has also been a dramatic loss of prehistoric cultural relics on account of this erosion (Mieth et al. 2002). Erosion damage is obvious on the south-west, north-east, and north slopes as well as right above the cliffs of Poike Peninsula. But also affected are near-cliff areas on Rano Kau, on the inner and outer slopes of Rano Raraku and several other volcano slopes of the inland.

Gullying has also dissected the youngest beds of areal deposition which, as has been shown for the north-western upper slope of Pua Katiki, continued up to the beginning of the twentieth century. Severe gullying is thus a phenomenon that did not start before the first half of the twentieth century. This is the reason why gullying and micropedimentation was not mentioned by discoverers and scientists up to the twentieth century. Drawings by Routledge (1919 [1998], pp. 192, 196) from the early twentieth century show the south-west coast of Poike Peninsula without the red micropediments above the cliffs that are so conspicuous today, and the north-western slope of Pua Katiki without any gullies.

There is no doubt, then, that gullying and micropedimentation are elements of the European import of land use, namely the keeping of large herds of grazing animals. Regular burning of the grassland and overgrazing in combination with vegetation damage and soil compaction along animal tracks triggered linear erosion. Originally track-wide rills rapidly expanded. The gullies today are up to 6 m deep, up to 50 m wide and several hundreds of metres long. Lateral erosion and the removal of walls between them created extended micropediments on their floors.

The formation of gullies and micropediments has thus been the outcome of barely four decades of sheep-grazing (from the 1930s to the 1960s). Comparison of today's situation with aerial photographs from 1981 shows that there has not been much change of the erosion forms since then. The slowing of erosion after the early 1960s is probably due to the increasing vegetation cover after the end of grazing and annual grass burning on Poike Peninsula. The numerous cattle tracks with their compacted soils, though, still cause local runoff concentration and linear erosion. Check 'dams' made of metal plates from cut-up oil drums set up on the gully heads of eastern Poike had at best no effect at all, or even accelerated gullying there (Mieth and Bork 2003b).

17. The Intensity of Recent Soil Erosion

Erosion balances were calculated for the gully systems in the north-west, north-east and south-west of Poike Peninsula (Mieth and Bork 2005). 40 years of gully formation were assumed, although it may have been concentrated on the 30 years of sheep grazing from the 1930s to the early 1960s. The erosion rates of between 190 and 650 t/ha/y are extremely high. Assuming an average erosion rate of 350 t/ha/y for the 20 per cent taken up by gullies and micropediments, soil loss on the whole peninsula is 70 t/ha/y (Mieth and Bork 2005). The rates of annual soil loss are thus 20 to 75 times higher than the rates of sheet erosion in prehistoric times induced by forest clearing, settlements and horticulture. The figures are also exceptional in a world-wide comparison (Bork et al. 2003). The other big difference is that the material eroded in prehistoric times was mostly redeposited on the slopes of the island, whereas the soil eroded by gullying and modern micropedimentation has been washed over the cliff and into the Pacific.

C. SUMMARY OF THE SOIL AND LANDSCAPE HISTORY OF EASTER ISLAND

18. Peculiarities of the Early Phase of Prehistoric Land Use

The detailed statements on the early phase of horticultural land use presented here are the outcome of our recent studies conducted on Poike Peninsula, but they are likely to apply to other parts of the island as well, e.g. the Rano Kau in the south-west. It appears that the first centuries of settlement were characterised by a sustainable subsistence economy adapted to the environment. The Polynesian discoverers found a mesophytic forest dominated by *Jubaea* palms, in which they cultivated a range of plants without any overall forest clearing. The positive effects of the forest vegetation were used and preserved for centuries. The kaolinite- and sesquioxide-rich topsoil between the trees of the palm forest was mechanically loosened up and homogenised by cultivation. Mulching of plant litter greatly increased the humus content of the cultivated layer and thus led to high soil fertility. The complete preservation of the tube systems of the palm roots in the areas of deposition is proof that mechanical soil treatment was careful. In this phase of sustainable horticulture erosion was avoided. The integration of various garden crops with the existing palm forest vegetation and only low and selective use of the trees themselves corresponds to the type of sustainable agroforestry described in the South Pacific by Hunter-Anderson (1998) and Kirch (2000).

It is not yet known whether there was any deliberate tree cultivation in those first centuries. As seeding and planting of ***Cocos nucifera*** and other trees (e.g. ***Artocarpus*** sp.) are common on other South Pacific islands, described as 'arboriculture' by Kirch (2000), it is quite likely that this was also the case on Easter Island. It is also a matter of speculation as to what extent the palm trees were used. It is likely that trunks and leaves were used for a range of purposes (Grau 1998; Orliac 2003). Some parts may also have been used for human consumption. Grau (1989), for instance, thinks that the 'hearts' of palm saplings may have been used for food. Eating the pulp of ***Jubaea*** nuts – similar in taste to the endosperm of ***Cocos nucifera*** – is still mentioned in the reports of the European discoverers. Cracked nutshells in cooking and refuse pits equally indicate that the palm fruits were consumed (Orliac 2003). Grau (1998, p. 122) writes that also the soft exocarp of the fruits is edible. In view of an average annual yield of 100 kg of nuts from a fully grown palm tree (Grau 1989) and the calculated figure of 16 million palms on the island this must have been an enormous food potential. The same is true for the sugar-rich juice of the palms, which on the South American continent is still collected from felled trunks of ***Jubaea chilensis*** and processed into 'palm honey' (Darwin 1999). There is as yet no evidence of it on Easter Island, but it is quite unlikely that such a valuable resource should have been overlooked for centuries. Even if the maximum yields of 400 litres per trunk common in Chile today (Grau 1998; Orliac 2003) were not reached on the island, simply because of the large number of trees, this should have been another abundant food resource (Bork and Mieth 2003). Food and non-food uses of the palm would not have been mutually exclusive.

From these data and inferences it appears that a reassessment of the ***Jubaea*** palm should be made in view of its economic function during the early culture phase of Easter Island. It seems that the early land-use phase was one of 'cold' processing of material and food resources. In the oldest anthropogenic layers of all the profiles studied there are no ***umu***, perhaps an indication that plants and animals were eaten raw then. The use of burned plant remains as fertiliser seems not to have been significant, in contrast to the later horticultural phases. Even tiny fragments of charcoal are hardly ever found in the oldest cultivated layers, which makes it difficult to date the beginning of the early phase of horticulture exactly. Its end, however, because of the disappearance of the palm forest, shows up very well in the stratigraphies.

19. The Causes of Deforestation: Natural Versus Human Factors

The former existence of a palm-dominated forest on Easter Island and the almost complete deforestation in prehistoric times is by now undisputed. There is still a controversy on the causes and timing of deforestation. There are two schools of thought, the one favouring the theory of palm forest disappearance by one or several climatic catastrophes, the other assuming that the forest ecosystem was destroyed by land use. As Easter Island lies at the western fringe of the cold Humboldt current, it has repeatedly been discussed whether the El Niño effect may not have led to long-lasting droughts in prehistoric times and thus to ecological catastrophes (Macintyre 1999; Nunn 2000; Genz and Hunt 2003). Recent climate modelling suggests, however, that the various El Niño effects compensate each other in the Southeast-Polynesian region (Mcintyre 2001), making the climatic catastrophe-hypothesis quite improbable. Flenley et al. (1991), Bahn and Flenley (1992) and Flenley (1993b) think that the demand for palm wood for technical purposes such as boat construction and transporting the ***moai***, as well as for firewood and open spaces for horticulture were the reasons behind forest clearing. The authors further assume that eating of the palm nuts by the Polynesian rat which had come with the settlers, may have been an additional handicap for the regeneration of palm vegetation.

Based on his palynological studies Flenley (1993b) dates the beginning of forest clearing to around AD 800, and the almost total extinction of the palm forest to the fifteenth century. For Bahn and Flenley (1992) the correlation of these date limits with the supposed beginning of settlement and the climax of the ***moai*** culture leaves no other conclusion as to causes of palm forest degradation than human interference. The effects of Late Holocene climatic fluctuations such as the Little Ice Age, postulated by McCall (1993), are not excluded by Bahn and Flenley but their potentially destructive effect on the palm vegetation is. From their pollen studies Bahn and Flenley (1992) can show that the palm forest survived much stronger climatic fluctuations than the Little Ice Age during the last 37,000 years.

Another author vehemently in favour of the climatic hypothesis is Hunter-Anderson (1998). By means of examples taken from other Pacific and continental regions she concludes that any other than sustainable land-use techniques would have been unthinkable on Easter Island. She also excludes any destruction of the forest due to the extraction of palm trunks for construction and as firewood, with the argument that the economy of the Pacific region was traditionally resource-preserving. She also thinks that any intended systematic forest clearing for the creation of garden plots is unlikely. Instead, she postulates the former existence of a soil- and plant-preserving agroforestry, which, from her conclusions, would have continued if the forest had not been destroyed by extreme climatic events. In her view the pollen studies by Flenley show that there had already been a retreat of the forest ***before*** the first settlement (Hunter-Anderson 1998, p. 89). Any negative effects the Polynesian rat might have had on the reproduction of palms she thinks unlikely; quite to the contrary, opening up of the exocarp may even have helped germination of the nuts.

Orliac and Orliac (1998, 2001) and Orliac (2000, 2003), from their anthracological studies, report a significant decrease of charcoal from woody plants in favour of burned grass in numerous fireplaces and refuse pits of the seventeenth century. The authors take this as evidence of an abrupt change from a forest- to a grass-dominated vegetation at that time. They caution the reader, however, that the frequency and composition of charcoal in archaeological structures is evidence that the species existed at a given time, but will yield no information on the frequency and distribution of the plant species. In their conclusions the authors also tend towards a climatically induced change of the ecosystem. They note a rapid undifferentiated retreat of the forest. Because of the climatically induced dying of the forest there would have been an oversupply of firewood for several decades, before a lack of firewood made grass a substitute (Orliac and Orliac 1998, p. 132). The hypothesis is contradicted by Orliac (2003), though. The youngest palm relics (charred endocarpon) dated are from the middle of the fifteenth century, and among several thousand pieces of charcoal from the seventeenth century she found merely four pieces of ***Jubaea***.

The latest research results, presented in this chapter, are unequivocal evidence of the anthropogenic destruction of the forest of Easter Island. The ***in situ*** relics of palm stumps and nuts within the burned layer, together with the absence of any larger pieces of burned wood in the soil profiles of Easter Island, are proof of slash-and-burn forest clearing, where the trunks were extracted for efficient use and the unusable remains were just as efficiently burned. The root tubes in the soil are circumstantial evidence of human forest clearing by fire. As described above, only the root tubes of the ***last*** palm-tree generation have been preserved by agglutination of the soil particles surrounding them – just the generation that was destroyed by fire. Thus only forest clearing followed by fire can have preserved them. In the central-Chile national park of La Campana – a region where ***Jubaea chilensis*** used to be cultivated – no such open root tubes could be found (Bork unpublished, 2003), which leaves fire as the only explanation for the agglutination of the soil particles. Direct heat has to be excluded, however, as the root tubes have been preserved down to a depth of several metres. A possible explanation might be physiological reactions of the fresh palm stumps during a fire by way of excretion from the roots, but there is clearly a need for more research.

The spatial and temporal retreat of the palm forest, in particular in the east of Easter Island, can be described in much detail from the stratigraphic studies undertaken. It appears that deforestation of once completely palm-covered Poike Peninsula took more than 200 years, from about AD 1250 to about AD 1450. There was no reforestation on the peninsula after forest clearing. The local regeneration of palm vegetation after a first phase of clearing on Rano Raraku is an exception. As dated, the newly grown palms lasted there for no more than 150 years. There is no evidence as to whether the stand resulted from spontaneous regrowth or was due to afforestation.

Our own radiocarbon datings of *Jubaea* relics are in good temporal agreement with the *Jubaea*- (*Paschalococos*) datings by Orliac (2003, pp. 195–6). Most of her datings of wood and nut fragments from archaeological structures fall between about AD 1200 and AD 1450. According to her there is almost no evidence of younger palms, which is confirmed by our own stratigraphic analyses. The dating of a palm nut from a fire site on Poike from the sixteenth to seventeenth century should rather be regarded as an exception and as evidence that a few specimens of the species could survive the forest clearing for a few more centuries, but not so the palm woodland as a whole.

Our own studies as well as those by Orliac (2003) have not yielded any *Jubaea* charcoal from the time *before* AD 1250. Neither our own stratigraphic studies nor Orliac's studies (2000) answer the question as to whether forest clearing in other parts of the island really began as early as AD 800–1000, as postulated by Flenley (1993b) from his palynological studies and only a very few radiocarbon datings. These datings are not only earlier than the main phase of clearing proved for Poike and inferred for the rest of the island; they are also close to the assumed beginning of settlement. As we have calculated, clearing of the island forest within an assumed period of 400 years would have taken an average of 500 workers for cutting and processing of the palms (Mieth et al. 2003a). Such a large workforce would only have been available a few centuries after the initial settlement. Therefore only spatially limited clearing is likely to have taken place before AD 1200.

The stratigraphic evidence of extended burning and the changes in the soils following them unequivocally show that, together with the areal clearing of the palm forest, also the extent of the shrub component of the mesophytic forest was much reduced. The anthracological evidence by Orliac (2000) of a still highly diverse composition of wood species in the fire sites of the seventeenth century should not be interpreted as evidence of a forest that was still species-rich. More likely branches were laboriously collected from small relict woodlots, which became fully exhausted by the middle of the seventeenth century so that grass had to be collected for making fire.

There may be a number of reasons why the forest was cleared. The studies in the east, north and south-west of Poike Peninsula show that immediately after clearing the areas were used for new settlements and as ceremonial and burial plots (Mieth et al. 2002; Mieth and Bork 2003a). No doubt population growth created a demand for cleared land and was the reason why new settlements sprang up in even the most remote parts of the island. The period of forest clearing on Poike Peninsula between AD 1250 and AD 1450 (possibly with some stands remaining until about AD 1600) is in good agreement with the time of increased expansion of settlement to areas higher up and farther away from the coast of Easter Island as identified by Stevenson et al. (2002) and Martinsson and Wallin (2002).

The intensification of horticulture – specifically the rapid transition from agroforestry to horticulture in the open – was possibly another central reason for clearing the forest. It is true that the soil profiles from lower slope positions in the east and south-west of Poike Peninsula suggest that there were only weak attempts at horticulture after forest clearing. The colluvia overlying the cultivated layers as well as occasional post-clearing planting pits at the base of the colluvium indicate, however, that at least for some time after the clearing intensive disturbance on the upper slopes must have existed, triggering the upslope-progressing sheet erosion and degradation of the soils. Keeping the areas open by horticulture – and thus prone to erosion – is one of a limited number of plausible reasons for the disturbances.

With increasing population an increased demand for timber is another likely reason for the gradual deforestation of the island. Extraction of timber was certainly more important than that of firewood, as can be deduced from the almost complete absence of burned pieces of trunks in the soil profiles and the archaeological structures (Orliac 2003, p. 191). Extracting palm juice is another possible explanation for systematic forest clearing, as the trees had to be felled to get the sugary juice (Bork and Mieth 2003). The wood could be put to other uses afterwards.

It is still not clear, however, how most of the trunks were used. Given the estimated number of 16 million palms before clearing began, their exclusive use for the construction of houses and boats is highly unlikely. Whether the transport of the stone statues called for large amounts of trunks may only be answered after more research into the transport techniques. Assuming that most of the ***moai*** transported from the quarry of Rano Raraku were produced within 300 years, not more than one or two statues would have had to be transported per year. Even if a few hundred trunks were used up, not more than at best a few thousand of them would have been used in a year, and fewer than one million for the total production period of the statues.

20. The Late Phase of Prehistoric Land Use: Interrelationships between Landscape and Cultural Development

In all probability the changes in the landscape and the soils had not only ecological, but also cultural consequences. Some considerations on such interrelationships will be presented below. To a large extent they will have to be speculations, and in some cases, only questions.

With the clearing of the palm forest of Easter Island the principle of a sustainable, soil-protecting subsistence economy was abandoned. A high turnover of the natural resources became the characteristic of the new type of land use. With high labour input and obviously also technical efficiency the forest was cleared within a few centuries. The 'cold' use of the resources was replaced by 'hot' processes. Fire sites and ***umu*** in large numbers dot the soil horizons and sediments after forest clearing. Some of them, installed immediately after the clearing, had sizes unknown from earlier times. In the east of Poike Peninsula a box-shaped ***umu*** 2.7 m wide and 1.7 m deep was found and dated to that time. Several thick layers of charcoal and numerous animal and vegetable food remains are evidence of several meals prepared there. The tradition of cooking food in such sunken ovens was continued well beyond the end of the last wood resources. A large number of ***umu*** occur in the colluvia of the last three centuries, as on the lower slope of Maunga Orito. For some time firewood could still be collected from isolated stands of woodland, but later had to be replaced by low-grade fuels.

Figure 10.14. Uncommon large prehistoric umu (fire place for cooking) in eastern Poike, constructed after woodland clearance. The trench-like feature is filled by several thick layers of charcoal, ashes and red burned clay and contains a broad variety of food relics (e.g. bones of chicken and rats, shells of various molluscs and sea-urchins, coral fragments, palm nuts).

Numerous planting pits from the post-clearing phase of horticulture have a high content of fire remains. This is taken as evidence that charcoal and ashes were increasingly used for soil improvement, at least supplementing the traditional method of mulching with organic material.

Forest clearing obviously brought a dramatic change of the environment. After the loss of the forest not only the plants once growing among the trees, but also the people who would mainly have lived and worked in their open shade, would have been increasingly exposed to the extremely intensive solar radiation, frequent strong downpours and gale-force winds. One of the reasons why the garden-plots and settlements in exposed slope positions above the cliff coast of Poike Peninsula were used for only a short time may have to do with the severe change of the local climate after forest clearing – a question hardly discussed in the literature. The increasing use of caves as places of settlement in the late prehistoric phase (Flenley and Bahn 2003) may have come in reaction to the changed climatic situation and may not have resulted from social conflicts and/or spiritual demands, as frequently assumed.

The Poike Peninsula case study shows that an unstoppable process of degradation set in after deforestation, in the course of which the fertile soils were eroded, resedimented and thus withdrawn from horticultural use for a long time. Even for the sweet potato, which is adapted to rather poor soil conditions (Kirch 2000), cultivation would no longer have been profitable. Garden plots, which in the course of spatial expansion had been set up on higher slopes away from the coast (Stevenson et al. 2002) probably had to be given up for that reason.

A third reason why ceremonial sites, burial grounds, settlements and garden plots were given up may have been the frequent runoff, soil erosion and sedimentation events of which there is evidence in the soil profiles of Poike Peninsula. Houses, *ahu*, burial sites, and gardens were repeatedly flooded and covered with mud. Stone structures became completely buried in sediment, perhaps within only a few years after they had been set up. The people who had not known such erosion events before, who could not explain them by cause and effect and who probably took recourse in religious or spiritual explanations, apparently took no technical or strategic countermeasures. At least the soil profiles of Poike Peninsula show that the people permanently abandoned the areas affected by erosion and sedimentation.

The retreat of settlements and horticulture from large parts of Poike Peninsula within a few years after forest clearing is in good agreement with the retreat of horticulture from the upper slopes of the volcanoes as described, but not explained by Stevenson et al. (2002). In the lower areas, to which the people had withdrawn, the nutrient-rich soil of the former palm forest and the humus-rich colluvium had partly been preserved on the concave, gently inclined lower slopes. But there too,

Figure 10.15. Gully in the north-east of Poike Peninsula, a result of overgrazing by sheep and frequent burning of the grassland in the first half of the twentieth century

Figure 10.16. Badlands in eastern Poike with volcanic bedrock exposed by severe soil erosion. Basaltic stones on the surface do not originate from this area, they are brought by man to this area and are relics of prehistoric settlements and cultural places which have been destroyed by erosion.

because of the absence of any protective tree cover, the surfaces were fully exposed to the extreme climatic conditions.

Cropland in the centre, in the south, north and west of the island now became protected by two extremely labour-intensive horticultural techniques new to Easter Island: stone mulching on the surface (Bork et al. 2004) and the construction of ***manavai***, little gardens surrounded with windbreaks of stone (Stevenson et al. 2002). The stones spread out on the soil surface were probably supposed to protect against transpiration and wind, dampen the temperature curve at the surface and give mechanical support to plant growth. Perhaps the gardeners of that time may also have applied these measures to reduce runoff, increase infiltration and reduce soil erosion. A recent island-wide analysis of stone mulching and the amount of work involved has shown that in the late prehistoric and perhaps also in the earliest historic phase more than a billion stones (mainly basalts of Maunga Terevaka) were taken from numerous small quarries and spread out on the garden land of Easter Island (Bork et al. 2004). The total area with an anthropogenic stone cover was about 76 km^2, with an average of 150,000 stones per hectare. A calculation of the amount of work necessary for stone mulching showed that more than 100 persons may have been necessary for about 400 years for cutting, transporting and distributing the stones (Bork et al. 2004). The third and last prehistorical phase of horticulture thus obviously was very labour-intensive.

At least from the fifteenth to sixteenth century onward, the clearing of the palm forest, because of the ecological and land-use change involved, had considerable detrimental effects on the population of Easter Island. The following ecological and economic consequences can either be reliably derived or, in some cases, at least inferred from the stratigraphic analysis:

- loss of the ***Jubaea*** palm and its usable parts (wood, nuts, palm juice, palm hearts)
- onset of extreme climatic conditions in the immediate environment of the people, leading to more difficult and insecure site conditions for the cultivation of plants at the remaining garden plots
- destruction of crops, houses and ceremonial places by frequent runoff events
- high competition for land, specifically over-use of the remaining horticultural sites
- problems of food supply due to the simultaneous loss of good soils, loss of nutrient-supplying species of the natural vegetation (mainly the ***Jubaea*** palms) and to decreasing yield (in part also because of the climatically unfavourable exposure of the cultivated plants), and at the same time

- high labour input and strenuous work for the stone-based horticultural techniques.

The factors identified in this recent study confirm and complete the theoretical model of the interrelationships of ecosystem, land-use and cultural changes on Easter Island as proposed by Bahn and Flenley (1992). Our studies indicate that there was already a dramatic change of economic and ecological conditions in the labour-intensive megalithic culture in the fifteenth or at the latest at the beginning of the sixteenth century. This is in contrast to other authors who place the collapse of the system in the seventeenth century or, without substantial proof, even exactly in the year 1680 (Esen-Baur 1989; Van Tilburg 1994).

The soil degradation which followed the clearing of the forest had a decisive influence on the ecological and economic situation. The key effect of soil change has so far either been overlooked or underestimated. As yet it can only be speculated, though, what role the changing vegetation and soils played in the collapse of the ***moai*** culture, or whether it triggered the rise of the birdman cult. The coincidence in time itself between the end of forest clearing and the final phase of ***moai*** production at the turn of the fifteenth to the sixteenth century suggests a causal relationship between ecosystem and cultural change or even disruption. Whether it was the scarcity of tree trunks for transporting the giant stone figures, the high labour input necessary for the stone-mulching system, increasingly difficult crop production and failed harvests because of lowered soil fertility and the scarcity of soils suitable for horticulture, or whether there were social tensions and diseases resulting from deteriorating nutrition that led to the collapse of the unique megalithic culture, will most likely never be found out. However, the new results presented here fully support the hypothesis that it was mainly the ***anthropogenic*** and not any natural change of the environment that profoundly altered the cultural and social situation on Easter Island.

21. Ecological Long-Term Effects of Historical Land Use

Both the late Polynesian and the early European phase of land use had profound ecological and long-lasting effects. By the turn of the fifteenth to the sixteenth century, or at the latest by the mid-1600s, the total clearing of the forest led to the severe depletion or even extinction of a number of woody plant species of the former mesophytic forest, among them the possibly endemic palm of Easter Island related to ***Jubaea chilensis*** on the mainland, the toromiro shrub (***Sophora toromiro***) and others of which only palynological, anthracological or carpological traces remain (Flenley 1993b, Orliac 2003). The extinction of several species of birds, probably dependent on the forest environment, has been proved by a small number of zooarchaeological studies (Steadman et al. 1994), which most likely give only an incomplete picture of the original faunal spectrum.

By the example of the Poike Peninsula it has been shown that the almost ubiquitous soil erosion triggered by forest clearing by the end of the thirteenth century and kept alive by the types of land use following it continued for more than 600 years and only came to an end when the intensive phase of European land use had already begun. A single and perhaps only brief disturbance of the surface may have been necessary to trigger erosion, possibly the one-time forest clearing itself. Once begun, however, the erosion continued as a dynamic process and only came to an end when the erosion scarp had almost reached the crater rim of the highest volcanic structure of the peninsula. Even long phases without any land use, after settled and horticultural areas had been abandoned from the fourteenth century onward, could not stop the erosion process. It even continued throughout the last centuries before European discovery, when the land-use intensity had been much reduced or had been restricted to the coastal areas.

By the beginning of the twentieth century sheet erosion was replaced by linear erosion triggered by European sheep raising. This phase of land use also had long-lasting effects. A few decades of intensive grazing were enough to start a degradational dynamism that is still active today. So far none of the measures of conservation management have succeeded in checking it. The first European phase of land use also had a long-lasting effect on the vegetation. On one hand it was the grazing and trampling by sheep that completely destroyed the remaining indigenous and endemic flora, and with it the final chance for regeneration of the typical vegetation of the island. At the same time, right up to the end of the twentieth century, there has been an introduction of allochthonous species for pasture improvement and forestry, but also for reasons of nature protection, leading to a massive anthropogenic transformation of the island flora (Feeser 2003). It also should not be overlooked that the existing prehistoric cultural remains are severely endangered by the rapid linear soil erosion.

22. Perspectives on Soil Protection

The results presented here have shown the sensitive reaction of the environment of Easter Island to various types of land use, in Polynesian as well as European times. Sustainable and non-sustainable phases of land use could clearly be differentiated. Human impact, not natural catastrophes, caused change within the system. The consequences of land use – or rather the misuse of the land – are largely irreversible. The ongoing linear soil erosion and with it the further growth of the micropediments will only be stopped once grazing by cattle and horses has been considerably reduced. The institutions responsible for the protection of nature and the antiquities of the island will have to face this conflict of land use.

A second important strategy of nature protection would be the regeneration of a protective vegetation cover by planting habitat-adapted shrubs, such as *Jubea chilensis*. This should be done in combination with other shrubs either known to have existed in Polynesian times on Easter Island or still present in other parts of southern Polynesia. First trials with ***Jubaea chilensis*** on Easter Island have shown that it grows much better there than on the Chilean mainland. This should be taken as encouragement for large-scale planting of this tree that was so characteristic of the early prehistoric environment. Planting shrubs in the terrain already affected by severe erosion, as tried in the 1970s in the east of Poike Peninsula, will certainly be unsuccessful. Continuing runoff and erosion dynamics, together with a nutrient deficit, high resistance of the little-weathered country-rock to root penetration and poor water supply there, as well as the considerable exposure to solar radiation and wind in those areas would kill all the plants in a short time.

23. Perspectives on Research

The goal of future research will be to test whether the detailed results obtained for the land-use history of Poike Peninsula are valid island-wide. The stratigraphic markers identified for the various phases of vegetation and land use could be used of for an island-wide interpretation of prehistoric and historic landscape history at a high temporal and spatial resolution. With less research effort than for the Poike Peninsula study, it should be possible to determine the phases of forest clearing in time and space as well as the intensity, succession, and changes of horticultural land-use practice. From these data, in combination with the analysis of former soil fertility, the horticultural and ecological carrying capacity of Easter Island can probably be reconstructed.

Such calculations, together with projections of the manpower needed for clearing the palm forest, mechanical soil treatment, stone mulching, the production of ***moai*** and the construction of ***ahu*** and the roads for transporting the ***moai***, may lead to more reliable estimates of population during the earlier phases of settlement. It should also be possible, from these data, to reconstruct the causes, sequence of events, and interrelationships involved in crossing the thresholds of ecological and economic capacity in prehistoric times.

In a joint approach by the disciplines of ecology, archaeology, and anthropology further evidence should be sought to identify the causes behind the dramatic and cultural change that affected the society of Easter Island at some time between the eleventh and thirteenth century. What really triggered the large-scale forest clearing and abandonment of a sustainable subsistence economy that had worked successfully for centuries? Were there further waves of settlement after AD 1000, which brought with them new technologies or new concepts of society? Did the large-scale clearing of the palm forest only become possible with the arrival of more manpower from the Polynesian region? Was there a need for new cultivated plants that caused the complete restructuring of horticulture? Did the farmers of Easter Island see a promise of higher yields by cultivating the sweet potato, imported at a later date, under the cool and humid conditions of the volcanic slopes, and did they hope that this undemanding plant might still be sufficiently productive with increasing nutrient deficiencies? Was the extreme flourishing of the ***moai*** culture at the time of forest clearing, as well as the enormous growth of the settled areas and number of ceremonial sites perhaps also due to new influences from abroad? Most likely, some of these questions will remain unanswered forever.

A particular research effort should be devoted to the extinct palm of Easter Island. Soil ecological studies should determine how the root tubes of the last palm generation could be preserved in the soils. Also the significance of the palm as possibly the most important material and food resource of the early Easter Island culture deserves more scientific attention, in particular as it seems to have been unique in the Polynesian cultural realm.

REFERENCES

Bahn, P. 1993a. 'The History of Settlement on Rapanui'. In S. R. Fischer (ed.), *Easter Island Studies: Contributions to the History of Rapa Nui in Memory of William T. Mulloy*, Oxbow Monograph 32 (Oxford: Oxbow Books), pp. 53–5.

Bahn, P. 1993b. 'A Brief History of Rapa Nui Archaeology up to 1956'. In S.R. Fischer (ed.), *Easter Island Studies: Contributions to the History of Rapa Nui in Memory of William T. Mulloy*, Oxbow Monograph 32 (Oxford: Oxbow Books), pp. 73–8.

Bahn, P. and J. Flenley. 1992. *Easter Island – Earth Island.* London: Thames and Hudson Ltd.

Baker, P.E. 1997. 'Petrological Factors Influencing the Utilization of Stone on Easter Island'. In C.M. Stevenson, G. Lee and F.J. Morin (eds), *Easter Island in Pacific Context*, South Seas Symposium, Proceedings of the Fourth International Conference on Easter Island and East Polynesia (Los Osos: Easter Island Foundation), pp. 279–83.

Bork, H.-R., H. Bork, C. Dalchow, B. Faust, H.-P. Piorr and T. Schatz. 1998. *Landschaftsentwicklung in Mitteleuropa*. Gotha/Stuttgart: Klett-Perthes

Bork, H.-R., H.R. Beckedahl, C. Dahlke, K. Geldmacher, A. Mieth and Y. Li. 2003. 'Die erdweite Explosion der Bodenerosionsraten im 20. Jh.: Das globale Bodenerosionsdrama – geht unsere Ernährungsgrundlage verloren?' *Petermanns Geographische Mitteilungen*, **147**: 16–25.

Bork, H.-R. and A. Mieth. 2003. 'The Key Role of Jubaea Palm Trees in the History of Rapa Nui: A Provocative Interpretation'. *Rapa Nui Journal*, 17(2): 119–22.

Bork, H.-R., A. Mieth and B. Tschochner. 2004. 'Nothing but Stones? A Review of the Extent and Technical Efforts of Prehistoric Stone Mulching on Rapa Nui'. *Rapa Nui Journal*, **18**(1): 10–14.

Coblentz, B.E. 1978. 'The effects of Feral Goats (*Capra hircus*) on Island Ecosystems'. *Biological Conservation*, **13**: 279–86.

Darwin, C. 1999 (reprint). *The Voyage of the Beagle*. Amherst: Prometheus Books

Diamond, J. 2005. *Collapse, How Societies Choose to Fail or Succeed.* New York: Viking, Penguin Group.

Diaz Vial, C. 1949. 'Asociación de Suelos de Isla de Pascua', Apertado de Agricultura Técnica IX (Santiago de Chile: Ministerio de Agricultura), pp. 115–34.

Dirección Meteorológica de Chile, Departamento Climatología y Meteoroloía Aplicada. 2002. 'Normales de temperatura media mensual (°C), periodo 1961–1990' (unpublished).

Dransfield, J., J.R. Flenley, S.M. King, D.D. Harkness and S. Rapu. 1984. 'A Recently Extinct Palm from Easter Island'. *Nature*, 312: 750–2.

Esen-Baur, H.-M. 1989. 'Megalithische Steinplastiken und monumentale Architektur'. In H.-M. Esen-Baur (ed.), *1500 Jahre Kultur der Osterinsel* (Mainz: Verlag Philipp von Zabern), pp. 87–108.

Etienne, M., G. Michea and E. Diaz. 1982. *Flora, Vegetacion y Potencial Pastoral de Isla de Pascua.* Santiago de Chile: Faculdad de Ciencias Agrarias, Veterinarias y Forestales.

Feeser, I. 2003. 'Untersuchungen zum Grünland der Osterinsel (Chile)'. Diplomarbeit am Botanischen Institut der Mathematisch-Naturwissenschaftlichen Fakultät der Christian-Albrechts-Universität zu Kiel (unpublished).

Fischer, H. 1999. *Schatten auf der Osterinsel, Plädoyer für ein vergessenes Volk.* Oldenburg: Bibliotheks- und Informationssystem der Universität Oldenburg

Fischer, S.R. and C. Love 1993. 'Rapanui: The Geological Parameters'. In S.R. Fischer (ed.), *Easter Island Studies. Contributions to the History of Rapa Nui in Memory of William T. Mulloy*, Oxbow Monograph 32 (Oxford: Oxbow Books), pp. 1–6.

Flenley, J.R. 1993a. 'The Present Flora of Easter Island and its Origins'. In S.R. Fischer (ed.), *Easter Island Studies. Contributions to the History of Rapa Nui in Memory of William T. Mulloy*, Oxbow Monograph 32 (Oxford: Oxbow Books), pp. 7–15.

Flenley, J.R. 1993b. 'The Palaeoecology of Easter Island, and its Ecological Disaster'. In S.R. Fischer (ed.), *Easter Island Studies. Contributions to the History of Rapa Nui in Memory of William T. Mulloy*, Oxbow Monograph 32 (Oxford: Oxbow Books), pp. 27–45.

Flenley, J.R. 1996. 'Further Evidence of Vegetational Change on Easter Island'. *South Pacific Studies*, **16**(2): 135–41.

Flenley, J.R. 1998. 'New Data and New Thoughts about Rapa Nui'. In C.M. Stevenson, G. Lee and F.J. Morin (eds), *Easter Island in Pacific Context*, South Seas Symposium, Proceedings of the Fourth International Conference on Easter Island and East Polynesia (Los Osos CA: Easter Island Foundation), pp. 125–8.

Flenley, J.R. 2001. 'Forest and Civilization on Easter Island'. In Y. Yasuda (ed.), *Forest and Civilization* (New Delhi), pp. 55–62.

Flenley, J.R. and S.M. King. 1984. 'Late Quaternary Pollen Records from Easter Island'. *Nature*, **307**: 47–50.

Flenley, J.R. and P. Bahn. 2003. *The Enigmas of Easter Island.* Oxford/New York: Oxford University Press.

Flenley, J.R., J.T. Teller, M.E. Prentice, J. Jackson and C. Chew. 1991. 'The Late Quaternary Vegetational and Climatic History of Easter Island'. *Journal of Quaternary Science*, **6**(2): 85–115.

Forster, G. 1784, reprinted 1983. *Reise um die Welt.* Frankfurt am Main: Insel Verlag.

Fosberg, F.R. 1963a. 'The Island Ecosystem'. In F.R. Fosberg (ed.), *Man's Place in the Island Ecosystem: A Symposium* (Honolulu: Bishop Museum Press), pp. 1–7.

Fosberg, F.R. 1963b. 'Disturbance in Island Ecosystems'. In J.L. Gressitt (ed.), *Pacific Basin Biogeography* (Honolulu: Bishop Museum Press), pp. 557–61.

Genz, J. and T.L. Hunt. 2003. 'El Niño/Southern Oscillation and Rapa Nui Prehistory'. *Rapa Nui Journal,* 17(1): 7–14.

Gill, G.W. 1993. 'William Mulloy and Rapanui'. In S.R. Fischer (ed.), *Easter Island Studies. Contributions to the History of Rapa Nui in Memory of William T. Mulloy,* Oxbow Monograph 32 (Oxford: Oxbow Books), pp. XI-XII

Grau, J. 1998. 'The *Jubaea* Palm, Key in the Transportation of Moai on Easter Island'. In C.M. Stevenson, G. Lee and F.J. Morin (eds), *Easter Island in Pacific Context,* South Seas Symposium, Proceedings of the Fourth International Conference on Easter Island and East Polynesia (Los Osos CA: Easter Island Foundation), pp. 120–24.

Grau, J. 2001. 'More about *Jubaea chilensis* on Easter Island'. In C.M. Stevenson, G. Lee and F.J. Morin (eds), *Pacific 2000,* Proceedings of the Fifth International Conference on Easter Island and the Pacific (Los Osos CA: Easter Island Foundation), pp. 87–90.

Heyerdahl, T. and E.N. Ferdon (eds). 1961. *Reports of the Norwegian archaeological expedition to Easter Island and the East Pacific, Vol 1: Archaeology of Easter Island,* Monographs of the School of American Research and the Museum of New Mexico. London: Allen and Unwin.

Honorato, R., A. Norero, J. Rodriguez and D. Suarez. 1991. 'Ambiente Agrofísico y Fertilidad de Suelos'. In H. Benedetti (ed.), *Estudio de la Productividad Silvoagropecuaria del Fundo Vaitea, Isla de Pascua* (Santiago de Chile: Pontificia Universidad Católica de Chile, Facultad de Agronomía), pp. 25–43.

Hunt, T. and C.P. Lipo. 2001. 'Cultural Elaboration and Environmental Uncertainity in Polynesia'. In C.M. Stevenson, G. Lee and F.J. Morin (eds), *Pacific 2000,* Proceedings of the Fifth International Conference on Easter Island and the Pacific (Los Osos CA: Easter Island Foundation), pp. 103–15.

Hunter-Andersson, R. 1998. 'Human vs Climatic Impacts at Rapa Nui: Did the People Really Cut Down All those Trees?' In C.M. Stevenson, G. Lee and F.J. Morin (eds), *Easter Island in Pacific Context,* South Seas Symposium, Proceedings of the Fourth International Conference on Easter Island and East Polynesia (Los Osos CA: Easter Island Foundation), pp. 85–99.

Kirch, P.V. 1984. *The Evolution of Polynesian Chiefdoms.* Cambridge: Cambridge University Press.

Kirch, P.V. 2000. *On the Road of the Winds: An Archaeological History of the Pacific Islands before European Contact.* Berkeley/Los Angeles/London: University of California Press.

Lee, G. 1992. *The Rock Art of Easter Island: Symbol of Power, Prayers to the Gods,* Monumenta Archaeologia 17. Los Angeles: The Institute of Archaeology, University of California.

Lee, G. 2001. *Te Moana Nui: Exploring Lost Isles of the South Pacific.* Los Osos CA: Easter Island Foundation/Bearsville Press.

Louwagie, G. and R. Langohr. 2002. 'Testing Land Evaluation Methods for Crop Growth on two Soils of the La Pérouse Area'. *Rapa Nui Journal,* **16**(1): 23–8.

MacIntyre, F. 1999. 'Is Humanity Suicidal? Are there Clues from Rapa Nui?' *Rapa Nui Journal*, **13**(2): 35–41.

MacIntyre, F. 2001. 'Enso, Climate Variability, and the Rapanui, Part II: Oceanography and Rapa Nui'. *Rapa Nui Journal*, **15**(2): 83–94.

Martinsson-Wallin, H. 2002. 'Sea, Land, and Sky as Structuring Principles in Easter Island Prehistory'. *Rapa Nui Journal*, **16**(2): 74–6.

Martinsson-Wallin. 2004. 'Archaeological Excavation at Vinapu (Rapa Nui)'. *Rapa Nui Journal*, **18**(1): 7–9.

McCall, G. 1993. 'Little Ice Age: Some Speculations for Rapanui'. *Rapa Nui Journal* 7: 65–70.

McCall, G. 1994. *Rapanui, Tradition and Survival on Easter Island*. Honolulu: University of Hawaii Press

Métraux, A. 1940, reprinted 1971. *Ethnology of Easter Island*. Honolulu: Bishop Museum Press.

Métraux, A. 1957. *Die Osterinsel*. Stuttgart: Verlag Kohlhammer.

Mieth, A., H.-R. Bork and I. Feeser. 2002. 'Prehistoric and Recent Land Use Effects on Poike Peninsula, Easter Island (Rapa Nui)'. *Rapa Nui Journal*, **16**(2): 89–95.

Mieth, A. and H.-R. Bork. 2003a. 'Diminution and Degradation of Environmental Resources by Prehistoric Land Use on Poike Peninsula, Easter Island (Rapa Nui)'. *Rapa Nui Journal*, 17(1): 34–41.

Mieth, A. and H.-R. Bork. 2003b. 'Land Degradation on Easter Island, Southern Pacific'. In Y. Li, J. Poesen and J. Zhang (eds), *Gully Erosion under Global Change* (Chengdu: Sichuan Publishing House of Science and Technology), pp. 185–99.

Mieth, A., H.-R. Bork, W. Markgraf, I. Feeser and K. Dierssen. 2003a. 'Bodenerosion – ein Schlüssel zum Verständnis der Kulturgeschichte der Osterinsel'. *Petermanns Geographische Mitteilungen*, **147**: 30–37.

Mieth, A. and H.-R. Bork. 2005. 'History, Origin and Extent of Soil Erosion on Easter Island (Rapa Nui)'. *Catena*, **63**: 244–60.

Mueller-Dombois, D. and F.R. Fosberg. 1998. *Vegetation of the Tropical Pacific Islands*. Heidelberg/New York/Berlin: Springer Verlag.

Nunn, P. 2000. 'Environmental Catastrophe in the Pacific Islands around A.D. 1300'. *Geoarchaeology* **15**(7): 715–40.

Orliac, C. 2000. 'The Woody Vegetation of Easter Island between the Early 14th and the Mid-17th Centuries AD'. In C.M. Stevenson and W.S. Ayres (eds), *Easter Island Archaeology: Research on Early Rapa Nui Culture* (Los Osos CA: Easter Island Foundation), pp. 211–20.

Orliac, C. 2003. 'Ligneaux et Palmiers de l'île de Pâques du XI ème au XVII ème siècle de notre ère'. In C. Orliac (ed.), *Archéologie en Océanie insulaire – peuplement, sociétés et paysages* (Paris: Éditions Artcom), pp. 184–99.

Orliac, C. and M. Orliac. 1995. *Easter Island, Mystery of the Stone Giants*. New York: Harry N. Abrams.

Orliac, C. and M. Orliac. 1998. 'The Disappearance of Easter Island's Forest: Over-exploitation or Climatic Catastrophe?' In C.M. Stevenson, G. Lee and F.J. Morin (eds), *Easter Island in Pacific Context*, South Seas Symposium, Proceedings of the Fourth International Conference on Easter Island and East Polynesia (Los Osos CA: Easter Island Foundation), pp. 129–34.

Orliac, C. and M. Orliac. 2001. 'Composition et Évolution de la Flore de L'Ile de Pâques du 14è au 19ès'. Rapport sur les Travaux de la Mission Archéologique Octobre – Decembre 2000, Ministère des Affaires Étrangères, France, Consejo de Monumentos, Corporación Nacional Forestal, Chile Museum d'Histoire Naturelle, Paris (unpublished).

Palmer, J.L. 1870. 'A Visit to Easter Island, or Rapa Nui, in 1868'. *The Journal of the Royal Geographic Society*, **40**: 165–80.

Ramírez, J.M. 2001. 'Cultural Resource Management of Easter Island: Utopia and Reality'. In C.M. Stevenson, G. Lee and F.J. Morin (eds), *Pacific 2000*, Proceedings of the Fifth International Conference on Easter Island and the Pacific (Los Osos CA: Easter Island Foundation), pp. 383–90.

Reichardt A. and I. Reichardt 2000. *Osterinsel – Ein Reiseführer*. Heidelberg: Max Kasparek Verlag .

Rohdenburg, H. 1977. 'Beispiele für Holozäne Flächenbildung in Nord- und Westafrika'. *Catena*, 4(1/2): 65–109.

Routledge, K. 1919, reprinted 1998. *The Mystery of Easter Island.* Kempton, Illinios: Adventures Unlimited Press.

Skottsberg, C. 1953. *The Natural History of Juan Fernandez and Easter Island, Vol. 2, Botany, 1920–1953.* Uppsala.

Skottsberg, C. 1956. *The Natural History of Juan Fernandez and Easter Island, Vol. 1, Geography, Geology, Origin of Island Life, 1920 – 1956.* Uppsala.

Steadman, D.W., P. Vargas Casanova and C. Cristino Ferrando. 1994. 'Stratigraphy, Chronology, and Cultural Context of an Early Faunal Assemblage from Easter Island'. *Asian Perspectives*, **33**: 79–96.

Stevenson, C.M. 1997. *Archaeological Investigations on Easter Island, Maunga Tari: An Upland Agricultural Complex.* Los Osos, CA: The Easter Island Foundation.

Stevenson, C.M. and C. Cristino. 1986. 'The Residential Settlement History of Easter Island's Southern Coastal Plain'. *Journal of New World Archaeology*, 7: 29–38.

Stevenson, C.M. and S. Haoa. 1998. 'Prehistoric Gardening Systems and Agricultural Intensification in the La Pérouse Area of Easter Island'. In C.M. Stevenson, G. Lee and F.J. Morin (eds), *Easter Island in Pacific Context*, South Seas Symposium, Proceedings of the Fourth International Conference on Easter Island and East Polynesia (Los Osos CA: Easter Island Foundation), pp. 205–13.

Stevenson, C.M. and S. Haoa 1999. 'Diminished Agricultural Productivity and the Collapse of Ranked Society on Easter Island'. In P. Wallin (ed.), *Lectures at the No Barriers Seminar 1999* (Oslo: The Kon-Tiki Museum, Institute for Pacific Archaeology and Cultural History), pp. 4–12.

Stevenson, C.M., J. Wozniak and S. Haoa. 1999. 'Prehistoric Agricultural Production on Rapa Nui'. *Antiquity*, **73**: 801–12.

Stevenson, C.M., Th. Ladefoged and S. Haoa. 2002. 'Productive Strategies in an Uncertain Environment: Prehistoric Agriculture on Easter Island'. *Rapa Nui Journal*, **16**(1): 17–22.

Sudzuki, F. 1979. 'Vegetación Arbórea y Erosión en la Isla de Pascua'. *Boletín Técnico* 45 (Santiago: Facultad de Agronomia, Universidad de Chile): 3–8.

van Tilburg, J.A. 1994. *Easter Island: Archaeology, Ecology and Culture*. London: British Museum Press.

Vogler, E. 1998. 'Die Europäer entdecken die Osterinsel'. In M. Esen-Baur (ed.), *1500 Jahre Kultur der Osterinsel* (Mainz: Verlag Philipp von Zabern), pp. 53–81.

Wallin, P., C.M. Stevenson and T. Ladefoged. 2005. 'Sweet Potato Production on Rapa Nui'. In C. Ballard, P. Brown, R.M. Bourke and T. Harwood (eds), *The Sweet Potato in Oceania: A Reappraisal*, Ethnology Monographs 19 (University of Pittsburgh), Oceania Monograph 56 (Sydney University), pp. 85–8.

Wozniak, J.A. 1998. 'Settlement Patterns and Subsistence on the North-west Coast of Rapa Nui'. In C.M. Stevenson, G. Lee and F.J. Morin (eds), *Easter Island in Pacific Context*, South Seas Symposium, Proceedings of the Fourth International Conference on Easter Island and East Polynesia (Los Osos CA: Easter Island Foundation), pp. 185–92.

Wozniak, J.A. 1999. 'Prehistoric Horticultural Practices on Easter Island: Lithic Mulched Gardens and Field systems'. *Rapa Nui Journal*, **13**: 95–9.

Wozniak, J.A. 2001. 'Landscapes of Food Production on Easter Island: Successful Subsistence Strategies'. In C.M. Stevenson, G. Lee and F.J. Morin (eds), *Pacific 2000*, Proceedings of the Fifth International Conference on Easter Island and the Pacific (Los Osos CA: Easter Island Foundation), pp. 91–102.

Yen, D.E. 1974. *The Sweet Potato and Oceania: An Essay in Ethnobotany*, Bernice P. Bishop Museum Bulletin 236. Honolulu: Bishop Museum Press.

Zizka, G. 1989. 'Naturgeschichte der Osterinsel'. In M. Esen-Baur (ed.), *1500 Jahre Kultur der Osterinsel* (Mainz: Verlag Philipp von Zabern), pp. 21–38.

Zizka, G. 1991. *Flowering Plants of Easter Island*, Palmarum Hortus Frankofurtensis 3. Frankfurt am Main: Palmengarten.

11

Know Your Soil: Transitions in Farmers' and Scientists' Knowledge in Germany

Frank Uekoetter

In the beginning, there was Liebig. That, at least, is the impression that one obtains from agricultural publications in the twentieth century. When German farming experts were discussing issues of plant nutrition and fertiliser use, and sought to squeeze in some historical information in order to give depth to the narrative, it was usually Liebig's name that shone the brightest. Of course, a good dose of patriotism was sometimes fuelling references to the German professor, and certainly admiration for Liebig's impressive career, from mediocre pupil to one of the most influential scientists of the nineteenth century (cf. Brock 1997). But most of all, a reference to Liebig seemed to be the perfect way to bestow more credibility upon one's own argument. While Liebig fought bitter controversies with other scientists of his time, retrospective references are almost exclusively of the complimentary kind.

However, the enduring appeal of Liebig to agricultural experts stands in stark contrast to the disagreements over what he stood for. Many experts cherish him as 'the founder of modern plant nutrition', as a textbook of 1951 noted (Schober 1951, p. 3). '[Liebig's] mineral theory destroyed the old humus theory', a publication of 1921 declared (Brandt 1921, p. 7). However, a third author invoked Liebig to underscore his argument for the complete collection of human faeces in the cities and their use as organic fertiliser (Deußen 1922, p. 826n). In the late 1920s, Max Reiser warned of the neglect of agriculture and destruction of soil fertility with a Liebig quote (Reiser 1929, p. 42). Johannes Görbing, who is today best known as a pioneer of organic farming methods, used a quotation of Liebig 'admonishing us to do our work with reverence to the Creation' (Görbing 1947, p. 205). And in 2003, an article on 'farm technology and sustainability' referred to Liebig in arguing for an economical use of minerals (Seufert 2003, p. 267). Obviously, Liebig, certainly a controversial figure in his own lifetime, has somehow turned into everybody's darling – something of a free-floating icon, hovering over the agricultural science profession, offering support to almost every faction in the field.

One could count these divergent readings of Liebig as mere curiosities if they were not so typical of the profession in general. Ever since Liebig took on Albrecht Thaer's humus theory, there have been multiple scientific opinions in the field (cf. Schling-Brodersen 1989 and Fussell 1971). Chemical, biological, bacteriological, geological and other approaches offered different perspectives on the soil, and the clashes between these approaches were as intense as they were inevitable. Interestingly, the competition between these approaches persists to the present day, and with soil fertility being the result of both chemical and biological processes, it seems unlikely that one approach will rise as unchallenged in the foreseeable future. However, the ambivalence of scientific knowledge did not prevent the chemical approach from emerging as the hegemonic one: from this point of view, soil fertility and plant nutrition were synonymous with putting the right minerals into the soil in the right quantities at the right time. 'The goal of agriculture is [...] the proper and purposeful transformation of nutrients into plant substance with the help of the soil as a mediator', a typical expression of this perspective ran (Jahn-Deesbach 1965, p. 35). To be sure, the chemical approach never managed totally to disprove or discredit the other approaches, but it did manage to push these alternative approaches to the margins, transforming them into a source of irritation for the farmer, rather than an inspiration for his daily work. Some of the problems of modern industrial agriculture, like the contamination of groundwater and the eutrophication of the landscape, are direct consequences of the hegemony of the chemical approach.

Justus von Liebig offered a stinging critique of agricultural experimental stations in his lifetime, claiming they were unable to conduct scientific research (cf. Schling-Brodersen 1989, p. 235n). However, these experimental stations would soon turn into key proponents of chemical conceptions of soil fertility, and as early as 1913, three out of four experimental stations were headed by a chemist, providing a head-start for agricultural chemistry that it would never really lose (Reichrath 1991, p. 118). Nevertheless, it is important to realise that there were always rivals to its hegemony. The early dominance of agricultural chemistry easily leads to the impression that its hegemony was a 'natural' phenomenon, and that is how it comes across in the current literature: as an incontestable scientific empire that rose in the nineteenth century and never showed any signs of weakness. However, it becomes clear upon closer investigation that the path of agricultural chemistry was far more complicated and twisted than that. Agricultural chemistry was involved in constant battles; and these battles reached their peak during the interwar years. The two decades between the world wars were a crucial time period that constitutes a period of transition and intellectual fermentation – a ***Sattelzeit***, to use Koselleck's famous term – in the history of plant nutrition and fertiliser use. By 1939, the tracks were laid down that German agriculture in the post-war era would generally follow.

1. THE MANY PERSPECTIVES ON THE SOIL: SOME THEORETICAL PERSPECTIVES

Nothing is more fundamental to farming than the soil. 'The soil is the most important part of a farm', Johannes Knecht noted in an agricultural textbook of 1949, adding that its proper cultivation was 'the supreme skill of farming' (Knecht 1949, p. 26). Therefore, it is safe to assume that there was a rich reservoir of ideas on how to foster the fertility of the soil in the agricultural population at large. After all, farmers were dealing with the soil in one way or another on a daily basis, and an attentive farmer would accumulate a whole host of experiences over time. In most cases, this knowledge was never put down in writing, and much of it was in fact tacit knowledge never put into words. Nevertheless, it is important to acknowledge the existence and persistence of these practical experiences to the present day because the farmers' experiences constitute an important part of the knowledge base of agriculture, even in the age of scientific agriculture. Scientific knowledge, readily accessible through tons of scientific and advisory literature, was only one contribution to the full body of agricultural knowledge, and for many farmers, it was not on a par with personal habits and experiences until well into the twentieth century.

While it is important to incorporate non-scientific forms of knowledge into a history of agricultural knowledge, it is equally important to realise that over the course of the twentieth century, these forms of knowledge came under increasing pressure as a result of the growing influence of scientific experts. This gradual devaluation of the farmers' own experiences is an important part of a history of agricultural knowledge, and needs to be studied carefully in its causes and consequences. However, in doing so, one should keep one's distance from nostalgia for the 'loss of traditional knowledge'. Countless reports show that farmers almost routinely mishandled the organic wastes that they produced, resulting in a significant loss of nutritional value as a fertiliser. 'Instead of treating manure in the way that its true value demands, and to take care of losses, the nutritional capital, won with great effort, is wasted without a thought', Johannes Schneider complained in his farming handbook of 1924 (Schneider 1924, p. 58). Similarly, Hans Schlange-Schöningen noted in one of his books that 'many farmers do not have a grasp of the losses they incur' through inadequate handling of organic wastes (Schlange-Schöningen 1931, p. 83). Clearly, 'traditional knowledge' can also be a synonym for laziness and misinformation.

Of course, the same can be true for scientific knowledge, and it is just as important to inquire more deeply into the rules and conducts of the relevant branch of science: Who supplied the farmers with knowledge? Who defined the 'good practice' of farming – and who did not? How did information flow from the researchers to the individual farmers – and what were the chances for feedback? How have farmers – a group with a legendary reputation for conservatism – come to accept advice on new farming methods so readily? Focusing on Liebig and his opponents, researchers have frequently overlooked the complex set of agricultural advisors who sought to bring scientific knowledge to the farming community. Standing between the scientific profession and the farmers, the community of agricultural advisors was where tensions and contradictions within the body of agricultural knowledge would come to the forefront. Therefore, the agricultural advisors deserve a prominent place in the history of agricultural knowledge in the twentieth century, making the system of agricultural knowledge production and diffusion even more complex.

All this is especially important given the background of recent arguments that modern society has come to be dominated by knowledge-dependent processes to such an extent that one could speak of a 'knowledge society'.1 For all its merits, it is important to realise that the theory of the 'knowledge society' includes an emphatic understanding of scientific research; characteristically, Nico Stehr argued in his seminal work on the knowledge society that scientific research is increasingly becoming the sole source of additional knowledge (Stehr 1994, p. 201). This focus on the increasing influence of scientific knowledge easily leads to a point of view that perceives every non-scientific kind of knowledge as a simple leftover from previous times that is just waiting to be enlightened by scientific knowledge (cf. Landwehr 2002, p. 17). However, non-scientific forms of knowledge persist even in today's farming practice, and may in fact fulfil a number of important functions. The rise of the agrarian knowledge society is not simply part of a Weberian process of rationalisation (cf. Stehr 1994, p. 16).

2. CRISIS AND RENEWAL: THE INTERWAR YEARS AS A TIME OF STRATEGIC DECISIONS

Historians have long argued that the First World War led to the disruption of a complex agricultural routine. Due to the lack of farm workers, and especially the almost complete confiscation of nitrate fertiliser for the production of explosives, farming yields declined significantly. Grain production fell from 27.1 million tons in 1914 to 17.3 million tons in 1918, while the potato harvest fell from 45.6 to 29.5

[1] Cf. the definition of Willke 1998, p. 162.

million tons (Mommsen 2002, p. 93). Therefore, the post-war years saw feverish efforts to restore soil fertility and to remedy the effects of the wartime *Raubwirtschaft* (exploitative economy) (Langenbeck 1921, p. 8). However, the supply of fertilisers had changed significantly since the 1910s. The invention of the Haber-Bosch process made atmospheric nitrogen available for fertiliser production, thus opening a practically limitless reservoir of the most critical plant nutrient (cf. Szöllösi-Janze 1998). At the same time, manure had become scarce as a result of wartime cutbacks on livestock and inadequate feed, which put the traditional role of organic wastes in jeopardy (cf. Kuhn 1920, p. 347). As a result, mineral fertiliser quickly became the method of choice: the almost unanimous hope of the scientific establishment was that with massive doses of mineral fertilisers, pre-war yields would return almost immediately. 'The exhaustion of our soils can be remedied only through an intensive, rational use of mineral fertiliser', a brochure of 1921 declared (Smalakies 1921, p. 7). Increasing the use of mineral fertilisers became the rallying cry of practically all advisors, and their efforts had the backing of both the scientific authorities and the bureaucracy. In fact, a memorandum of the Prussian ministry of agriculture expressed the urgency of the matter in a memorandum of 1920 as follows: 'The necessary amounts of mineral fertiliser are available, and can be purchased and put into the soil. If that does not happen, people will starve.' [2]

However, the promise of a quick return to pre-war conditions turned out to be deceptive. Per-acre yields stayed below pre-war figures for much longer than expected, and the massive propaganda (to use a contemporary term that was not yet tainted) in favour of mineral fertiliser came to appear dubious to many farmers. Mineral fertiliser was the most significant investment for many farmers, and with farmers struggling to keep out of debt during the 1920s, this issue was by no means unimportant (Schlange-Schöningen 1931, p. 23). In addition, the massive use of mineral fertiliser led to acidification of many soils, which in turn inhibited the growth of certain plants (Schellenberger 1924, p. 17). 'The acid condition of the soil constitutes a threat that has attracted too little attention so far', a publication of 1923 warned (Niggl 1923, p. 5). To be sure, farmers could use lime to restore a neutral condition, but this implied additional costs, took another two or three years with reduced yields, and provided the farmers with the irritating experiences that it took a chemical substance to remedy a problem caused by a chemical substance (Wagner 1929, p. 4). In fact, the sheer prospect of ruining the most essential basis of agricultural production must have been disturbing, if not traumatic, for many farmers. One expert later noted that the most important consequence of this episode was the 'psychological effect': 'Agricultural chemistry lost credit in the eyes of many

---

[2] Bundesarchiv R 3602 No. 606, Denkschrift des preußischen Landwirtschaftsministers zur Frage der Volksernährung, November 1, 1920, p. 1.

people' (Merkenschlager 1933, p. 123). In short, agricultural chemistry experienced a deep crisis of confidence, and that opened the door for potential rivals.

One of the consequences of acidification was that it significantly reduced the microbiological activity of the soil (Wiessmann 1926, p. 529). From the viewpoint of the agricultural chemistry establishment, this was particularly unfortunate, since the biological perspective on the soil was experiencing something of a boom in the 1920s. While bacteriology was instrumental for the rise of the medical sciences in the late nineteenth century, it had played a marginal role in the field of agriculture due to the dominance of agricultural chemistry (Glathe 1935, p. 182). As a result, scientific knowledge about bacteriological processes in the soil was generally scarce; Felix Löhnis, the most important agricultural bacteriologist of the 1920s, once spoke of a 'still quite unenlightened field' (Löhnis 1926a, p. 247). Löhnis had been a researcher at the University of Leipzig until 1914 when he joined the United States Department of Agriculture as an expert in soil bacteriology. In 1925 he returned to Leipzig to fill a newly created chair of agricultural bacteriology, a step which he hoped would encourage other universities to follow suit (Löhnis 1926b, p. 174). Löhnis was optimistic about the prospects for his field, not least because he saw the contemporary situation as generally conducive: 'In the present time, with its urgency to save labour and capital as much as possible, the rational use of soil bacteria working for free deserves close attention' (Löhnis 1926a, p. 251).

However, agricultural bacteriology soon encountered a number of significant problems. Perhaps the most important problem was essentially home-made: it focused on the investigation of the basics of soil bacteriology and failed to provide farmers with answers to questions of practical relevance. Without a thorough understanding of the fundamentals of soil bacteriology, Löhnis argued (1927, p. 242), the business of giving advice to farmers would stand on very shaky ground. To be sure, Löhnis would sometimes refer to practical farmers like Schultz-Lupitz and depict agricultural bacteriology as a response to the wishes of the farming community, but there was never any systematic attempt to get back to these farmers with specific rules of good practice (Löhnis 1926a, p. 247; Löhnis 1928, p. 818). In fact, even the question whether there was a usable method of soil investigation based on bacteriology was absurd for him (Löhnis 1927, p. 242). In a telling episode, he noted that more information on the effect of tillage on bacterial life would be desirable. However, lacking sufficient research on this issue, he urged farmers to find out the right approach for themselves 'and then wait patiently until scientific research has found an explanation for their practical successes' (Löhnis 1926a, p. 251). Apparently, it never occurred to Löhnis that the merits of a discipline that only explained retrospectively what people were already doing were somewhat limited.

Therefore, the prospects for the field were already dim when Löhnis died unexpectedly in 1931, essentially ending all hopes for a boom of agricultural bacteriology in Germany. However, even these small beginnings were met with considerable hostility from the establishment of the agricultural sciences. The meeting minutes of the committee on general fertiliser affairs with the Prussian ministry of agriculture provide a case in point. When a professor from the agricultural university of Berlin called for more attention to humus issues ('plants do not simply grow from mineral fertiliser') and expressed his regret about the recent death of Löhnis and the probable discontinuation of his chair, his statements met with strong opposition; at the following meeting, the professor spoke of the 'polemic' he had encountered, and stressed that he was using mineral fertiliser on his own farm, thus seeking to avoid the impression of being a fundamentalist opponent of the chemical approach.[3] Clearly, the agricultural chemistry establishment was monitoring the upswing of research in agricultural bacteriology with a good degree of scepticism, if not hostility.

However, agricultural bacteriology was not the only competitor that the agricultural chemists encountered in the 1920s. After Rudolf Steiner's lectures on agriculture in 1924, a group of ardent anthroposophists sought to develop a new kind of organic farming that refrained from the use of mineral fertiliser altogether. This 'biodynamic agriculture' (***biologisch-dynamische Landwirtschaft***) quickly drew criticism from the agricultural science establishment to an extent that is hard to explain on first glance. For example, the director of the Federal Bureau of Agriculture and Forestry (***Biologische Reichsanstalt für Land- und Forstwirtschaft***) stated in a letter of 1935 that in his eyes, organic farming was '99 percent humbug'.[4] When the German chemists' journal *Chemiker-Zeitung* published a scathing attack on organic farming in 1934, the article was supplemented by an editorial note complaining of the 'heresies' in the field of soil fertility being proposed 'with a fanaticism that is reminiscent of the dark ages of medieval ignorance' (*Chemiker-Zeitung* **58** (1934): 245). After the Nazis' seizure of power in 1933, an article with the emblematic title 'Hands Off Our Reliable Fertilising Methods!' even called upon the new regime to crack down on these 'charlatans' (Flieg 1933, p. 715).[5]

In his influential book ***Subsidised Nonsense*** (Die subventionierte Unvernunft) of 1985, Hermann Priebe wondered why a farming method that occupied less than one per cent of Germany's arable land was creating such a stir, and it seems worthwhile to ask the same question for the interwar years (Priebe 1985, p. 27). 'There is no doubt among the experts that only a small amount of ash will remain

---

[3] Bundesarchiv R 3602 No. 67, Niederschrift der 80. Sitzung über allgemeine Düngerangelegenheiten im Preußischen Landwirtschaftsministerium am 13. Mai 1931, p. 14; Niederschrift der 81. Sitzung über allgemeine Düngerangelegenheiten im Preußischen Landwirtschaftsministerium vom 13. Mai 1932, p. 23.

[4] Bundesarchiv R 3602 No. 608, the director of the Biologische Reichsanstalt to the Reichsgesundheitsamt, January 10, 1935.

[5] Similarly Schnelle and Nolte 1933, p. 223.

from this straw fire', one expert wrote (Flieg 1933, p. 715). But if that was so, why did these experts not simply relax and wait until the anthroposophists had worked their way into bankruptcy? Why did the agricultural science establishment react with so much vigour, and even go to the extreme of trying to solve the dispute with the help of the Nazi regime? To a certain extent, these attacks were a response to the way that anthroposophists were marketing their products. Organic farmers often focused on the public at large, rather than other farmers, in their publicity efforts, suggesting a link between the use of mineral fertiliser and cancer (Scharrer 1934, p. 245). The fear of cancer turned out to be a powerful marketing tool, and many consumers sought organic farming products as a result. Numerous farmers were selling their products as 'grown without mineral fertiliser' in order to achieve better revenues, even though they had followed the conventional methods of fertilising; certification of organic farming products, and even clear definitions of the 'do's' and 'don'ts' of organic farming, were still several decades away, thus the door was open for a lot of fraud under the organic farming label (Flieg 1933, p. 715). Of course, all this was bound to evoke no sympathy for organic farming in the agricultural science establishment, and yet it would be misleading to depict the massive attacks on organic farming as simply a response to a challenge from outside. In its critique of organic farming, the scientific establishment was also pursuing a number of strategic goals. In the light of their interests, and against the background of the crisis of confidence of the 1920s, one could even argue that the clash over organic farming was a golden opportunity to hammer home a few important points.

On the most general level, the critics of organic farming sought to prevent the formation of a competing network of expertise. The campaign did not fully succeed in this regard, as the continued existence of organic farming up to the present shows, but it did succeed in confining organic farming methods to a small fringe. Even more, the critics succeeded in preventing any meaningful exchange between conventional and organic farming: every practitioner knew that he could turn to either system of expertise, but not to both at the same time. Even on a scientific level, contacts were rare, and were met with great scepticism; up to the 1980s, seeking a dialogue with organic farming experts put a scientist's professional reputation at risk (Gerber et al 1996, p. 596). As a second welcome side-effect, the critique of organic farming led to a general scepticism towards biological concepts of soil fertility; in effect, the controversy canonised a chemical understanding of soil fertility. The true importance of this impact becomes clear when one looks at the general character of the field. The agricultural chemistry establishment was a chaotic conglomeration of research institutions from completely different backgrounds: university departments and institutes sponsored by the farmers (*Landwirtschaftskammern*) coexisted happily with institutes run by the producers of mineral fertilisers. The importance of the latter becomes apparent when one considers the major producers of mineral fertiliser in Germany: potash fertilisers came from state-owned mines, heavy industry sold phosphate from the production of steel in Thomas converters, and IG Farben supplied nitrogen fertilisers – the three most powerful forces of the German economy, united by their interest in increasing fertiliser sales! Against this background, the lack of a truly independent branch of agricultural chemistry is by no means surprising.

However, the complex mixture of industrial, agricultural, and scientific concerns also implied a considerable weakness of this research network: rules of good conduct were notoriously difficult to establish. There was no supreme authority that could define certain rules, and neither was there a place – with Peter Galison (1995), one might also speak of 'trading zones' – where the profession could meet and develop rules of this kind. Therefore, the conflict with organic farming provided a perfect opportunity to establish a few important rules: don't get in touch with organic farming experts, don't seek an understanding of soil fertility that moves beyond chemistry, and don't bother too much about organic fertilisers. Of course, every expert in the field was aware of the fact that the vast majority of farmers used manure on a regular basis – but at the same time, the general tendency in the field was to marginalise the study of manure as a fertiliser, and to focus on mineral fertiliser instead. Over time, this turned into a highly consequential bias: mineral fertiliser came to be regarded as the 'true' fertiliser that commanded the lion's share of attention – while organic wastes were relegated to the status of a marginal topic. As a result, manure did not attract major attention from the agricultural science establishment until long after the Second World War – and then, unexpectedly, manure became an issue not in the form of the asset that it actually was, but, strangely enough, in the form of an environmental problem.

3. TRUST IN NUMBERS: THE BEGINNINGS OF A HAZARDOUS GAME

While agricultural chemistry was competing with other schools of soil fertility, it was simultaneously struggling to resolve a second problem that was instrumental in its rise to dominance: the issue of dosage. The conventional literature provided ample information about the characteristics of the individual nutrients, the types of fertilisers, and their proper handling. However, as farmers became more reliant on mineral fertiliser during the interwar years, they were no longer content with general information. Faced with a crisis of confidence and the significant economic risk of high fertiliser expenditures, they were asking for more precise directions: what amount of what type of fertiliser should they use for their crops? From the point of view of agricultural scientists, this posed a tricky dilemma: they could not ignore this demand without risking their own jurisdiction over fertiliser use – but at the same time, they had no scientifically proper method to meet it.

The traditional method of field experiments was obviously inadequate for this purpose. The method of field experiments as described by the influential fertiliser expert Paul Wagner, the long-time head of the Darmstadt Agricultural Experiment Station, called for five plots of land with different inputs of nutrients: one plot got nitrogen, phosphate, and potash fertiliser, three plots got two of these nutrients, and one plot was left without fertiliser (Wagner 1920, p. 32). That way, experiments stayed away from quantification altogether and sought qualitative results only: they showed whether the individual nutrients contributed significantly to plant growth, but did not allow any conclusion on the amount needed. Lacking an established method for producing reliable figures, many advisors tried to avoid the question by pointing the farmers to their own intuition. For example, Gustav Höppner mentioned 'self-observation and self-examination' as the best source of information for farmers (Höppner 1911, p. 114). 'The general guideline for the choice and use of mineral fertilisers is the local and individual, as opposed to the schematic, approach', a handbook of 1930 declared (Hartmann 1930, p. 27). Even publications that offered precise figures by way of advice diligently warned of a schematic approach to fertiliser use and called upon the farmers to take the specific local conditions into account.[6]

It is interesting to note that the method of soil analysis, nowadays the common method in this field, initially met with almost unanimous scepticism. For example, Höppner noted that a chemical analysis would only produce usable results in those rare cases where one nutrient was missing totally or almost totally, which implied no superiority of the method over Wagner's approach (Höppner 1911, p. 107). Another publication of the same year mentioned field experiments as the only reliable method, depicting chemical and other methods as potentially 'misleading' (Gerlach 1911, p. 3). In 1924, an author explained that his own reservations were due to the fact that chemical investigations would only demonstrate the presence of certain minerals without revealing whether they were actually soluble and thus available to the plant (Schneider 1924, p. 56). In fact, even experts conducting these analyses were eager to tap other sources of information as well: one researcher supplemented his chemical analyses by handing a three-page questionnaire to the farmers in order to consider the wide range of influential factors as fully as possible (König 1929, pp. 11–13). A handbook of 1926 concluded after a discussion of soil investigation methods, 'Valuable as all these physical and chemical methods may be, personal observation, trained through constant practice, is standing above them all' (Stremme 1926, p. 40).

---

6 E.g. Arbeitsgemeinschaft der deutschen Stickstoff-Industrie für das landwirtschaftliche Beratungswesen 1937, p. 57.

However, this hesitancy gradually disappeared, and that paved the way for today's system of soil analysis: a laboratory of the state-funded agricultural research institute (***Landwirtschaftliche Untersuchungs- und Forschungsanstalt***) analyses the soil samples that farmers need to submit at least every six years, and for those farmers who submit an overview of future crops to the institute beforehand, a computer program with the charming name DungPro produces a print-out with detailed information on the necessary nutrients (cf. Jacobs and Remmersmann 2003). This routine procedure is all the more remarkable since the initial uncertainties of the method have by no means disappeared. In fact, a recent handbook on plant nutrition and fertilising praised soil sampling for its speed and low costs but also mentioned 'the findings' low precision as a disadvantage' (Schilling 2000, p. 294). The causes of error are numerous: acres are usually not uniform in their mineral content, a fact that soil samples usually ignore in favour of an arithmetic average; the method does not account for the concentrations in different depths; the solvent needs to extract within a matter of hours what plants would extract over the span of an entire year; and the method inevitably needs to ignore the impact of future weather conditions (Schilling 2000, pp. 294, 299). All in all, the uncertainties are staggering.

However, scientific precision is not necessarily the most important aspect for the selection of a certain method. In the everyday consultation business, time soon came to dominate the choice of methods. The experience of Eilhard Alfred Mitscherlich with soil investigations in Eastern Prussia provides a case in point. Faced with the challenge of writing some 2,000 reports within a brief span of time, he came to use a simple form that included exact figures and postponed detailed, non-quantitative comments to personal meetings during the winter (Mitscherlich 1931, pp. 11–14). The farmers' persistent demand for exact figures met with a growing readiness of scientists to supply this kind of information, and that created a dynamic of its own that gradually pushed all scientific doubt to the margins. Interestingly, Mitscherlich took a much more sceptical view of this method two decades later: 'It is certain that one will never be able to tell the farmer how much of a certain fertiliser he should give on the basis of chemical soil analyses' (Mitscherlich 1952, p. 7). However, objections of this kind did little to stop the rise of this method.

Today, the advisory literature generally ignores all the uncertainties that chemical soil analyses include. 'Soil analyses are the indispensable tool for optimising fertiliser use', a recent article in a popular farming journal declared (Jacobs and Remmersmann 2003, p. 32). The environmental contingencies of this approach become apparent when one realises that even if the figures were precise, the approach would inevitably ignore a whole host of factors. Condition and size of the humus layer, bacterial activity, density of the soil – these and other parameters naturally have an influence on soil fertility, but they do not enter into the calculations of DungPro, which simply sees the soil as a temporary depot of minerals. In fact, even a severe problem like erosion will remain undetected within this system of knowledge – a worrisome situation in a country where erosion is usually a long-term process without conspicuous events of the dust bowl kind. Erosion is not simply an environmental problem in German agriculture. It is also something that happened to the body of agricultural knowledge.

4. BOYS AND THEIR TOYS: COPING WITH EXPERT DEPENDENCY

If agricultural chemistry emerged from the conflicts of the interwar years as the dominant discipline, and gradually won a prominence that, through erosion and other environmental problems, puts at risk the most essential precondition of agricultural production, this obviously brings up the question of the farmers' reaction. Many farmers have essentially surrendered command over their fields to a software program nowadays, and it is difficult to imagine that this happened without major repercussions. Yet precisely that seems to be the case: after the discussions of the interwar years, there has never been significant protest from Germany's farmers against the intrusion of scientific knowledge in this field. This is all the more remarkable since there is plenty of evidence to show that groups usually resist tooth and nail any attempt at usurpation of their productive knowledge. Therefore, it seems that a closer investigation of the farmers' motives is called for.

The most fundamental cause of this paradoxical situation is an immense degree of trust that the farmers have developed towards farm advisors in general.[7] Without the strong sense of identity among farmers, and a popular juxtaposition of 'we the farmers' against 'the rest of society', the trusting relationship between the farmers and the scientific advisors would certainly be unthinkable.[8] However, the key factor was certainly the ultimate success of the relationship from a short-term production perspective. If few farmers nowadays remember the crisis of confidence of the 1920s, this is due to the fact that while mineral fertiliser did not increase per-acreage yields as quickly as promised, it did become clear in the longer run that it boosted agricultural productivity significantly. Specifically, the massive rise of per-acre yields after the Second World War underscored the legitimacy of the agricultural advisors, and in fact provided them with an aura of indispensability. The great sense of trust becomes apparent in the recent announcement of a software program that allows farmers to customise their fertiliser usage plans: the project was commissioned by Hydro Agri, a major producer of mineral fertiliser. Apparently, no one felt that entrusting the calculation of fertiliser demand to a producer of fertiliser was a risky undertaking (cf. Remmersmann 2004).

---

[7] For the recent interest in trust as a historical phenomenon, see Frevert 2003.

[8] For a telling expression of this line of reasoning, see Planck 1985.

However, trust is only one part of the explanation. The other part is distraction – the preoccupation of today's farmers with other issues. It is by no means coincidental that the increasing dominance of scientific advisors went parallel with the mechanisation of agricultural production: the former was possible only because from the farmers' standpoint, it was embedded in the latter. As in the field of fertiliser use, the early relationship between the farmers and their machines went through a period of estrangement. In the 1920s, one author reported that many farmers still saw technology as 'a necessary evil' (Endres 1926, p. 256). Similarly, Hans Schlange-Schöningen (1927, p. 42) mocked the 'tractor-fancy' as a showcase of the exaggerations of modern farming. But soon machines became a routine part of agricultural work, and ultimately a source of pride. A new type of farmer emerged: the technologist with a penchant for up-to-date machinery and do-it-yourself repairs. In the 1990s, agricultural machinery dealers were selling (as opposed to installing) about 50 percent of the total number of spare parts because the farmers wanted to install these parts themselves (Thiede 1992, p. 65n). Of course, saving money was part of the motivation, but so was a fancy for technology. Interestingly, the advisory literature no longer pushes the use of the latest machinery but rather calls for caution in this respect. 'Those who are buying the latest technology for reasons of prestige are usually producing with excessive costs', a recent handbook on farm economy noted (Mohn 1995, p. 139). However, the persistent rise of horsepower figures of farm tractors since the Second World War provides a telling documentation of the farmers' priorities.

The links between the growing prominence of scientific experts and the mechanisation of farm production were functional as well as ideological. On a functional level, machines required a lot of attention and time for maintenance, thus diminishing the farmers' chance to monitor other parts of their enterprises as diligently as before. For example, while farmers traditionally learned a lot about their soils by walking behind the plough, farmers using tractors for tillage devoted much of their attention to the proper operation of the engine. Thus, an expert who offered detailed information on proper fertiliser use could appear not as an intruder into a core area of farming knowledge but as a welcome aide who brought clarity into an area which one could not deal with thoroughly for lack of time.

However, the impact of mechanisation on the farmers' identity was probably even more important. If a powerful tractor and up-to-date equipment was the farmers' new fetish, it was consistent that they became uninterested in the issue of fertiliser; after all, chemicals and odorous wastes are about the least prestigious farming tools that one can imagine. As a result, farmers readily ceded authority on this dull subject to a trustworthy expert; for them, the key issue was not autonomy of decision, but prestige. Naturally, it would take a major incentive, or incident, for these farmers to redirect their attention to the soil.

5. CONCLUSIONS

One of the surprising features of today's system of agricultural production is its remarkable resilience to criticism. Reading Hermann Priebe's ***Subsidised Nonsense*** almost twenty years after its publication, one is struck by the enduring validity of much of what he wrote. In fact, it seems that there are few fields where the environmental movement has made as little headway as in the field of agriculture. Obviously, chastising farmers for their environmental toll, or urging them to convert to organic farming, has been less effective in the field of agriculture than analogous criticism of industrial corporations. In fact, it seems that this kind of criticism raised tempers far more than it raised awareness of environmental problems.

In light of the previous discussion, an important reason for this dismal situation is now clear: the environmentalists' critique ignores the fact that the environmental problems of industrial agriculture are intrinsically connected to a knowledge system that farmers cannot easily abandon. Having surrendered jurisdiction over soil fertility and plant nutrition to outside experts over the course of the twentieth century, the key problem is not a lack of good intentions on the side of the farmers but a lack of knowledge systems. Under an advisory regime that ignores any non-numerical input, the farmer has little choice but to accept the DungPro print-out and follow its directions – and the preoccupation with the latest machinery keeps most farmers from thinking about their fateful situation. And if a farmer chooses to distrust the experts' recommendations, that leaves him with the even more dismal task of deciding autonomously. In their book on the history and future of the soil, John Seymour and Herbert Girardet (1985) repeatedly note that farmers are heeding scientific advice while at the same time being uneasy about it. This may look like irrational behaviour at first glance – until one considers the available alternatives. If farmers are heeding their advisors' recommendations, this is not due to the aura of scientific expertise, or to ignorance about the narrow approach that the experts are taking. In many cases, farmers simply trust for lack of a choice.

Interestingly, even the reforms of recent years have done little to change this situation. Renate Künast, the German minister of agriculture who came into office in early 2001 in the wake of the BSE scandal, devoted much energy to the promotion of organic farming, hoping for a 20 per cent market share in the foreseeable future. Aside from the still open question whether there is really a sizable market for organic farming products – prices for organic farming products have dropped significantly in recent years, indicating an overproduction even with the present market share[18] – this approach has tended to decrease pressure on conventional

[18] Cf. Budde 2003. In 2000, 3.2 per cent of Germany's farmland was under cultivation pursuant to the rules of organic farming. (Schmidt and Jasper 2001, p. 111.)

farmers to improve their environmental record, and in fact tends to depict the system of conventional farming as a dated model that farmers can only abandon, but not change in an environmentally responsible way. To be sure, some reformers have developed a complex network of financial incentives and legal prohibitions that seek to spread certain practices and ban others. However, there seems to be a strange reluctance to put the issue of agricultural knowledge centre stage: who is defining the 'good practice' of farming, according to which criteria, with what overall goal? Not least, a reform of the agricultural knowledge system would have to consider ways to incorporate the farmers' experiences in a productive way. At present, it seems likely that the new system of agricultural knowledge will marginalise the experiences of the farmers in the same way that the old one did. Farmers will be no less helpless in dealing with the new environmental canon than they are now under the regime of DungPro.

In his monograph on the knowledge society, Nico Stehr notes that a knowledge society differs from previous types of society in that its structure is to a greater extent the result of social action. 'It is a society where "secondary" nature is vastly more important than "primary" nature', Stehr notes (1994, p. 218). An environmental historian will certainly read such a statement with a sense of alarm: what happens if the rules of the knowledge society are at odds with the dynamics of nature? From such a point of view, the agrarian knowledge society appears like a runaway steam train – a system of knowledge with a vast impact on nature and little room for feedback. The prospect of this situation should be all the more reason to put the current system of agricultural knowledge under scrutiny – and to start an open discussion on the knowledge base of modern agriculture. The history of agricultural knowledge in the twentieth century shows that this will demand a lot from all parties involved, but also that such a discussion may easily gain momentum once it has started. After all, agricultural knowledge has always been pluralistic, even at times when agricultural chemistry seemed uncontested at first glance. Promoting a plurality of opinions, and of approaches, is an important contribution to the quest for an agriculture that is both productive and sustainable. The call is for scientists to communicate with both farmers and their perceived rivals in other scientific branches – and to combine these divergent contributions into the vision of a new agriculture. In other words, the agricultural knowledge society will need to turn into an agricultural learning society. And researchers should not be discouraged by the fact that there is probably no Liebig quotation to legitimate this endeavour.

REFERENCES

Arbeitsgemeinschaft der deutschen Stickstoff-Industrie für das landwirtschaftliche Beratungswesen. 1937. *Düngungsratschläge für den Bauernhof*, 2nd edition. Berlin: Arbeitsgemeinschaft der deutschen Stickstoff-Industrie für das landwirtschaftliche Beratungswesen.

Brandt, O. 1921. *Gemeinverständliche Einführung in die Stickstoffrage*. Hannover: Engelhard & Co.

Brock, William H. 1997. *Justus von Liebig: The Chemical Gatekeeper*. Cambridge/New York: Cambridge University Press.

Budde, Franz-Josef. 2003. 'Biobauern trifft's gleich mehrfach'. *Landwirtschaftliches Wochenblatt Westfalen-Lippe*, no. 38, 18 September 2003: 11.

Deußen, E. 1922. 'Kritisches zu unserer Düngernot, nebst einem Beitrag zum Phosphorgehalt der Jauche'. *Chemiker-Zeitung*, 46: 826–8.

Endres, W. 1926. 'Maschinentechnische Hilfsmittel des Landwirts'. In F. Christoph (ed.), *Die Technik in der Landwirtschaft* (Berlin-Schmargendorf and Leipzig: Andermann), pp. 256–88.

Flieg, O. 1933. 'Hände weg von unseren bewährten Düngemethoden!' *Die Umschau. Illustrierte Wochenschrift über die Fortschritte in Wissenschaft und Technik*, 37: 713–5.

Frevert, Ute (ed.). 2003. *Vertrauen. Historische Annäherungen*. Göttingen: Vandenhoeck & Ruprecht.

Fussell, George E. 1971. *Crop Nutrition: Science and Practice before Liebig*. Lawrence: Coronado Press.

Galison, Peter L. 1995. 'Context and Constraints'. In Jed Z. Buchwald (ed.), *Scientific Practice: Theories and Stories of Doing Physics* (Chicago and London: University of Chicago Press), pp. 13–41.

Gerber, Alexander, Volker Hoffmann and Michael Kügler. 1996. 'Das Wissenssystem im ökologischen Landbau in Deutschland. Zur Entstehung und Weitergabe von Wissen im Diffusionsprozeß'. *Berichte über Landwirtschaft*, 74: 591–627.

Gerlach, Max. 1911. *Langjährige Feldversuche über die Wirkung der Kalisalze auf Sandböden* (Landwirtschaftliche Lehrhefte no. 5). Berlin: Verlagsgesellschaft für Ackerbau.

Glathe, H. 1935. 'Vorkommen und Tätigkeit von Mikroorganismen im Boden'. In F. Löhnis et al (eds), *Handbuch der landwirtschaftlichen Bakteriologie. Band II: Dünger- und Bodenbakteriologie* (2nd rev. edition, Berlin: Borntraeger), pp. 159–629.

Görbing, Johannes. 1947. *Die Grundlagen der Gare im praktischen Ackerbau, vol. 1*. Hannover: Landbuch-Verlag.

Hartmann, Friedrich. 1930. *Kalkfibel in Frage und Antwortstil*. Coburg: Hofbuchhandlung A. Seitz.

Höppner, Gustav. 1911. *Die Kunstdüngemittel und ihre Anwendung in der modernen Landwirtschaft* (Löbe's Landwirtschaftliche Bibliothek vol. 28/29). Leipzig: Reichenbach'sche Verlagsbuchhandlung.

Jacobs, Günter and Theo Remmersmann. 2003. 'Erst untersuchen, dann düngen'. *Landwirtschaftliches Wochenblatt Westfalen-Lippe*, no. 31, 31 July 2003: 32–4.

Jahn-Deesbach, W. 1965. 'Aufgaben und Probleme der Gründüngung im modernen Ackerbau'. In Ruhr-Stickstoff AG (ed.), *Bodenfruchtbarkeit ohne Stallmist?*, Landwirtschaftliche Schriftenreihe Boden und Pflanze no. 12 (Bochum: Ruhr-Stickstoff AG), pp. 27–59.

Knecht, Johannes. 1949. *Das Jahr des jungen Landwirts. Ein Lehr- und Handbuch für den landwirtschaftlichen Berufsschüler und Landwirtschaftslehrling*, 2nd edn. Stuttgart/ Ludwigsburg: Ulmer.

König, J. 1929. *Die Ermittlung des Düngerbedarfs des Bodens*. Berlin: Parey.

Kuhn. 1920. 'Landwirtschaftliche Produktion im Zeichen der Düngernot'. *Mitteilungen der Deutschen Landwirtschafts-Gesellschaft*, **35**: 346–52.

Landwehr, Achim. 2002. 'Einleitung: Geschichte(n) der Wirklichkeit'. In Achim Landwehr (ed.), *Geschichte(n) der Wirklichkeit. Beiträge zur Sozial- und Kulturgeschichte des Wissens* (Augsburg: Wißner), pp. 9–27.

Langenbeck, E. 1921. *Zur Berufswahl des Landwirts* (Flugschriften der Deutschen Landwirtschafts-Gesellschaft no. 17). Berlin: Deutsche Landwirtschafts-Gesellschaft.

Löhnis, F. 1926a. 'Bodenbakterien und Bodenfruchtbarkeit'. *Die Technik in der Landwirtschaft*, 7: 247–51.

Löhnis, F. 1926b. 'Bodenbakterien und Bodenfruchtbarkeit'. *Vorträge gehalten anläßlich der sechsten sächsischen Landwirtschaftlichen Woche in Dresden vom 25.-29. Januar 1926 und des Vortragskursus für praktische Landwirte in Leipzig vom 5.-7. Januar 1926* (Arbeiten aus dem Gebiete der sächsischen Landwirtschaft no. 14, Dresden: Verlag der Landwirtschaftskammer für den Freistaat Sachsen), pp. 173–83.

Löhnis, Felix. 1927. 'Die Aufgaben der Biologie des Bodens'. *Fortschritte der Landwirtschaft*, **2**: 241–2.

Löhnis, F. 1928. 'Untersuchungen über Wirkung und Wert des Wirtschaftsdüngers'. *Fortschritte der Landwirtschaft*, **3**: 817–21.

Merkenschlager, F. 1933. 'Zeitkrise und die sogenannte "Biologisch-dynamische Düngung"'. *Natur und Kultur. Monatsschrift für Naturwissenschaft und ihre Grenzgebiete*, **30**: 121–5.

Mitscherlich, Eilhard Alfred. 1931. *Ein Leitfaden zur Anwendung der künstlichen Düngemittel*, 2nd rev. edition. Berlin: Parey.

Mitscherlich, Eilhard Alfred. 1952. *Erkenntnisse bei der Pflanzendüngung. Ein Festvortrag zur Feier des 250jährigen Bestehens der Deutschen Akademie der Wissenschaften zu Berlin* (Deutsche Akademie der Wissenschaften zu Berlin, Vorträge und Schriften issue 44). Berlin: Akademie-Verlag.

Mohn, Reiner. 1995. 'Gesetzmäßigkeiten wirtschaftlichen Entscheidens und Handelns'. In Erwin Reisch, Gerhard Knecht and Julius Konrad (eds), *Betriebslehre* (7th edition, Stuttgart: Ulmer), pp. 103–42.

Mommsen, Wolfgang J. 2002. *Die Urkatastrophe Deutschlands. Der Erste Weltkrieg*. Stuttgart: Klett-Cotta.

Niggl, Ernst. 1923. *Bedeutung und Anwendung der Kalkdüngung*. Berlin: Verlag des Vereins Deutscher Kalkwerke.

Planck, Ulrich. 1985. 'Die Landwirtschaft in der Industriegesellschaft und die Industrialisierung der Landwirtschaft'. *Zeitschrift für Agrargeschichte und Agrarsoziologie*, **33**: 56–77.

Priebe, Hermann. 1985. *Die subventionierte Unvernunft. Landwirtschaft und Naturhaushalt.* Berlin: Siedler.

Reichrath, Susanne. 1991. *Entstehung, Entwicklung und Stand der Agrarwissenschaften in Deutschland und Frankreich*. Frankfurt: Peter Lang.

Reiser, Max. 1929. 'Die Wiesen- und Weidewirtschaft im Allgäu unter Berücksichtigung der Güllewirtschaft'. In Verein der Thomasmehlerzeuger (ed.), *Unseren Freunden vom Thomasmehl* (Berlin: Verlag für Bodenkultur), pp. 32–42.

Remmersmann, Theo. 2004. 'Der Computer hilft bei der Düngeplanung'. *Landwirtschaftliches Wochenblatt Westfalen-Lippe*, no. 4 , 22 January 2004: 28–30.

Scharrer, K. 1934. 'Die "biologisch-dynamische" Düngung im Lichte der Agrikulturchemie'. *Chemiker-Zeitung*, **58**: 245–7, 267–70.

Schellenberger. 1924. *Zeitgemäße Vorschläge für die Düngung der landwirtschaftlichen Kulturpflanzen* (Arbeiten aus dem Gebiete der sächsischen Landwirtschaft no. 7). Dresden: Verlag des Landeskulturrats Sachsen.

Schilling, Günther. 2000. *Pflanzenernährung und Düngung*. Stuttgart: Ulmer.

Schlange-Schöningen, Hans. 1927. *Rationalwirtschaft und Nationalwirtschaft. Betrachtungen eines praktischen Landwirtes*. Berlin: Reinhold Kühn.

Schlange-Schöningen, Hans. 1931. *Das Wirtschaftsjahr des praktischen Landwirts*. Berlin: Parey.

Schling-Brodersen, Ursula. 1989. *Entwicklung und Institutionalisierung der Agrikulturchemie im 19. Jahrhundert. Liebig und die landwirtschaftlichen Versuchsstationen* (Braunschweiger Veröffentlichungen zur Geschichte der Pharmazie und der Naturwissenschaften 31). Stuttgart: Deutscher Apothekerverlag.

Schmidt, Götz, and Ulrich Jasper. 2001. *Agrarwende oder die Zukunft unserer Ernährung.* München: Beck.

Schneider, Johannes. 1924. 'Düngung'. In Johannes Schneider (ed.), *Praktisches Handbuch der Landwirtschaft* (Reutlingen: Enßlin & Laiblins Verlagsbuchhandlung), pp. 53–94.

Schnelle, G. and O. Nolte. 1933. 'Ueber die sogenannte Reformbewegung für Umstellung der Volkskost'. *Fortschritte der Landwirtschaft*, 8: 220–3.

Schober, Karl. 1951. *Düngerlehre. Ein Handbuch für praktische Landwirte und Studierende.* Wien and Heidelberg: Bohmann.

Seufert, Hermann. 2003. 'Nachhaltige Landtechnik im Sinne Liebigs'. In Der Präsident der Justus-Liebig-Universität Gießen (ed.), *Justus Liebig (1803–1873). vol. 1: Seine Zeit und unsere Zeit. Chemie - Landwirtschaft – Ernährung* (Gießen: Justus-Liebig-Universität), pp. 267–76.

Seymour, John, and Herbert Girardet. 1985. *Fern vom Garten Eden. Die Geschichte des Bodens. Kultivierung - Zerstörung – Rettung*. Frankfurt: Wolfgang Krüger Verlag.

Smalakies, Arthur. 1921. *Die Bedeutung der Kunstdünger für den Wiederaufbau der deutschen Landwirtschaft.* Berlin: Elsner.

Stehr, Nico. 1994. *Arbeit, Eigentum und Wissen. Zur Theorie von Wissensgesellschaften.* Frankfurt: Suhrkamp.

Stremme, Hermann. 1926. *Grundzüge der praktischen Bodenkunde.* Berlin: Borntraeger.

Szöllösi-Janze, Margit. 1998. *Fritz Haber 1868–1934. Eine Biographie.* München: Beck.

Thiede, Günther. 1992. *Die Grüne Chance. Landwirte zwischen Tradition und Fortschritt.* Frankfurt: DLG-Verlag.

Wagner, Paul. 1920. *Anwendung künstlicher Düngemittel* (Thaer-Bibliothek no. 100). 7th edition, Berlin: Parey.

Wagner, Paul. 1929. 'Das Thomasmehl als Frühjahrsdünger'. In Verein der Thomasmehlerzeuger (ed.), *Unseren Freunden vom Thomasmehl* (Berlin: Verlag für Bodenkultur), pp. 3–6.

Wiessmann, H. 1926. 'Die Bodenacidität und ihre Bedeutung für das Pflanzenleben'. *Zeitschrift für angewandte Chemie*, **39**: 525–30.

Willke, Helmut. 1998. 'Organisierte Wissensarbeit'. *Zeitschrift für Soziologie*, 27: 161–77.

ARCHIVAL SOURCES

Bundesarchiv R 3602 No. 67.

Bundesarchiv R 3602 No. 606.

Bundesarchiv R 3602 No. 608.

Index

Index

Index

Index

Index

Index

Index

Index

Index

H

I

Index

Index

N

Index

Index

Index

Index

Index

Index

T

Index

Index